Springer Tracts in Modern Physics

Volume 284

Springer Tracts in Modern Physics provides comprehensive and critical reviews of topics of current interest in physics. The following fields are emphasized:

- Particle and Nuclear Physics
- Condensed Matter Physics
- Light Matter Interaction
- Atomic and Molecular Physics

Suitable reviews of other fields can also be accepted. The Editors encourage prospective authors to correspond with them in advance of submitting a manuscript. For reviews of topics belonging to the above mentioned fields, they should address the responsible Editor as listed in "Contact the Editors".

More information about this series at http://www.springer.com/series/426

Roman Kogler

Advances in Jet Substructure at the LHC

Algorithms, Measurements and Searches for New Physical Phenomena

Springer

Roman Kogler
Institute for Experimental Physics
University of Hamburg
Hamburg, Germany

ISSN 0081-3869 ISSN 1615-0430 (electronic)
Springer Tracts in Modern Physics
ISBN 978-3-030-72857-1 ISBN 978-3-030-72858-8 (eBook)
https://doi.org/10.1007/978-3-030-72858-8

This Springer imprint is published by the registered company Springer Nature Switzerland AG
The registered company address is: Gewerbestrasse 11, 6330 Cham, Switzerland

Dedicated to Jennifer, Louis and Linda.
Without you none of this would matter.

Preface

This book has been written as part of my habilitation at the University of Hamburg. It is intended to serve graduate students and researchers to get familiar with the state-of-the-art of jet substructure at the LHC. I have attempted to provide enough details such that this book can serve as a work of reference for experienced researchers as well.

The versatility of this branch of particle physics is reflected in multi-faceted developments of algorithms and their numerous applications in measurements and searches for new phenomena. While this versatility is the reason for the success of this field, the inexperienced researcher may feel lost when first exposed to this wealth of information. I advise researchers starting in this area to read the whole book without stopping at passages which may not be immediately clear at first. After having obtained a good overview of the possibilities jet substructure methods offer, re-reading the book will help to deepen the knowledge and reinforce connections. I hope that the level of detail given will help more experienced researchers to find solutions to their problems or connections to other areas. I have tried to provide as many relevant references as possible, to aid the interested reader in finding all information needed to delve deeply into every topic discussed in this book.

My scientific work, leading to this book, would not have been possible without the support of Prof. Johannes Haller (University of Hamburg), who has encouraged me to proceed with my habilitation at the Institute for Experimental Physics. I am grateful to Dr. Alberto Orso Maria Iorio (INFN Napoli), who has led the physics analysis group "Beyond Two Generations" within the CMS Collaboration together with me from 2017 to 2019. The work with this group has inspired me and led to exciting new developments. I would also like to thank my CMS collaborators for the excellent working atmosphere and stimulating environment. I have profited from the scientific exchange at numerous conferences and workshops, most notably at the annual BOOST conference (International Workshop on Boosted Object Phenomenology, Reconstruction and Searches in HEP).

I am thankful for financial support by the German Bundesministerium für Bildung und Forschung (BMBF) and the Deutsche Forschungsgemeinschaft (DFG, German Research Foundation) under Germany's Excellence Strategy—EXC 2121 "Quantum Universe". I am grateful to Prof. Thomas Müller (KIT Karlsruhe) for making the contact with Springer, initiating that this document could be published as part of the "Springer Tracts in Modern Physics" and to the publisher for the professional partnership when creating this book. I am indebted to Prof. Johannes Haller, Prof. Michael Spannowsky (IPPP and the University of Durham), Dr. Arne Reimers (University of Zürich), Dr. Dennis Schwarz (HEPHY Vienna), Dr. Anna Benecke (UC Louvain) and Andrea Malara (University of Hamburg) for proof-reading the book.

Hamburg, Germany Roman Kogler

Contents

Acronyms

ANN	Artificial neural network
aNNNLO	Approximate next-to-next-to-next-to-leading-order
AO	Angular ordering
BDT	Boosted decision tree
BEH	Brout-Englert-Higgs
BEST	Boosted event shape tagger
BSM	Beyond the standard model
CA	Cambridge/Aachen
CHS	Charged hadron subtraction
CL	Confidence level
CM	Centre-of-mass
CMSTT	CMS top tagger
CNN	Convoluted neural network
CR	Control region
DDT	Designed decorrelated tagger
DNN	Deep neural network
ECAL	Electromagnetic calorimeter
ECF	Energy correlation function
EFP	Energy flow polynomial
EM	Electromagnetic
EMD	Energy mover's distance
EW	Electroweak
FSR	Final state radiation
ggF	Gluon-gluon fusion
HCAL	Hadronic calorimeter
HL-LHC	High luminosity LHC
HOTVR	Heavy object tagger with variable R
HP	High purity
HTT	HEPTopTagger
HVT	Heavy-vector triplet
IRC	Infrared and collinear
ISR	Initial state radiation

JER	Jet energy resolution
JES	Jet energy scale
JMR	Jet mass resolution
JMS	Jet mass scale
LAR	Liquid Argon
LC	Light cone
LCW	Local cell weighting
LHC	Large Hadron Collider
LL	Leading logarithmic
LO	Leading order
LP	Low purity
LQ	Leptoquark
LSP	Lightest supersymmetric particle
LV	Leading vertex
MB	Minimum-bias
MD	Mass-decorrelated
MDT	Mass drop tagger
ML	Machine learning
MLP	Multilayer perceptron
mMDT	Modified mass drop tagger
MPI	Multiple parton interaction
MPV	Most probable value
NLL	Next-to-leading-logarithmic
NLO	Next-to-leading order
NN	Neural network
NNLL	Next-to-next-to-leading-logarithmic
NNLO	Next-to-next-to-leading-order
PDF	Parton distribution function
PDG	Particle Data Group
PF	Particle flow
pQCD	Perturbative quantum chromodynamics
PUPPI	Pileup per particle identification
QCD	Quantum chromodynamics
QED	Quantum electrodynamics
RG	Renormalisation group
RNN	Recurrent neural network
ROC	Receiver operating characteristic
RPV	R-parity violating
SCET	Soft-collinear effective theory
SD	Shower deconstruction
SISCone	Seedless infrared-safe cone
SM	Standard model (of particle physics)
SR	Signal region
SUSY	Supersymmetry
TCC	Track-CaloClusters

TRT	Transition radiation tracking
UFO	Unified flow object
VBF	Vector-boson fusion
VLQ	Vector-like quark
VR	Variable *R*
WIMP	Weakly interacting massive particle
WTA	Winner-takes-all
XCone	Exclusive cone

Chapter 1
Introduction

Abstract The standard model of particle physics is a remarkably successful theory. It has been tested by countless measurements and has withstood all attempts at its falsification. Jet substructure techniques offer new possibilities to test this exceptional theory and search for unknown effects.

The standard model (SM) of particle physics is one of the most successful theories humankind has developed. Its core elements have been formulated in the 1960–70s to explain phenomena at energy scales of a few GeV. Nowadays, more than 50 years later, the SM has been probed by many experiments and numerous measurements spanning energy ranges from GeV to multiple TeV. Its validity has been confirmed in each of these measurements, with cross sections spanning many orders of magnitude. The SM has also proven to be extremely successful in predicting new particles and processes, spectacular discoveries accompany its development.

In our modern picture, the SM of particle physics is governed by the Lagrange density $\mathcal{L}_{\mathrm{SM}}$ with left-handed doublets ψ_L and right-handed singlets ψ_R of the quark and lepton fields. It is one of the largest successes in science that this theory could be devised by following symmetry considerations only. This leads to a structure of $\mathcal{L}_{\mathrm{SM}}$ which is remarkably plain, but leads to a vast phenomenology. Since any description of nature is only a valid physical theory if it can be falsified, the success of the SM is founded on its predictions of observable processes.

With the start of data acquisition at the CERN LHC in 2010, a new era of particle physics has begun. The energy range of particle collisions has been extended by about an order of magnitude relative to earlier experiments. In conjunction with a dataset of unprecedented size, this has led to a wealth of measurements and searches exploring regions not accessible before. A tremendous experimental effort by the ATLAS and CMS Collaborations has climaxed in the last addition to the SM. The Higgs boson H has been discovered in 2012 [1, 2], about 50 years after the prediction of its existence. So far, all studies of this new particle suggest that it has the properties predicted by the minimal version of the Brout-Englert-Higgs (BEH) [3–5] mechanism incorporated in the SM [6–10]. The experimental programme at the LHC has also brought forth numerous measurements of SM processes and

R. Kogler, *Advances in Jet Substructure at the LHC*, Springer Tracts in Modern Physics 284, https://doi.org/10.1007/978-3-030-72858-8_1

precision measurements of fundamental parameters of the theory. Taken together, this has resulted in SM consistency tests with remarkable precision, confirming the validity of the theory up to scales probed by the LHC. There is an ever increasing degree of scientific confidence in the SM.

While the SM is exceptionally successful, no explanation exists for a variety of fundamental questions. These include the unlikeness of the Higgs boson mass from its naive quantum-mechanical expectation, the nature of dark matter, the masses of neutrinos, the number of fermion generations, to name a few. Especially intriguing is the question of the nature of the Higgs boson and the dynamics of the underlying scalar field, responsible for the spontaneous electroweak (EW) symmetry breaking. Central to this mechanism is the top quark, the SM particle with the largest coupling to the Higgs field. With its centre-of-mass energy $\sqrt{s} = 13\,\text{TeV}$, the LHC is the ideal machine to study these questions and probe the SM at the EW scale $v \simeq 246\,\text{GeV}$ and above.

Broadly, there exist three different approaches to probe for effects from unknown physics: direct searches, precision measurements and measurements of very rare processes. At the LHC, direct searches have been focussing on finding traces of particles predicted by beyond the standard model (BSM) theories, often with very specific signatures. So far, these direct searches have not found any evidence for BSM effects, even though hundreds of analyses have been carried out [11]. Precision measurements allow for the possibility to find small deviations from SM expectations, giving indirect hints of BSM effects. Important milestones of the experimental programme at the LHC are precision measurements of Higgs boson properties, such as its mass, its coupling to vector bosons and fermions, and its parity. The programme also includes measurements of the W boson mass, the top quark mass, the effective weak mixing angle and cross sections of W, Z and jet production. Taken together, these measurements have the potential to reveal unknown effects through comparisons with precision calculations and global fits of SM parameters [12–16]. The last possibility, the observation of very rare processes, allows to probe for BSM physics through quantum corrections entering the calculations of rare decays or production mechanisms. Examples are $B^0_{(s)} \to \mu^+\mu^-$ [17, 18], four-top [19] or di-Higgs [20] production at the LHC.

With the end of the data acquisition period from the end of 2015–2018, called Run 2, LHC analyses are being completed using the full 13 TeV data corresponding to about 140 fb^{-1}. It becomes apparent that none of the three approaches has resulted in evidence for unknown effects. In this sense, particle physics at the LHC is at a crossroads. The next data acquisition period of the LHC experiments, Run 3 , is planned to start in 2022 and continue until the end of 2025. The total amount of data collected by the end of 2024 will correspond to about 300 fb^{-1}. However, the sensitivity of direct searches for BSM effects improves only little when doubling the amount of data. In terms of discovery significance, the mass reach improves only by 10–20% for analyses dominated by the statistical precision of the data and even less for analyses dominated by systematic uncertainties. Precision measurements face the difficulty of accurately understanding even small experimental effects and correctly estimating corresponding systematic uncertainties, a process which takes years to complete due to the complexity of the LHC experiments.

This leaves three choices for our immediate hopes to find BSM effects: either measure rare processes, becoming accessible only now with the large amount of data, find novel ways to perform searches for BSM effects, or devise complementary strategies for precision measurements. This book describes techniques related to the internal structure of jets, offering the possibility to advance on all three of these topics. The versatility of jet substructure techniques results from the presence of jets in virtually every process in high energy pp collisions. The substructure of these jets contains information about their origin and the underlying dynamics. For example, jet substructure allows to distinguish quark from gluon jets, remove the effects from uncorrelated radiation from multiple simultaneous pp collisions, identify jets originating from collimated decays of W, Z, H bosons or top quarks, or reduce the influence of non-perturbative effects and thus improve the precision of measurements. Since either one or more of these aspects are important for every analysis of LHC data, jet substructure techniques have percolated into all physics analysis groups of the ATLAS and CMS Collaborations.[1] The physics of jet substructure has enhanced the physics potential of the LHC and will continue to do so in the years to come.

The aim of this book is to review the most important developments in the field of jet substructure from an experimentalists' point of view. This encompasses theoretical and algorithmic developments, analytic calculations, advances in modelling, experimental studies of substructure observables and commissioning of new tools, measurements of substructure observables, measurements making use of them, as well as direct searches for BSM effects with jet substructure. It is my intention to put these developments into a larger context, showing the relevance of this relatively young field to particle physics as a whole. The versatility of jet substructure leads from scientific gains in precision studies of Quantum Chromodynamics (QCD), to the determination of fundamental parameters of the SM, to searches for new physical phenomena at the highest energy scales. If there are BSM effects observable at the LHC, jet substructure techniques will play an important role to discover them.[2]

[1] Jet substructure has also become a field of study in heavy ion collisions, which are not discussed in this book.

[2] While some aspects of LHC physics may not have direct connections to jet substructure techniques, the advancements in this field result in new developments of reconstruction algorithms, influencing all stages of data analyses.

This book summarises the status of this field until about the end of 2020. Regular meetings of theorists and experimentalists in the context of the BOOST conference series,[3] have led to a fruitful exchange, triggering developments that shaped the field. While several publications and studies on jet substructure existed already prior to the first BOOST conference, it marks the beginning of an exciting scientific undertaking. The three reports from this conference series provide valuable material [21–23]. The interested reader is also referred to recent comprehensive reviews on theoretical [24, 25] and experimental [26] aspects of jet substructure.

[3]Conferences in this series are dedicated to the physics of jet substructure, and are called *International Workshops on Boosted Object Phenomenology, Reconstruction and Searches in HEP*. They are organised annually, where the first one was held in 2009 at the SLAC National Accelerator Laboratory at Menlo Park (USA).

Chapter 2
Phenomenology of Jet Substructure

Abstract Jet substructure techniques are devised to analyse the internal radiation pattern of jets, thereby identifying their origin or revealing the dynamics of the strong force. These techniques are intricately connected with jet finding algorithms, either by modifying the clustering process, reversing it, or by storing information during the clustering. In this chapter, an overview is given of the currently known methods, algorithms and observables. The kinematics of heavy particle decays is discussed in detail to gain an understanding how individual jets can capture the full information from these decays. Theoretical methods used to calculate jet substructure observables are introduced, as well as models to simulate the rich substructure of jets.

2.1 Introduction to Jet Substructure

Jets are collimated sprays of particles, produced in abundance in high energy particle collisions.[1] They are ubiquitous in particle collider experiments and indispensable in studies of the underlying dynamics and interactions. Jets have played a central role in the discovery and property measurements of many fundamental particles like the gluon (g) [27–30] and the top quark (t) [31, 32]. They have provided key insights into the structure of the strong force and were indispensable in the study of H boson couplings to third-generation quarks [33–36]. Because of their large production rate at the LHC, jets feature prominently in searches for new particles and precision measurements of SM properties. However, important information on the underlying particle dynamics is not only carried by the four-momenta of jets, but also by their internal structure. Investigations of this jet substructure reveal a wealth of physical processes and pose interesting theoretical and experimental challenges. While relatively young, the field of jet substructure has become an important field of research over the last decade and will gain further importance with the future data taking periods at the LHC.

[1] The text in this subsection has been taken from [26] and has been written by the author. It has been adjusted to fit this book.

R. Kogler, *Advances in Jet Substructure at the LHC*, Springer Tracts in Modern Physics 284, https://doi.org/10.1007/978-3-030-72858-8_2

With the advent of the LHC it was realised that decays of hypothetical, very heavy resonances can lead to highly Lorentz-boosted heavy SM particles, W, Z, H bosons and top quarks [37–41]. Since these particles feature the largest branching fractions into hadrons, final states with fully-hadronic decays have high sensitivity in LHC analyses. The large boost leads to very collimated decays, where particle masses of $\mathcal{O}(100)$ GeV are not large enough for the outgoing quarks to be sufficiently separated relative to each other to be resolved into individual jets. It is the small opening angle between the decay products that leads to fully-merged particle decays. This chapter describes techniques for measuring jets as proxies for hadronic decays of W, Z, H bosons and top quarks, as well as the discrimination of quark and gluon jets.

Since the first evidence for jets in e^+e^- collisions at SPEAR [42], jets have had a significant impact on the research programme of every particle collider since DORIS through the LHC, and beyond to the design of future colliders. There is no single, universal definition of a jet—which particles belong to a jet depend on the algorithm used to combine particles into jets. In the beginning of jets from the mid 1970's, there were no jet clustering algorithms; information from the whole event was used instead of localised energy flows. The sphericity tensor [43] was typically used to obtain a jet axis for events with a back-to-back dijet topology. Quantitative statements about data were obtained from event shapes, like the sphericity or thrust [44–46]. Sphericity is a measure for the isotropy of the particles produced and thrust is a measure of the directed energy flow along an axis that maximises this flow in an event. These event shapes can be used to characterise how compatible events are with the assumption of two oppositely directed, collimated jets. A clear theoretical advantage of these event shapes is that they are calculable in perturbative Quantum Chromodynamics (pQCD). This was realised early on and the calculability, together with experimental data, ultimately resulted in the confirmation of the parton model and, with data from experiments at higher $\sqrt{s}$, the discovery of the gluon in three jet events at PETRA [27–30].

When studying the dynamics of quark and gluon scattering, it became necessary to perform quantitative analyses and calculations that go beyond event shapes. For these to be possible, it was realised that it is mandatory to define a deterministic set of rules on how particles are combined into jets. A schematic drawing depicting this problem is shown in Fig. 2.1. While the sphericity axis is uniquely defined and easily calculable, the direction and magnitude of the jet axes depend on which particles should be combined into a given jet, and how the particles are combined to obtain the axes. An intuitive definition for a jet algorithm consists of summing the momenta of all particles within a cone with fixed size [47]. Naive cone algorithms are not infrared and collinear (IRC) safe—the requirement that the resulting jets be insensitive to arbitrarily low energy particles and collinear splittings. IRC safety is a useful theoretical requirement for making calculations in pQCD and is also a convenient language for describing the experimental robustness to noise and detector granularity.

There exist many variants of cone-type algorithms, developed in the attempt to solve the IRC unsafety of naive cone jet algorithms. This stems from the necessity of an initial axis, which was eventually solved with the formulation of the SISCone algo-

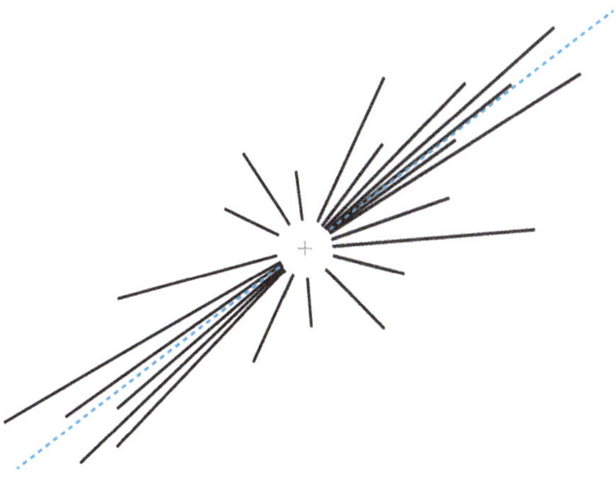

Fig. 2.1 Schematic drawing of particles emerging from the hard scattering of a high energy particle collision. The sphericity axis is shown as dashed line. Published in [26]

rithm [48]. Although this algorithm is IRC safe, it is not widely used today because it was found that sequential recombination algorithms have several advantages over cone-type algorithms. First used by the JADE Collaboration [49, 50], the initial version of a recombination algorithm defined for e^+e^- collisions was improved in several steps [51, 52], to finally arrive at the longitudinally-invariant k_T-clustering algorithm for hadron-hadron collisions [53, 54]. A generalisation of this algorithm leads to three classes, distinct only by the sign of the exponent k of the transverse momentum $p_{\mathrm{T},i}$ in the inter-particle distance measure d_{ij}. The original k_T algorithm, with $k = 1$, clusters soft and collinear particles first, the Cambridge/Aachen algorithm (CA) [55, 56], with $k = 0$, prioritises particles in the clustering solely by their angular proximity, and the anti-k_T algorithm [57], with $k = -1$, combines the hardest particles first. The size of the catchment area of a jet is regulated by the jet distance parameter R, often referred to as jet radius. The proposal of the latter algorithm is also responsible for the disappearance of cone-type algorithms in experimental studies. When it was realised that the anti-k_T algorithm results in nearly perfect conical jets the LHC collaborations made a transition to this algorithm. Today, almost all studies involving jets performed at the LHC use this algorithm. Even when analysing the substructure of jets with advanced grooming or tagging techniques, the initial step often consists of building an ensemble of particles, clustered with the anti-k_T algorithm.

So far, it has not been specified what the term particle refers to when using particles as input to jet clustering. In fact, in jet physics, the term particle is often used generically for different sorts of objects, whose ensemble comprises the input to a given jet algorithm. Three different ensembles are commonly used. The partonic final state includes all particles resulting from the parton shower before the hadronisation starts (which is unphysical). This also include photons when these were created in the hard interaction or emitted from charged particles during the parton shower. The ensemble on the particle level, also called hadron level, consists of hadrons and their decay products, including photons and leptons. The detector level input consists of calorimeter clusters, reconstructed particle tracks or combinations thereof. This ensemble is the only one available in real collision data recorded by particle physics experiments. Jet algorithms using these different ensembles as input result in parton, particle or detector level jets, respectively. Ideally, in any given event, the jets obtained on parton, particle and detector level are as similar as possible. Realistically, agreement can not be achieved, but a close correspondence ensures the possibility to study the underlying partonic dynamics with the use of jets. It is this correspondence, paired with calculability in pQCD, which makes jets indispensable tools at high energy particle colliders. For a theoretical introduction to jets, see [24, 25, 58, 59].

Soon after their discovery, it was realised that not only the kinematics of jets but also their internal structure carry information. The parton shower and subsequent hadronisation leads to a characteristic multiplicity, as well as angular and momentum distributions of hadrons inside jets, which depend on the parton that initiated the shower. For example, the probability of a $q \rightarrow qg$ splitting is proportional to the colour factor $C_F = 4/3$ at leading order in QCD, while the probability of $g \rightarrow gg$ is proportional to $C_A = 3$. The larger value of C_A results in a larger multiplicity of hadrons and in broader jets. This lead to the suggestion of measuring jet shapes, defined as the fractional transverse momentum profile of particles within a concentric inner cone, smaller than the jet cone of the original jet, and pointed to their usefulness for distinguishing quark from gluon jets [60]. Experimental results from LEP [61–64], Tevatron [65, 66] and HERA [67–69] confirmed this and can be considered the starting point of physics with jet substructure in particle physics.

At the LHC, jet substructure is used to identify highly boosted heavy SM particles in fully hadronic decays, as well as light quark and gluon jets. An example of a jet with substructure from a two-prong decay is shown schematically in Fig. 2.2. The difficulty lies in identifying the underlying process that led to the final state, for example distinguishing $W \rightarrow q\bar{q}'$, $Z \rightarrow q\bar{q}$ or $H \rightarrow b\bar{b}$ from QCD splittings like $q \rightarrow qg$, $g \rightarrow gg$ or $g \rightarrow q\bar{q}$. Numerous algorithms have been suggested to identify specific decays, which are part of a class of jet substructure taggers. The idea behind many of these algorithms is related to event shapes in e^+e^- collisions. By defining N axes within a jet, it is possible to check for the compatibility of a fully-merged N-prong decay. How these axes are found typically differs from algorithm to algorithm, and some techniques do not even explicitly require axes. Popular concepts are an

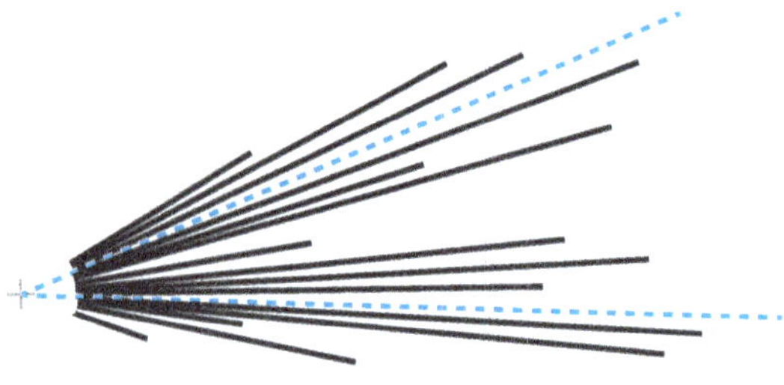

Fig. 2.2 Schematic drawing of particles clustered into a single jet. Two subjet axes are shown as dashed lines. Published in [26]

exclusive jet clustering using the particles inside a jet as input, or the maximisation of the projection of the jet constituents' momenta onto the desired number of axes, as illustrated in Fig. 2.2.

Since the opening angle between the quarks depends on the momentum of the parent particle and its mass, larger jets ($R \sim 1$) than normally employed in LHC analyses ($R \sim 0.4$) are used to reconstruct boosted heavy particle decays. A larger distance parameter is chosen to capture the full kinematics of the decay already at moderate momenta of 200–400 GeV. The drawback of jets with large areas is unwanted contributions from the underlying event and from multiple proton-proton collisions in the same or adjacent bunch crossings (pileup). These lead to a worsening of the resolution in quantities used to identify the substructure of jets, like the jet mass. Jet grooming and pileup removal algorithms have been developed to mitigate these effects. Grooming algorithms aim at removing soft and wide-angle radiation, therefore not only reducing the effects from the underlying event but also reducing the sensitivity to the details of fragmentation. Pileup removal algorithms are designed to identify and subtract contributions from a different interaction vertex, by eliminating uncorrelated radiation from jets. A combination of these techniques often leads to the best overall performance and it is an ongoing effort to understand the interplay of pileup removal, grooming and tagging algorithms.

The theoretical and algorithmic developments have been made possible thanks to advances in experimental methods. New technologies, like silicon pixel detectors, high-resolution tracking detectors in conjunction with strong magnetic fields, highly granular calorimeters with low electronic noise and lightweight materials for detector structures with little dead material inside the active detector volume have enabled increasingly precise jet measurements and studies of internal jet structure. Modern particle detectors at the LHC are equipped with many layers of high-resolution tracking detectors, strong and very homogeneous magnetic fields and finely segmented calorimeters with an excellent energy resolution. With these technologies, the ATLAS and CMS detectors[2] are equipped to track and reconstruct individual particles produced in high energy collisions. On average about 60% of a jet's momentum is carried by charged hadrons, photons account for about 25% of the total jet momentum and the remaining 15% can be attributed to long-lived neutral hadrons [70]. With increasing jet energy, the particle multiplicity increases, and also the fraction of the jet's momentum carried by soft particles. For example, on average 50% of the momentum of a 50 GeV jet is carried by particles with a momentum less than 5% of the jet's momentum. It is therefore crucial to ensure that particles with energies down to $\mathcal{O}(100\,\text{MeV})$ can be reconstructed in order to retain the full information on a jet's kinematics and internal structure.

As important as the reconstruction of the total jet energy is the measurement of the jet constituent multiplicity and their angular distributions. While charged particles

[2]The ALICE and LHCb detectors are also well-equipped to perform jet substructure studies. While these experiments do not have access to boosted massive particles due to their data rate (ALICE) or acceptance (LHCb), they are performing many interesting QCD studies with jet substructure. This review will be focused on ATLAS and CMS, but the future of jet substructure will involve key contributions from all four LHC experiments.

can be efficiently reconstructed as tracks, neutral particles only develop showers in the calorimeters and the possibility to resolve two separate showers depends on the granularity of the calorimeter and the lateral shower development. Hence, it becomes more difficult to separate two adjacent particles in dense environments, such as high momentum jets, and the situation is aggravated by the presence of hadronic showers from charged hadrons. Often it is impossible to build one calorimeter cluster per neutral particle. A way to improve the angular resolution in jet substructure analyses is to combine measurements from the tracking detectors and calorimeters. Using combined detector measurements as input to jet algorithms, for example using a particle flow approach [71–75], results in improved resolutions of jet substructure observables, compared to using only tracks or only calorimeter clusters.

An important aspect of experimental analyses at the LHC is the calibration of jets, necessitated by the non-compensating nature of hadron calorimeters, suppression of electronic noise, tracking inefficiencies, dead material in front of calorimeters, the influence of pileup and other effects. While the calibration of the total jet energy scale is an important aspect in all analyses using jets, the precise knowledge of the jet mass scale and the detector response to jet substructure observables and jet tagging algorithms is specific to jet substructure analyses. Calibrating the jet energy scale results in a change of the magnitude of the jet's four-momentum, where the jet mass scale comprises an additional degree of freedom that can not be constrained by the typical methods of balancing a jet with a well-calibrated reference object. The jet mass scale is usually calibrated using jets from fully-merged, highly boosted $W \to q\overline{q}'$ decays, facilitating a calibration of the peak position in the jet mass distribution. Measurements of the jet mass distribution from light quark and gluon jets, as well as from fully-hadronic highly-boosted W, Z and t decays allow for precise tests of the modelling of perturbative and non-perturbative effects in jet production. Similar measurements can also be used to study the detector response to jet substructure observables and their modelling in simulation. A mis-modelling of variables used for tagging, either in the detector simulation or on the level of the underlying physics, can result in a wrong estimation of the tagging efficiency or the misidentification rate, with important consequences for measurements. In order to overcome this limitation, measurements of tagging efficiencies and misidentification rates are performed in samples enriched with the particle decays in question. While these measurements do not help to understand the cause of the mis-modelling or to improve the description of jet substructure distributions, they can be used to correct the efficiencies in simulation. It is these measurements that have enabled the use of jet substructure taggers in numerous physics analyses since the beginning of data taking at the LHC. The increased statistical precision from a data sample corresponding to about 150 fb^{-1} per experiment at a centre-of-mass energy of 13 TeV can now be used to improve our understanding of the detector response to jet substructure algorithms, the underlying physics and the performance differences of taggers.

2.2 General Considerations

2.2.1 Heavy Particle Decays

Detailed knowledge of the decay properties of heavy particles in the SM is crucial for devising and understanding jet substructure applications. Here and elsewhere, heavy refers to masses at the order of the W boson mass or higher. Thus, heavy SM particles refer to the electroweak gauge bosons W and Z, the Higgs boson H, and the top quark t.

All of these particles have hadronic and leptonic decay modes. The top quark is the only fermion in this list and takes a special role, since its decay is governed by the W boson decay, $t \to Wb$. The boson decays in the SM are described by two-body kinematics,[3] while the top quark features a three-body decay. However, since in this case the narrow width approximation for the W boson is accurate with sufficient precision, the top quark decay can be treated as two subsequent two-body decays.

The total width Γ of a particle can be expressed as a sum over all partial widths, $\Gamma = \sum_i \Gamma_i$. This relation can be used to express the branching fraction $\mathcal{B}_{X\to Y}$ of a particle X to final state Y, i.e. the probability that the particle decays into a given final state, which is given by

$$\mathcal{B}_{X\to Y} = \frac{\Gamma_{X\to Y}}{\Gamma} . \qquad (2.1)$$

For all heavy SM particles, the largest branching fraction is given for hadronic decays, $\mathcal{B}_{X\to\mathrm{had}}$, making the hadronic decay channel indispensable in searches for new physics, but also for SM measurements.

The lifetime τ of a particle is given by the inverse of the total width, $\tau = 1/\Gamma$. The heavy SM particles have lifetimes smaller than $\mathcal{O}(10^{-20}\,\mathrm{s})$, making all decays prompt (i.e. not observable in an experimental setup) and thus their presence can be inferred only from the measurement of the decay particles.

2.2.2 W Boson Decay

The W boson has been discovered in 1983 by the UA1 and UA2 experiments at CERN [76, 77]. Its mass has been measured at LEP and the Tevatron with a value of $m_W = 80.385 \pm 0.015$ GeV [78–80]. When neglecting fermion mass effects, the partial widths of the W^+ boson can be obtained by counting arguments. The partial width at leading order (LO) for the decay into a pair of fermions $W \to f\bar{f}'$ can be readily calculated when neglecting fermion mass effects,

[3] The only exceptions are the $H \to WW^*$ and $H \to ZZ^*$ processes, with subsequent decays of the EW gauge bosons, which are four body decays. These decays are not discussed in detail here.

$$\Gamma_{W\to f\bar{f}'} = C\frac{G_F M_W^3}{6\sqrt{2}\pi}\,, \tag{2.2}$$

where G_F is the Fermi constant and M_W is the mass of the W boson. The colour factor C is 1 for decays into leptons and 3 for decays into quarks, thus one obtains at tree level

$$\frac{\Gamma_{W\to\text{hadrons}}}{\Gamma_{W\to\text{leptons}}} = \frac{\mathcal{B}_{W\to\text{had}}}{\mathcal{B}_{W\to\text{lep}}} = \frac{6}{3}. \tag{2.3}$$

The W boson decays twice as often to hadrons as to leptons. Higher order corrections and fermion masses can affect the numerator and denominator in (2.3) differently, leading to small deviations from this result. Known corrections include one-loop quantum electrodynamic (QED) and EW corrections for massless and massive fermions [81–88], one-loop QCD corrections for massive quarks [89, 90], QCD corrections up to four loops for massless [91–93] quarks, where the two- and three-loop corrections include quadratic quark mass effects [94], and mixed EW/QCD corrections [95]. Numerical results for the calculated partial widths including all known corrections are $\Gamma_{W\to\text{leptons}} = 680.34 \pm 0.05$ MeV and $\Gamma_{W\to\text{hadrons}} = 1409.4 \pm 0.8$ MeV, resulting in a total width of $\Gamma_W = 2089.7 \pm 0.8$ MeV [78]. The predicted branching fraction of $\mathcal{B}_{W\to\text{had}} = 67.45 \pm 0.04\%$ is about one percent larger than the LO result. These predictions agree very well with the combination of the LEP and the Tevatron measurements, $\Gamma_W = 2085 \pm 42$ MeV and $\mathcal{B}_{W\to\text{had}} = 67.60 \pm 0.27\%$ [78–80]. It is noteworthy that the decay $W^+ \to c\bar{b}$ is Cabibbo-suppressed with a factor of $|V_{cb}|^2$, which is about $1.7 \cdot 10^{-3}$ [78]. This results in $\mathcal{B}_{W\to cb} \approx 5 \cdot 10^{-4}$, and thus the contribution from b quarks to the decay of the W boson is small enough to be neglected in all practical uses of jet substructure.

The angular distribution of the fermions from the W boson decay depends on the W boson polarisation. For W^+ decays, the angular distribution is at Born level [96]

$$\frac{1}{\sigma}\frac{d\sigma}{d\cos\theta^*} = f_+\frac{3}{8}\left(1+\cos\theta^*\right)^2 + f_-\frac{3}{8}\left(1-\cos\theta^*\right)^2 + f_0\frac{3}{4}\sin^2\theta^*\,. \tag{2.4}$$

The decay angle θ^* is defined in the W rest frame and is the angle between the charged lepton (or the quark) and the W flight direction in the laboratory rest frame. The fractions f_+ and f_- refer to transversely polarised W^+ bosons with helicities $+1$ and -1, respectively. The fraction of longitudinally polarised W^+ bosons is given by f_0. For W^- bosons the fractions f_- and f_+ are interchanged in (2.4). Since the quark charge is impossible to reconstruct experimentally, only the absolute values are accessible for hadronic decays. The angular distribution can then be written as [97]

$$\frac{1}{\sigma}\frac{d\sigma}{d|\cos\theta^*|} = f_\pm\frac{3}{4}\left(1+|\cos\theta^*|^2\right) + f_0\frac{3}{2}|\sin\theta^*|^2\,, \tag{2.5}$$

where $f_\pm = f_- + f_+$ has been used. The helicity composition of W bosons depends strongly on their production mechanism. For example, in certain BSM scenarios only

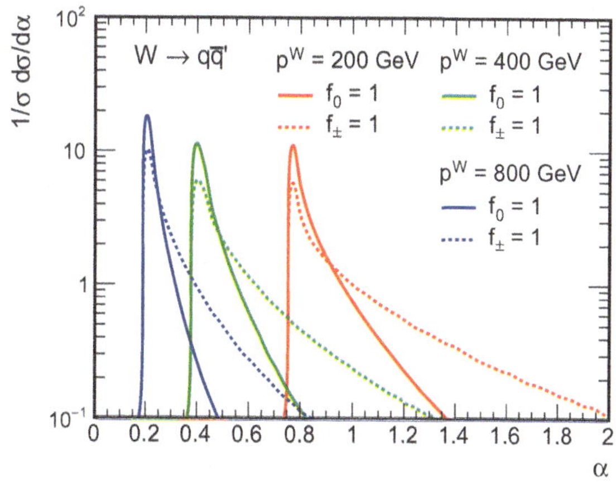

Fig. 2.3 Distribution of the opening angle α between the two quarks from W decays, calculated in the laboratory rest frame. Distributions are shown for longitudinal (solid lines) and transversal (dashed lines) W boson polarisations, for different momenta p^W

longitudinally polarised W bosons are produced via the decay of a heavy resonance. The helicity fractions for W bosons produced via SM processes at the LHC depends on the transverse momentum of the W boson and receives non-negligible QCD corrections [98–101]. At high transverse momenta, the W^- bosons are predominantly left-handed and W^+ bosons are mostly right-handed [102–104].

The opening angle α between two quarks from the W boson decay obtained in the laboratory rest frame is the figure of merit for jet substructure applications, as it controls the degree of collimation. The distribution of α is shown in Fig. 2.3 for longitudinally and transversely polarised W bosons with three different simulated values of the momentum. For transverse polarisations, the distributions have more pronounced tails due to decays collinear to the W boson flight direction. This can lead to differences in identification efficiencies in jet substructure analyses, due to lost decay products. Additionally, the shapes of jet substructure observables which take angular correlations into account can be different for different polarisation states.

2.2.3 *Z Boson Decay*

The Z boson was discovered in 1983 by the UA1 and UA2 collaborations [105, 106]. With a mass of $m_Z = 91.1875 \pm 0.0021$ GeV [107], it is about 10 GeV heavier than the W boson. At LO in EW perturbation theory, the decay width of the Z boson into a fermion-antifermion pair is

$$\Gamma_{Z\to f\bar{f}} = C\frac{G_F M_Z^3}{6\sqrt{2}\pi}(|g_{V,f}|^2 + |g_{A,f}|^2)\,, \tag{2.6}$$

where fermion masses have been neglected. The mass of the Z boson is given by M_Z, and $g_{V,f}$ and $g_{A,f}$ are the vector and axial-vector couplings of the Z boson to the fermion f, respectively. Due to electroweak unification, the coupling of the Z boson to fermions is very different from the one of the W boson. Inserting $g_{V,f}$ and $g_{A,f}$, one obtains the following branching fractions at LO: $\mathcal{B}_{Z\to\nu\bar{\nu}} = 20.4\%$, $\mathcal{B}_{Z\to\ell^+\ell^-} = 10.2\%$ and $\mathcal{B}_{Z\to\text{had}} = 69.2\%$. Again, the branching fraction to hadrons is dominant, with more than two thirds of all Z bosons decaying hadronically. Note that there is a sizeable branching fraction of 15.2% into $b\bar{b}$, which often results in a non-negligible background in analyses targeting the $H \to b\bar{b}$ decay. Owing to the Z bosons' importance in electroweak precision tests, higher order corrections to the partial widths have been calculated in EW two-loop order [108–111], including mixed EW/QCD one-loop diagrams [112–116], leading three- and four-loop corrections in the large top quark mass limit [117–124], factorisable final state two-loop QED and four-loop QCD radiation [125–127], and mixed EW/QCD one-loop non-factorisable vertex corrections [128–133]. Including all corrections, the branching fractions become $\mathcal{B}_{Z\to\nu\bar{\nu}} = 20.105 \pm 0.004\%$, $\mathcal{B}_{Z\to\ell^+\ell^-} = 10.092 \pm 0.002\%$ and $\mathcal{B}_{Z\to\text{had}} = 69.804 \pm 0.059\%$ [111], in excellent agreement with experimental measurements [15].

Similar to the W boson case, the angular distribution in the centre-of-mass frame of the decay fermions can be calculated [134],

$$\begin{aligned}\frac{1}{\sigma}\frac{d\sigma}{d\cos\theta^*} &= f_-\frac{3}{8}\left(1+\cos^2\theta^* - \frac{2(c_L^2-c_R^2)}{c_L^2+c_R^2}\cos\theta^*\right)\\ &+ f_+\frac{3}{8}\left(1+\cos^2\theta^* + \frac{2(c_L^2-c_R^2)}{c_L^2+c_R^2}\cos\theta^*\right)\\ &+ f_0\frac{3}{4}\sin^2\theta^*\,, \end{aligned} \tag{2.7}$$

where the angle θ^* is defined equivalent to (2.5). The couplings to left- and right-handed chiral states are given by c_L and c_R, respectively. Since the Z boson couples to both, left- and right-handed fermions, in principal the decay angle distributions of the outgoing quarks differs for left- and right-handed Z polarisation states. However, due to the inability of distinguishing quarks from anti-quarks, only absolute values $|\cos\theta^*|$ can be measured and the decay angle distribution becomes identical for left- and right-handed polarisations,

$$\frac{1}{\sigma}\frac{d\sigma}{d|\cos\theta^*|} = f_\pm\frac{3}{4}\left(1+|\cos\theta^*|^2\right) + f_0\frac{3}{2}|\sin\theta^*|^2\,, \tag{2.8}$$

where the dependence on c_L and c_R drops out. It follows that the decay angle distribution is identical to the one for W bosons, (2.5). Thus, similar as for W bosons, transversely polarised Z bosons show a more pronounced tail towards larger opening angles α when compared to longitudinally polarised Z bosons. The only difference between the α distribution from W and Z boson decays arises from the mass difference m_W and m_Z, resulting in a shift towards somewhat larger values of α in the case of Z bosons at a given momentum (the shift is between 0.1 and 0.01 radians for momenta between 200 and 2000 GeV). Similar as for W bosons, the differences in the angular distributions between longitudinal and transverse polarisations can lead to differences in efficiencies in jet substructure analyses, thus depending on the production mechanism of the Z boson.

2.2.4 Higgs Boson Decay

The Higgs boson is the youngest member of the family of SM particles. It has been predicted already in 1964 [3, 4], but its existence has been verified only in 2012 by the ATLAS and CMS Collaborations [1, 2]. It has a mass of $m_H = 125.09 \pm 0.24$ GeV [135] and is therefore the second-heaviest particle of the SM. While it has not been measured with a precision comparable to the one achieved for the W and Z bosons, all evidence points to it being the SM Higgs boson [7]. Its total and partial decay widths have been calculated in the SM up to, and including four-loop massless QCD corrections and two-loop electroweak corrections, see [136–138] for a complete discussion and references therein. Its branching fractions are shown in Fig. 2.4. The numerical values of the branching fractions together with their theoretical uncertainties are given in Table 2.1. The total width, assuming no invisible decays,[4] is $\Gamma_H = 4.10 \pm 0.06$ MeV, for $m_H = 125.09$ GeV.

The fully hadronic decay of the H boson comprises of direct decays into two quarks, decays into two gluons through loop effects, and decays into pairs of W and Z bosons with subsequent hadronic decays. In total, the hadronic branching fraction is $\mathcal{B}_{H\to\text{had}} = 80.25 \pm 0.86\%$. Especially relevant for jet substructure applications is the $H \to b\bar{b}$ decay with a branching fraction of $58.09 \pm 0.73\%$. Here, the presence of two b quarks facilitates experimental identification using the long lifetime of B hadrons. The second largest hadronic decay is $H \to WW^* \to 4$ quarks, with a branching fraction of $9.79 \pm 0.15\%$. This decay can in principle be distinguished from QCD branchings due to its four-prong nature, with two quarks having an invariant mass around the W boson mass. Equivalent considerations can be made for the $H \to ZZ^* \to 4$ quarks decay, however, this decay has only a branching fraction of $1.29 \pm 0.02\%$. The third largest branching fraction for hadronic decays originates from the loop-induced $H \to gg$ decay, with a rate of $8.18 \pm 0.42\%$. However, at high boosts, this decay is nearly indistinguishable from QCD branchings and therefore has not been specifically targeted in substructure analyses so far.

Due to the spin-0 nature of the Higgs boson, it has only a single polarisation state and exhibits an isotropic decay in the centre-of-mass frame. Therefore different production mechanisms do not introduce an angular dependence of the decay fermions.

---

[4]Except for invisible decays in the SM, namely $H \to ZZ^* \to \nu\bar{\nu}\nu\bar{\nu}$ with a branching fraction of 0.11%.

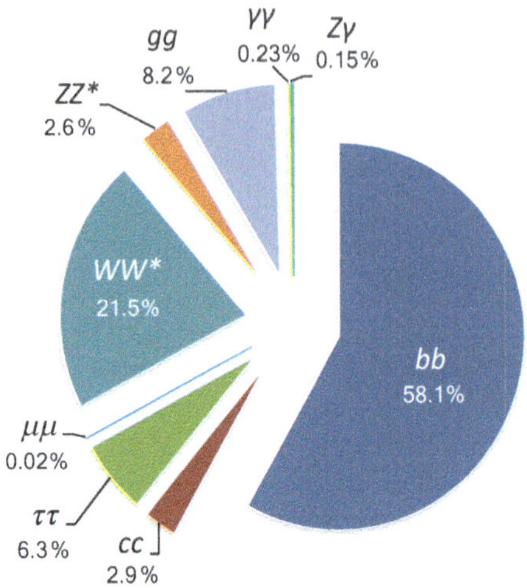

Fig. 2.4 Higgs decay branching fractions in the SM for $m_H = 125.09$ GeV. Taken from [139]

Table 2.1 Branching fractions of the H boson in the SM together with total theoretical uncertainties [137, 138], obtained for $m_H = 125.09$ GeV

Decay to two fermions	Decay to two bosons
$\mathcal{B}_{H\to b\bar{b}} = 58.09 \pm 0.73\%$	$\mathcal{B}_{H\to WW^*} = 21.52 \pm 0.33\%$
$\mathcal{B}_{H\to\tau^+\tau^-} = 6.26 \pm 0.10\%$	$\mathcal{B}_{H\to gg} = 8.18 \pm 0.42\%$
$\mathcal{B}_{H\to c\bar{c}} = 2.88^{+0.16}_{-0.06}\%$	$\mathcal{B}_{H\to ZZ^*} = 2.64 \pm 0.04\%$
$\mathcal{B}_{H\to s\bar{s}} = 0.025 \pm 0.001\%$	$\mathcal{B}_{H\to\gamma\gamma} = 0.227 \pm 0.005\%$
$\mathcal{B}_{H\to\mu^+\mu^-} = 0.0217 \pm 0.0004\%$	$\mathcal{B}_{H\to Z\gamma} = 0.154 \pm 0.009\%$

2.2.5 *Top Quark Decay*

The top quark was discovered in 1995 by the CDF and D0 experiments at Fermilab [31, 32]. With a mass of $m_t = 173.34 \pm 0.76$ GeV [140] it is the heaviest particle in the SM. The dominant decay is $t \to Wb$, where the Wd and Ws decays are Cabibbo suppressed by the squares of the CKM matrix elements V_{td} and V_{ts}. Assuming validity of the SM and unitarity of the CKM matrix implies a value of $V_{tb} > 0.999$ [78] and therefore a branching fraction $\mathcal{B}_{t\to Wb}$ close to unity. The total width is $\Gamma_t = 1.34 \pm 0.01$ GeV, calculated including finite b quark mass effects, finite width effects [141, 142], one and two-loop QCD corrections [141, 143–146], and one-loop EW corrections. The large width implies a short lifetime of the top quark $\tau = 1/\Gamma_t$, which is about an order of magnitude smaller than the typical formation time of hadronic bound states. In other words, the top quark can be treated as a free quark in production and decay processes [147].

The top quark decay is a weak process and hence results in polarised decay particles. The angular distribution of the longitudinal and transverse W bosons from the decay of polarised top quarks is at leading order [148]

$$\frac{1}{\Gamma}\frac{\mathrm{d}\Gamma}{\mathrm{d}\cos\theta*} = \frac{1}{2}\left(1 + \frac{m_t^2 - 2m_W^2}{m_t^2 + 2m_W^2}\pi_t\cos\theta^*\right), \tag{2.9}$$

where the summation over W polarisations leads to the coefficient of the $\pi_t\cos\theta^*$ term. The angle θ^* is defined between the top quark spin orientation and the W boson flight direction in the top quark rest frame. The parameter π_t denotes the magnitude of the top quark polarisation, where $\pi_t = 0$ corresponds to unpolarised top quarks and $\pi_t = 1$ stands for fully polarised top quarks. For top quark pair production at the LHC, which proceeds mainly through the strong interaction, the top quarks are predicted to be unpolarised. Only a negligible value of $\pi_t = 0.003$ is generated by the weak interaction [149]. These predictions are in good agreement with current measurements [150, 151].

The helicity fractions of the W^+ boson from the $t \to bW^+$ decay are at leading order $f_0 = 1/(2 + y^2)$, $f_- = 2y^2/(2 + y^2)$ and $f_+ = 0$, where $y^2 = m_W^2/m_t^2$ [152]. Numerically, this gives $f_0 = 0.7$ and $f_- = 0.3$, and consequently a dominant fraction of the W bosons is longitudinally polarised. Calculated at NNLO, the helicity fractions become $f_0 = 0.687 \pm 0.005$, $f_- = 0.311 \pm 0.005$, and $f_+ = 0.0017 \pm 0.0001$ [153], in excellent agreement with experimental observations [154–157]. Note that these fractions are defined in the top quark rest frame. When calculating them in the laboratory rest frame a dependence on the p_T of the W boson is introduced [96], as shown in Fig. 2.5. At low p_T, the helicity fractions differ considerably from the calculation in the top quark rest frame, with $f_- \approx f_+ \approx 0.3$. At high p_T, the helicity fractions in the laboratory rest frame approach the helicity fractions obtained in the top quark rest frame.

2.2.6 *Kinematic Considerations*

The decay of a short-lived particle X with mass M decaying to two secondary particles a and b with masses m_a and m_b is best described in the centre-of-mass (CM) frame, where the total energy $\sqrt{s}$ is known, $\sqrt{s} = M$. Since the two decay particles will be back-to-back in the CM frame (see Fig. 2.6 (left)), their momenta fulfil $\mathbf{p}_a^* = -\mathbf{p}_b^*$ and $p^* = p_a^* = p_b^*$, where $p^* = |\mathbf{p}^*|$ is the magnitude of momenta. Starred quantities refer to energies and momenta expressed in the CM frame. It follows that $M = E_a^* + E_b^* = \sqrt{s}$, and the magnitude p^* can be expressed in terms of the particle masses,

$$p^* = \frac{1}{2M}\sqrt{(M^2 - (m_a - m_b)^2)(M^2 - (m_a + m_b)^2)}\,. \tag{2.10}$$

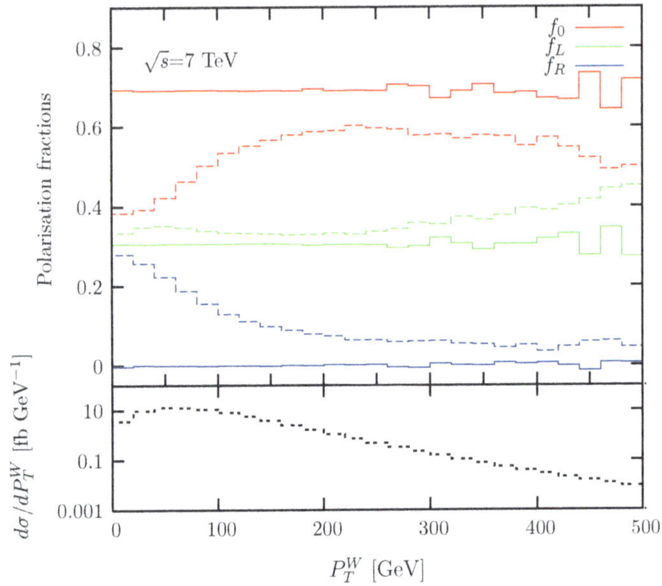

Fig. 2.5 Polarisation fractions for W^+ bosons from top quark decays, in the top quark rest frame (solid) and laboratory rest frame (dashed), as a function of W^+ boson p_T in the laboratory rest frame. The fractions f_L and f_R correspond to f_- and f_+ as used in the main text, respectively. Taken from [96]

Fig. 2.6 Kinematics of a two-body decay in the CM frame (left) and in the laboratory rest frame (right)

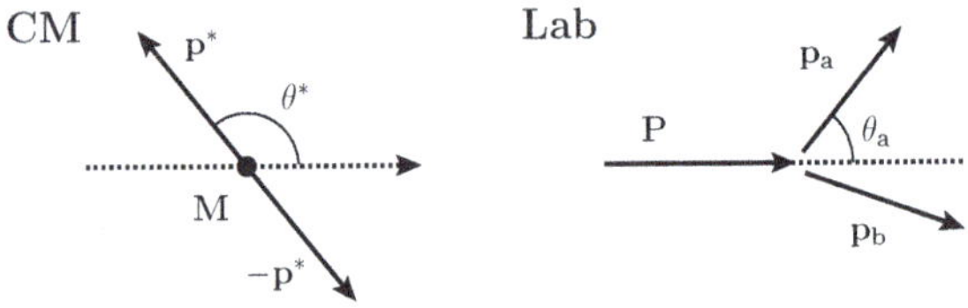

The decay angle θ_a^* in the CM frame denotes the angle of particle a with respect to the parent particle's direction of flight, as shown in Fig. 2.6. Relating this to the corresponding decay angle in the laboratory rest frame, one obtains

$$\tan\theta_a = \frac{\sin\theta_a^*}{\gamma\left(\beta/\beta_a^* + \cos\theta_a^*\right)} \tag{2.11}$$

where $\gamma = E/M$, $\beta = P/E$ and $\beta^* = p^*/E^*$, and $P = |\mathbf{P}|$ denotes the magnitude of the parent particle's momentum. The minimum of θ_a is obtained when $\theta^* = \pi/2$, and hence $\tan\theta_a = \beta_a^*/(\gamma\beta)$. This expression can be simplified if the masses of the daughter particles are negligible compared to M, which results in $p^* = E^* = M/2$ and $\beta^* = 1$. Consequently, one obtains

$$\tan\theta_{a,\min} = \frac{1}{\gamma\beta} = \frac{M}{P} \quad \rightarrow \quad \theta_{a,\min} \approx \frac{M}{P}, \tag{2.12}$$

where the last relation is obtained by taking only the first term of the Taylor expansion of $\tan\theta_{a,\min}$. Since $\theta^* = \pi/2$ it follows that $\theta_a = -\theta_b$, and thus the minimum value of the opening angle $\alpha = |\theta_a - \theta_b|$ can be approximated by

$$\alpha_{\min} \approx \frac{2M}{P}. \tag{2.13}$$

The minimum opening angle between the particles a and b observed in the laboratory rest frame decreases as $1/P$. Equation (2.13) holds irrespective of the chosen coordinate system, as no specific direction is needed for the derivation of $\alpha_{\min}$.

In a realistic experimental environment the relationship in (2.13) needs to be modified to preserve its usefulness. The reason is the coordinate system implied by the beam axis and the detector geometry. Typically, at the LHC one chooses a right-handed coordinate system, where the x-axis points to the centre of the LHC ring, the y axis points upwards, perpendicular to the LHC plane, and the z-axis points along the anti-clockwise beam direction. Angular distances are measured using the azimuthal angle ϕ, measured from the positive x-axis in the x-y plane, and the pseudorapidity defined as $\eta = -\ln[\tan(\theta/2)]$, where θ is the polar angle measured from the positive z axis in the y-z plane. The advantage of using η instead of θ is that differences in η are invariant under longitudinal Lorentz-boosts. Hence, the opening angle α is replaced by the angular distance $\Delta R = \sqrt{\Delta\phi^2 + \Delta\eta^2}$ in the azimuth-pseudorapidity plane, where $\Delta\phi = \phi_a - \phi_b$ is the distance in azimuthal angles and $\Delta\eta = \eta_a - \eta_b$ is the distance in pseudorapidity between particles a and b. This results in an angular distance measure invariant under boosts in the beam direction, which is also the reason why ΔR is used in jet finding algorithms (see Sect. 2.3). For centrally produced particles X ($\eta \approx 0$), with $P > 2M$ and $M \gg m_a, m_b$, the difference between α and ΔR is small. For increasing $|\eta|$, the value of ΔR becomes increasingly larger than α at fixed values of P. In the limiting case of $P = P_z$, it follows that $\Delta\phi = \pi$ and therefore $\Delta R > \pi$. The reason for this behaviour is that $\Delta\eta$ is not affected by longitudinal boosts, so only the size of the transverse component P_T is responsible for decreasing values of ΔR in the laboratory rest frame with respect to the CM frame. Hence, replacing P with P_T in equation (2.13) results in ΔR being invariant under variations of η for fixed values of P_T. This results in

$$\Delta R \approx \frac{2M}{P_T}, \tag{2.14}$$

which can be found extensively in the literature to approximate the angular distance between the decay products of a two particle decay. One should note that this is only an approximate relation, which in fact gives a lower bound on ΔR. The $1/P_T$ behaviour can also be altered by additional kinematic requirements on the decay particles. The relation (2.14) is only valid for small values of ΔR since the expansion in (2.12) has to hold. Consequently, $P_T \gg M$ needs to be fulfilled to ensure its validity. Practically, $P_T > 2M$ is sufficient to ensure that (2.14) gives a reliable lower bound on ΔR.

2.2.7 Kinematics of Vector and Higgs Boson Decays

The hadronic two-body decays $W/Z/H \rightarrow q\bar{q}^{(\prime)}$ are the most important applications of the above considerations in jet substructure analyses. In the following, results are presented from numerical calculations using realistic decay angle distributions. Quark mass effects are included, albeit small. The considerations are based on the kinematics of the plain $1 \rightarrow 2$ process with coloured quarks in the final state, not including radiation or hadronisation effects.

In realistic applications there exist a minimum detection threshold for the quarks from the boson decay. This threshold is introduced due to the inability to distinguish soft and wide-angle radiation from uncorrelated radiation in hadron-hadron collisions, such as contributions from the underlying event or pileup. Typically, p_T thresholds on the reconstructed quarks are applied, either directly or indirectly, when using jet substructure observables. The effect such a p_T threshold has on the detection efficiency depends on the decay angle distribution, and thus on the nature and polarisation state of the heavy particle. In Fig. 2.7 (left) the detection efficiency is shown for different quark p_T thresholds for longitudinal (W_L) and transverse (W_T) polarisations of the W boson, as a function of p_T of the W boson. For W_L bosons, the efficiency is larger than 90% for $p_T > 200$ GeV, even for quark p_T thresholds as large as 30 GeV, and quickly reaching the plateau at almost 100%. In contrast to this, the efficiency is much smaller for W_T, where it is 70–80% at low values of p_T, with only a moderate rise as a function of p_T. The efficiency never reaches 100%, even at very high values of p_T. The reason for this are decays collinear to the boson flight direction, i.e. $\theta^* \approx 0$ and $\theta^* \approx \pi$, which only occur for W_T states (see (2.5)). In this case the flight direction of one of the quarks will be anti-parallel to that of

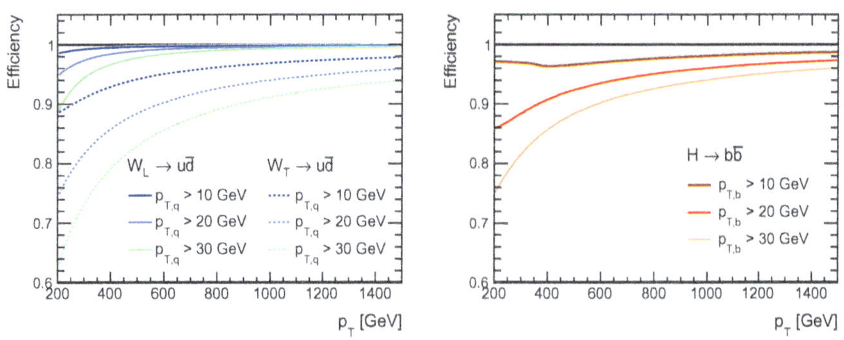

Fig. 2.7 Relative occurrence (efficiency) of both quarks from a W (left) and H (right) boson decay having a p_T larger than indicated, as a function of the boson p_T. For W bosons, results for longitudinal (W_L) and transverse (W_T) polarisations are shown

the W boson, resulting in a small value of quark p_{T} and a large angular distance ΔR between the two quarks. Consequently, when requiring the decay quarks to have a minimum p_{T}, a cut-off at large values of the ΔR distribution is introduced, and vice-versa. Since the decay angle distributions $\mathrm{d}\sigma/\mathrm{d}|\cos\theta^*|$ are identical for W and Z bosons, the results from Fig. 2.7 (left) are also valid for Z bosons, where the small correction due to the mass difference $m_W - m_Z$ leads to a negligible change. In Fig. 2.7 (right) the detection efficiency is shown for $H \to b\bar{b}$ decays. Since the distribution $\mathrm{d}\sigma/\mathrm{d}|\cos\theta^*|$ is flat for spin-0 particles, there are more collinear decays than in the W_{L} case, but less than for W_{T} bosons. This is reflected in the detection efficiency, which lies between the efficiencies obtained for W_{L} and W_{T} decays.

The angular distance ΔR between the two decay quarks is expected to decrease proportional to $1/p_{\mathrm{T}}$ of the parent particle. While the relation (2.14) gives the minimum value of ΔR, the tails of the ΔR distribution are more difficult to obtain, especially if detection thresholds on the quarks are included. The scaling of the ΔR distribution as a function of p_{T} can be readily estimated, though. Since any given decay angle θ^* will transform proportional to $1/\gamma\beta$, see (2.11), any given point of the ΔR distribution will scale as $1/p_{\mathrm{T}}$. Once the ΔR distribution is known for a given value value of p_{T}, this scaling can be used to calculate an expected value of ΔR as long as no additional kinematic requirements on the decay quarks are imposed. The relation

$$\int_0^{\Delta R_{90}} \mathrm{d}\Delta R \, \frac{\mathrm{d}\sigma}{\mathrm{d}\Delta R} \Big/ \int_0^{\infty} \mathrm{d}\Delta R \, \frac{\mathrm{d}\sigma}{\mathrm{d}\Delta R} = 0.9 \tag{2.15}$$

defines the 90th percentile ΔR_{90}, which denotes the value below which 90% of the values of ΔR can be found. For example, in the case of W_{L} decays, $\Delta R_{90} = 0.46$ for $p_{\mathrm{T}} = 500\,\mathrm{GeV}$ and hence 90% of all possible values of ΔR lie in the interval $[(160\,\mathrm{GeV})/p_{\mathrm{T}}, (230\,\mathrm{GeV})/p_{\mathrm{T}}]$. Since these values depend on the nature and polarisation of the decaying boson, the replacement of $2M$ with ρ_{90} in (2.14),

$$\Delta R_{90} = \frac{\rho_{90}}{p_{\mathrm{T}}^{\alpha}}, \tag{2.16}$$

can help to estimate the expected range of ΔR for a given boson decay. The parameter α has been introduced to allow for modifications from the $1/p_{\mathrm{T}}$ behaviour once p_{T} thresholds on the quarks are introduced. An overview of the ρ_{90} values for W, Z and H bosons and different polarisation states is given in Table 2.2.

The effect a p_{T} detection threshold has on the ΔR distribution is shown in Fig. 2.8, where both quarks from the W_{T} (left) and H (right) boson decay are required to have a minimum transverse momentum, $p_{\mathrm{T},q} > 20\,\mathrm{GeV}$. For comparison, the expected $1/p_{\mathrm{T}}$ scaling for the ΔR_{90} percentile is shown, as obtained for the ρ_{90} value determined at $p_{\mathrm{T}} = 1500\,\mathrm{GeV}$. While this function describes the behaviour of the ΔR distribution well if no kinematic requirements on the decay quarks are imposed, significant deviations from the $1/p_{\mathrm{T}}$ scaling are seen once a detection threshold is introduced. Especially at small values of p_{T}, the naive $1/p_{\mathrm{T}}$ scaling overestimates the angular distance significantly. When allowing for deviations of the form $1/p_{\mathrm{T}}^{\alpha}$,

Table 2.2 Numerical values of ρ_{90} and α for the calculation of the 90th percentile ΔR_{90}, as given in (2.16). The top part is obtained without a kinematic requirement on the decay quarks, the bottom part is obtained for $p_{\mathrm{T},q} > 20\,\mathrm{GeV}$

	W_{L}	W_{T}	Z_{L}	Z_{T}	H
	no p_{T} selection, $\alpha = 1$				
ρ_{90} [GeV]	230	420	255	460	550
	$p_{\mathrm{T},q} > 20\,\mathrm{GeV}$				
ρ_{90} [GeV]	160	67	180	74	150
α	0.95	0.76	0.95	0.76	0.83

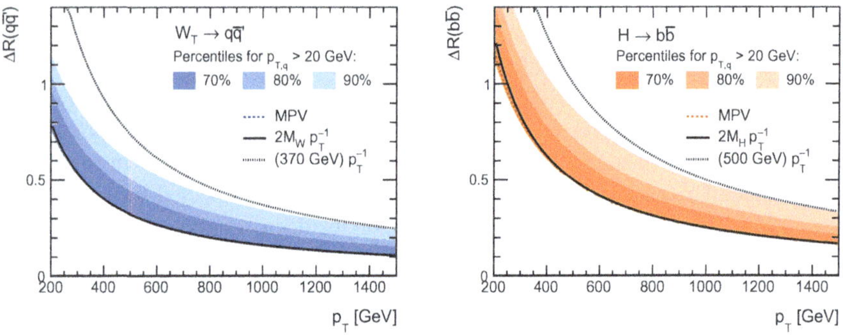

Fig. 2.8 Angular distance ΔR between the two quarks from the decay of transversely polarised W (left) and a H bosons (right), as a function of the boson p_{T}. The transverse momenta of the two quarks from the decay are required to be $p_{\mathrm{T},q} > 20\,\mathrm{GeV}$. The fraction of events contained within a given interval in ΔR are shown by shaded areas, the most probable value (MPV) is depicted by a dashed line. For comparison, also shown are the naive expectations $2M_W/p_{\mathrm{T}}$ and $2M_H/p_{\mathrm{T}}$ (solid lines), and the functions $(370\,\mathrm{GeV})/p_{\mathrm{T}}$ and $(500\,\mathrm{GeV})/p_{\mathrm{T}}$ (dotted lines)

see (2.16), parametrisations can be found that describe the shapes of the percentiles well. An example is given in the bottom part of Table 2.2, where the parameters for the ΔR_{90} percentiles are given for the requirement $p_{\mathrm{T},q} > 20\,\mathrm{GeV}$. Another observation from Fig. 2.8 is that the minimum value of ΔR can be accurately predicted by (2.14), which also coincides with the MPV. Only for H decays with $p_{\mathrm{T}} < 300\,\mathrm{GeV}$ the approximation (2.14) starts to deviate from the realistic shape, since for small values of p_{T} the approximation of small angles in (2.12) is not valid any more.

2.2.8 *Kinematics of Top Quark Decays*

In the SM, the top quark decays to bW with a branching fraction larger than 99.9%. The subsequent hadronic decay of the W boson leads to the decay chain $t \to bW \to bq\bar{q}'$, with three quarks in the final state. The narrow width approximation, where the top quark and the W boson are on-shell, is sufficiently accurate to study the kinematics of this decay for substructure related applications. Thus, the decay can be modelled by two consecutive two-body decays, where the polarisation states of the top quark and the W boson need to be taken into account.

At the LHC, $t\bar{t}$ production results in unpolarised top quarks, which leads to a uniform decay distribution in the decay angle in the CM frame. This transforms into a uniform energy distribution of the b quarks and W bosons in the laboratory rest frame, for mono-energetic top quarks. The minimum and maximum energies are obtained for $\theta^* = 0$ or π, thus the distributions of the b quark and W boson energies, E_b and E_W, are flat within the kinematically allowed boundaries,

$$\gamma(E'_{b,W} - \beta p^*) < E_{b,W} < \gamma(E'_{b,W} + \beta p^*), \tag{2.17}$$

where p^* is given in (2.10) and $E'^2_{b,W} = p^{*2} + m^2_{b,W}$. The minimum of E_b becomes smaller for increasing top quark momenta p and approaches m_b for large values of p, above 1500 GeV. At $p = 550$ GeV the minimum of E_b is about 10 GeV. The result is that even at very high values of p very soft b quarks will emerge from the top quark decay, which are not detectable. However, since the E_b distribution is flat, the relative occurrence of very soft b quarks will decrease with increasing p.

The decay angle distribution of the two quarks from the W boson decay follows from an admixture of longitudinal and transverse W polarisation states. Convoluted with the flat distribution of E_W, this leads to a peak in the quark momentum distribution and a steeply falling tail with a maximum value slightly above p. Overall, the distributions of the light quark momenta are softer than the distribution of the b quark momentum. Consequently, when introducing a detection threshold on p_{T} of the b and light quarks, a larger loss in detection efficiency originates from the light quarks compared to the b quark, as shown in Fig. 2.9. When requiring $p_{\mathrm{T},q} > 20$ GeV for all three quarks, the detection efficiency is about 80% at top quark $p_{\mathrm{T}} = 400$ GeV and increases to 90% for $p_{\mathrm{T}} > 800$ GeV. Even at very high p_{T} of 1500 GeV the efficiency is only 95%, owing to decays with $\theta^* \approx 0$ or π. The detection efficiency is very sensitive to the exact value of the quark p_{T} threshold. For example, at $p_{\mathrm{T}} = 400$ GeV,

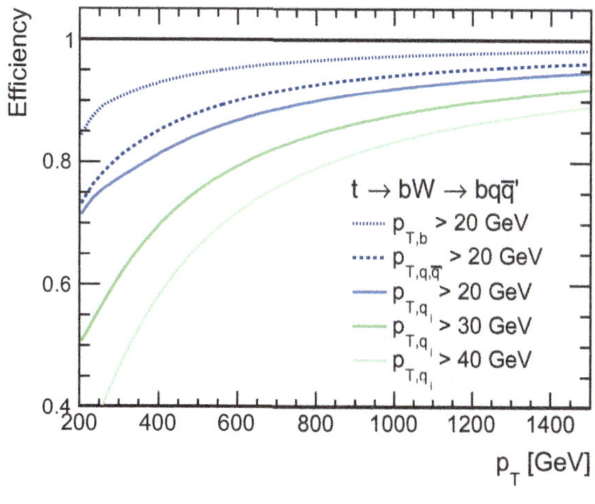

Fig. 2.9 Relative occurrence (or efficiency) of the b quark (dotted), the two quarks from the W decay (dashed) or all three quarks (solid lines) having a p_{T} larger than indicated, as a function of top quark p_{T}

Table 2.3 Numerical values of ρ and α for top quark decays for the approximation in (2.19). Shown are the values for the minimum value that $\Delta R_{\max}$ can take, and the 70th, 80th and 90th percentiles of $\Delta R_{\max}$. All values are obtained for a quark threshold of $p_{\mathrm{T},q} > 20\,\mathrm{GeV}$

	min	70%	80%	90%
ρ [GeV]	180	190	200	210
α	0.92	0.84	0.83	0.815

a requirement of $p_{\mathrm{T},q} > 30\,\mathrm{GeV}$ leads to a drop of the efficiency to 70%, and with $p_{\mathrm{T},q} > 40\,\mathrm{GeV}$ the efficiency is only 60%.

Similar to the case of vector boson and H decays, a p_{T} threshold on the quarks leads to a change in the distribution of the angular distance between the quarks. Since there are three quarks involved, the relevant quantity is the maximum angular distance

$$\Delta R_{\max} = \max\left[\Delta R(b, q), \Delta R(b, \overline{q}'), \Delta R(q, \overline{q}')\right], \tag{2.18}$$

which represents a proxy for fully merged final states. Obviously, approximation (2.14) will not yield an accurate prediction for $\Delta R_{\max}$. Surprisingly, it still gives a relatively good estimate for the MPV of $\Delta R_{\max}$, as can be seen in Fig. 2.10. However, the $1/p_{\mathrm{T}}$ scaling does not give an accurate description of the minimum value $\Delta R_{\max}$ can take, and also cannot be used to predict a given percentile of the $\Delta R_{\max}$ distribution. A better phenomenological approximation is obtained by modifying the scaling with an exponent α, similar to (2.16),

$$\Delta R_{\max} = \frac{\rho}{p_{\mathrm{T}}^{\alpha}}. \tag{2.19}$$

Numerical values for ρ and α are given in Table 2.3, for the three percentiles shown in Fig. 2.10. The values given approximate the shape of the three percentiles to within 1% in the range $200 < p_{\mathrm{T}} < 1500\,\mathrm{GeV}$.

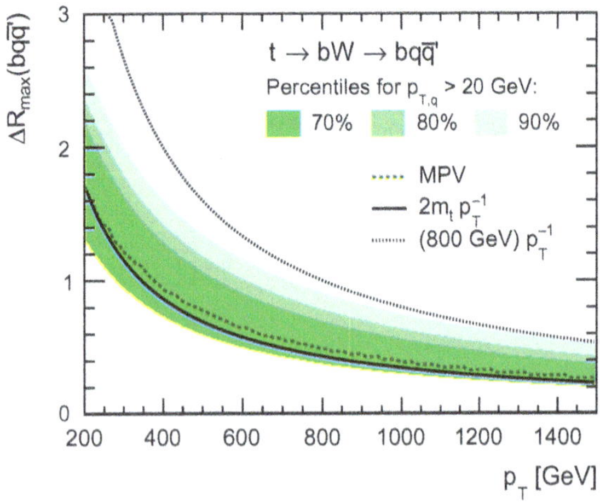

Fig. 2.10 Maximum angular distance $\Delta R_{\max}$ of the three quarks from the hadronic top quark decay, as a function of the top quark p_{T}. The transverse momenta of the three quarks b, q and $\overline{q}'$ are required to be $p_{\mathrm{T},q} > 20\,\mathrm{GeV}$. The fraction of events contained within a given interval in $\Delta R_{\max}$ are shown by shaded areas, the MPV is depicted by a dashed line. For comparison, also shown are the expressions $2m_t/p_{\mathrm{T}}$ (solid line) and $(800\,\mathrm{GeV})/p_{\mathrm{T}}$ (dotted line)

The MPV of the ΔR_{max} distribution is about twice as large as the corresponding MPV of the ΔR distribution for W boson decays, over the full p_{T} range. The same is true for the 70th, 80th and 90th percentiles, showing the need for larger jets in analyses targeting top quark decays, compared to vector boson or H decays.

2.3 Jet Algorithms

The colour charges of quarks and gluons lead to collimated sprays of particles as their footprint in high energy collisions. The first detailed proposal of measuring these jets in e^+e^- collisions has been made more than 40 years ago [47], and since then jets have become an indispensable tool in particle physics. However, there is no unique way of defining a jet. Instead, many jet definitions have been invented, tested, modified, re-invented or discarded. Nowadays, with the advent of the LHC, only a small set of jet algorithms is in use. These are the three sequential clustering algorithms k_{T} [53, 54], anti-k_{T} [57] and CA [55, 56], which have proven to be practical in numerous theoretical and experimental works. They all have in common that they fulfil the criteria of the Snowmass accord [158]. It was a long struggle to devise jet definitions fulfilling all the Snowmass criteria simultaneously, being IRC safe [159], while computationally feasible [160]. The reader is referred to [59, 161–163] for a comprehensive overview of the developments that led to modern clustering algorithms. With the development of the FastJet [164] package, jet algorithms got a common framework which resulted in a new standard for jet physics at the LHC with numerous applications [165–174]. With these developments the situation became satisfactory for jet algorithms themselves, fulfilling all experimental and theoretical requirements.

The much younger field of jet substructure is adolescent with developments in progress and new ideas still shaping the field. Some of the algorithms in use in jet substructure analyses have been developed by maximizing the sensitivity of a search for a given signal model, or by optimising the efficiency versus the misidentification rate for the hadronic decay of a heavy particle. While in some cases the first developments lacked simplicity, experimental feasibility, or calculability in all orders in perturbation theory, these already showed the huge potential of jet substructure for data analysis at the LHC. The ongoing developments have guided us and helped to gain an understanding of the possibilities that jet substructure offers. These may be a gain in performance, a deeper understanding of the underlying partonic dynamics and hadronisation processes, or resilience against experimental effects. The use of new developments in substructure techniques have been shown to improve the sensitivity and physics potential of LHC analyses. These developments have been facilitated largely by the availability of extensions to the FastJet package, which is now the standard for new developments in this area. The common framework allows for an easier comparison of different methods and a faster integration time of new developments in analyses.

Often new algorithmic developments are closely related to advances in calculations, which can also prompt the introduction of novel variables. Work is ongoing in order to classify existing and novel methods and evaluate their usefulness, where the largest scientific gain is obtained by methods well understood theoretically as well as experimentally.

Methods for substructure analyses usually build up on existing jet algorithms. In the following, sequential recombination algorithms are shortly reviewed because of their importance for jet substructure. Traditional cone algorithms [175–178], where the seedless infrared-safe cone (SISCone) [48] is an example of an IRC safe algorithm, play an insignificant role in this field and are not considered. Instead, two recent developments, the XCone[5] [179, 180] and the Georgi [181–183] algorithms, are described briefly.

2.3.1 Sequential Clustering Algorithms

All sequential recombination algorithms start with an input list of particles, also denoted as pseudojets.[6] The clustering continues the processing until the input list is empty. The distance measure d_{ij} between pseudojets i and j, and distance $d_{i\mathrm{B}}$ between pseudojet i and the beam axis are defined as

$$d_{ij} = \min\left(p_{\mathrm{T},i}^{2k}, p_{\mathrm{T},j}^{2k}\right) \frac{\Delta R_{ij}^2}{R^2}, \tag{2.20}$$

$$d_{i\mathrm{B}} = p_{\mathrm{T},i}^{2k}. \tag{2.21}$$

Here, $p_{\mathrm{T},i}$ is the transverse momentum of pseudojet i and $\Delta R_{ij}^2 = (y_i - y_j)^2 + (\phi_i - \phi_j)^2$ is the geometric distance in rapidity y and azimuth ϕ between the pseudojets i and j. The value of the parameter k defines the class of the jet algorithm: $k = 1$ for the k_T algorithm [53, 54], $k = 0$ for the CA algorithm [55, 56] and $k = -1$ for the anti-k_T algorithm [57]. Given a list of pseudojets, the algorithm proceeds with the following steps:

1. Compute the distances d_{ij} for all possible pairs of pseudojets and the beam distances $d_{i\mathrm{B}}$, using (2.20) and (2.21).
2. Find the smallest d_{ij} and the smallest $d_{i\mathrm{B}}$. If $d_{ij} < d_{i\mathrm{B}}$, combine pseudojets i and j to pseudojet l. Remove i and j from the list and add l to the list of pseudojets. If $d_{i\mathrm{B}} < d_{ij}$, call pseudojet i a jet and remove it from the input list.
3. Repeat the steps above until no pseudojets are left.

---

[5] The name is derived from exclusive cone algorithm.

[6] Following the FASTJET [164] terminology, a pseudojet denotes an entity entering the jet clustering. This can be coloured partons, stable particles, reconstructed detector objects or combined objects from a previous clustering iteration.

Despite the fact that the k_T, anti-k_T and CA algorithms are identical except for the choice of the parameter k, each of them exhibits a very different behaviour.

The k_T algorithm starts with clustering soft and collinear objects, harder objects are clustered at later iteration steps. This leads to irregular jet boundaries with on average larger jet areas [184] than those of other algorithms with the same value of R [185].

The anti-k_T algorithm starts with the hardest object and accumulates all objects within a distance smaller than R into a jet. If there is no harder jet within the jet's vicinity with a distance smaller than R, the resulting jet is circular in the $y - \phi$ plane and has an area of exactly πR^2.[7] Anti-k_T jets also exhibit the smallest amount of back-reaction [59] among the three clustering algorithms discussed here. Because of these features anti-k_T jets became the standard choice for jet analyses at the LHC. Since the softest particles are clustered last in the anti-k_T algorithm, it is unsuited for substructure taggers which include decomposition steps.

The CA algorithm is insensitive to the objects' transverse momenta and builds jets using geometrical proximity in the $y - \phi$ plane as the only criterion. Like the k_T algorithm, the resulting jets are somewhat irregular, but the angular hierarchy makes the clustering sequence very useful for substructure techniques, where different angular scales can reveal the underlying dynamics.

2.3.2 Variable R Algorithm

The optimal choice of the distance parameter R used in an analysis depends on the physics case under consideration. In general, perturbative and non-perturbative effects influence jet observables. The relative size of these contributions depends on the choice of R. The transverse momentum of a jet with a given value of R is modified by perturbative effects proportional to $\ln R$, while hadronisation effects lead to a change proportional to R^{-1}, and corrections due to the underlying event grow as R^2 [186]. The same hierarchy holds for the jet mass, albeit modified by two powers of R (see Sect. 2.4.1). The average squared jet mass is modified by perturbative corrections proportional to $\alpha_s R^2 p_T^2$ [161], whereas hadronisation corrections grow linearly with $R p_T$ and the underlying event affects the jet mass proportional to $R^4 p_T$ at leading order [186].

This leads to a predicament for jet substructure analyses. At low values of p_T of the decaying object, a large value of R must be chosen to combine all decay products in a single jet. At low values of p_T the non-perturbative effects on the jet substructure are still manageable. At large values of p_T the influence of non-perturbative effects on the jet mass and other substructure observables is much larger, resulting in a performance loss of substructure techniques. A possible solution to this dilemma is

[7] The jet axis will change between the different clustering steps because of combining particles i and j, resulting in a slight deviation from the exact value of πR^2.

a modification of the distance parameter, replacing it by an effective radius

$$R \to R_{\mathrm{eff}} = \frac{\rho}{p_{\mathrm{T}}} \tag{2.22}$$

in (2.20). This is known as the Variable R (VR) algorithm [187]. It results in jets with a dynamically adjusted radius, decreasing as $1/p_{\mathrm{T}}$. The parameter ρ is a constant controlling the slope of R_{eff}. It needs to be chosen according to the specific physics case under study. A minimum and maximum cut-off value, R_{min} and R_{max}, can be chosen for robustness against experimental effects by using

$$R_{\mathrm{eff}} = \begin{cases} R_{\mathrm{min}} & \text{for } \rho/p_{\mathrm{T}} < R_{\mathrm{min}}\,, \\ R_{\mathrm{max}} & \text{for } \rho/p_{\mathrm{T}} > R_{\mathrm{max}}\,, \\ \rho/p_{\mathrm{T}} & \text{else}\,. \end{cases} \tag{2.23}$$

The VR algorithm can be run for all values of k (i.e. in k_{T}, anti-k_{T} or CA mode) and is IRC safe. It leads to an improved resolution of substructure variables when averaged over a large range in p_{T}. Albeit its advantages, the VR algorithm has not played an important role in the development of substructure techniques so far. The reason is a known shortcoming of the VR algorithm, which is the clustering of additional radiation into jets in QCD multijet production, resulting in a higher jet p_{T} on average and an increased rate once a p_{T} selection is applied [187]. This can be overcome by a modification of the algorithm using a vetoed clustering (see the HOTVR algorithm in Sect. 2.4.3).

2.3.3 XCone

Another way of dealing with the transition from the resolved regime of well-separated jets to the boosted regime of overlapping jets with substructure from N-prong decays is the recently developed XCone algorithm [179, 180]. The algorithm is based on a minimisation of the event shape variable N-jettiness [188], defined as

$$\tilde{\mathcal{T}}_N = \sum_i \min\left[\rho_{\mathrm{jet}}(p_i, n_1), \ldots, \rho_{\mathrm{jet}}(p_i, n_N), \rho_{\mathrm{beam}}(p_i)\right]\,. \tag{2.24}$$

The sum runs over the the four-momenta p_i of all input particles. The value of $\tilde{\mathcal{T}}_N$ is calculated for a set of N normalised light-like axes $\{1, n_1, \ldots, n_N\}$, where $\rho_{\mathrm{jet}}(p_i, n_j)$ is a distance measure between particle i and axis j, and $\rho_{\mathrm{beam}}(p_i)$ is a distance measure to the beam. The minimum value of $\tilde{\mathcal{T}}_N$ with respect to all possible axes,

$$\mathcal{T}_N = \min_{n_1, n_2, \ldots, n_N} \tilde{\mathcal{T}}_N\,, \tag{2.25}$$

assigns each entity in the list of input particles to one of the N jet regions or to an unclustered beam region. Together with a suitable choice for ρ_{jet} and ρ_{beam}, this defines an IRC safe exclusive jet algorithm. A suitable choice of the measures ρ_{jet} and ρ_{beam} results in approximate circular jet boundaries, an example is given by the XCone default [179],

$$\rho_{\text{jet}}(p_i, n_j) = \frac{2\cosh y_j}{R^2} n_j \cdot p_i \,, \tag{2.26}$$

$$\rho_{\text{beam}}(p_i) = p_{\text{T},i} \,, \tag{2.27}$$

where y_j denotes the rapidity of axis n_j and R is the distance parameter, similar as in clustering algorithms. The presence of the dot product $n_j \cdot p_i$ makes the beam measure linear in n_j and p_i. The linearity in the jet axis n_j implies that the total three-momentum of the jet is aligned with the axis direction, and the linearity in p_i leads to factorisation properties of $\mathcal{T}_N$ making higher-order perturbative calculations technically feasible [189–195]. Additionally, jets defined by this choice have an active area of πR^2 to within 1% over a wide rapidity range. This similarity to the anti-k_{T} algorithm is beneficial in an experimental context, where the jet energy calibration is usually derived for the leading anti-k_{T} jets in an event.

From the substructure point of view, the interesting feature about the XCone algorithm is that exactly N jet regions are defined by the choice $\mathcal{T}_N$, regardless of how close some of the axes might be to each other. This results in a smooth interpolation between the boosted and the resolved regime, which is shown in Fig. 2.11a for the signal efficiency for boosted top quark reconstruction [180], where a comparison to traditional resolved and boosted analyses based on anti-k_{T} jets is made. The background misidentification rate shown in Fig. 2.11b is approximately constant at about 10% for the XCone algorithm, which is slightly higher than the values obtained with the two traditional approaches. However, a visible improvement is observed in the signal significance shown in Fig. 2.11c. Similar results have been obtained for

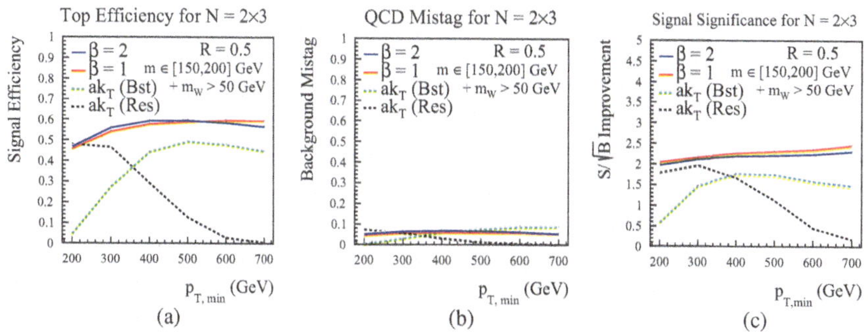

Fig. 2.11 Performance of the XCone algorithm for the reconstruction of boosted top quarks as a function of the top quark p_{T}, compared to a resolved (Res) and boosted (Bst) analysis based on anti-k_{T} jets. Shown are **a** the signal efficiency, **b** background misidentification rate and **c** the signal significance gain. Taken from [180]

two-prong $H \to b\bar{b}$ decays [180]. The N-jettiness factorisation theorem provides a basis for precision calculations (see e.g. [196–198]) which can also be performed for jet substructure observables [195], making this algorithm an interesting choice for future studies.

2.3.4 The Georgi Algorithm

A very different approach to jet finding, based on maximising a fixed function of the total four-momentum, has been suggested by H. Georgi [181] for e^+e^- collisions. The algorithm is based on a jet function depending on the total energy and the mass squared divided by the energy. One particular choice is given by

$$J_\beta = E - \beta P_\mu P^\mu / E \,, \tag{2.28}$$

where $\beta > 1$ and P denotes the total four-momentum, obtained by adding the jet constituents' four-momenta. The jet function monotonically increases in energy and decreases in jet mass squared. It is closely related to parton shower kinematics and can be modified to obtain Lorentz-invariant jet finding [199]. Maximising (2.28) by subsequently adding particles to the jet, leads to conical a jet with a cone size that goes to $1/\sqrt{\beta}$ for large values of β. Removing all particles associated to the jet from the input list and iterating the procedure leads to non-overlapping, IRC safe jets.

For hadron-hadron collisions, the energy E in (2.28) is replaced by the transverse energy E_T, resulting in the J_{E_T} algorithm [182]. The jets feature a shrinking jet cone size as the jets are closer to the beam direction. A first analytical calculation at NLO on parton level [200] established a similar cross section for inclusive jet production in hadron-hadron collisions as for the k_T algorithm. In fact, it can been shown that maximising the jet function, minimising N-jettiness (XCone), and stable cone finding (SISCone) can be reduced to a more fundamental optimisation problem [201], explaining the similar behaviour of the cross section.

A promising feature of the J_{E_T} algorithm for substructure applications is its global strategy, which leads to a better reconstruction efficiency for hadronically decaying W bosons into a single jet, compared to the anti-k_T algorithm [182]. A modification to the algorithm has been proposed [183], where an additional term related to Fox-Wolfram moments [202, 203] is added to the jet function. This results in a better performance in terms of signal acceptance versus background efficiency for reconstructing boosted hadronically decaying W bosons when compared to strategies using sequential recombination algorithms together with filtering [40] and pruning [204]. These studies should be extended, exploiting also other substructure techniques, in order to establish the usefulness of this algorithm.

2.4 Identifying Particle Decays with Jet Substructure

Once the dynamics of an event in a high energy collision have been determined using a suitable jet algorithm, analysing the substructure of these jets reveals valuable information on the particles produced and their decays. Jet substructure is ubiquitous in the identification of boosted heavy SM particles in their hadronic decays, but can also be used in the search for exotic decays of BSM particles.

2.4.1 *Jet Mass*

The jet mass is the most important observable for identifying jets from heavy particle decays. It is defined as the square root of the Lorentz-invariant product $P_\mu P^\mu$, where the four-momentum P_μ is obtained by summing the four-momenta of all jet constituents. While partons can be massless, jets always have mass due to perturbative radiation and hadronisation effects. For a massless parton the mass generated by the collinear $1 \rightarrow 2$ splitting can be approximated by [59]

$$m^2 = (p_1 + p_2)^2 \approx p_{\mathrm{T},1} p_{\mathrm{T},2} R_{12}^2 \tag{2.29}$$

where p_i are the four-momenta of the massless particles 1 and 2 with transverse momenta $p_{\mathrm{T},i}$ and R_{12} is the geometrical distance in η and ϕ between them. A jet acquires mass by a sequence of splittings, followed by non-perturbative hadronisation. The partonic cross section is proportional to $1/m$ and thus features singularities in the collinear ($R_{12} \rightarrow 0$) and soft ($p_{\mathrm{T},i} \rightarrow 0$) regime for fixed-order calculations. The singularities can be overcome by resumming no-splitting probabilities, resulting in Sudakov form factors similar as in predictions for inclusive observables, like event shapes in e^+e^- collisions. However, a similar precision as obtained for inclusive observables is more difficult to achieve for jet substructure observables because of non-global logarithms which complicate matters [205–207]. A resummation of leading logarithms from subsequent collinear splittings results in a Sudakov peak in the jet mass distribution, which position in m increases roughly linear with jet p_T. There are additional contributions from the hadronisation and the underlying event, which can be estimated analytically similarly to the changes in the jet p_T (Sect. 2.5.2).

The leading power corrections for m^2 scale with $p_\mathrm{T} R$ for hadronisation corrections and with $p_\mathrm{T} R^4$ for contributions from the underlying event. Hence, the hadronisation corrections dominate by three powers of R for $R < 1$. Note that this approximation is only valid to the right of the Sudakov peak and breaks down in the vicinity of the peak [208]. An example is shown in Fig. 2.12, where results from a resummed calculation of the m/p_T distribution with two different choices of the hadronisation scale are compared to results from PYTHIA 8 and HERWIG++. The results from the analytical calculations agree well with predictions from parton showers, giving confidence in the calculations and showing that the modelling of non-perturbative

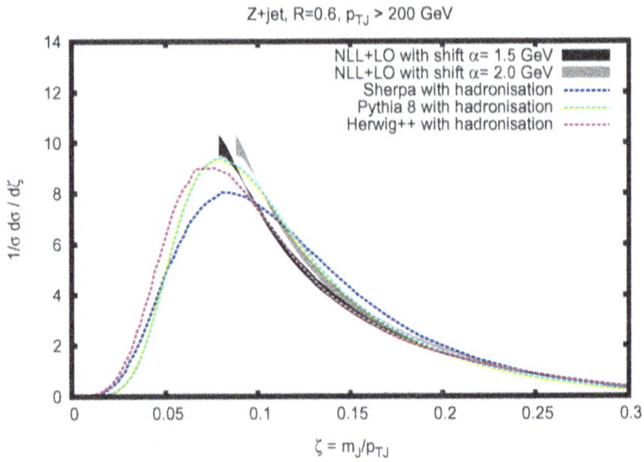

Fig. 2.12 Comparison of the scaled jet mass distribution m/p_T between analytical, resummed calculations with two different choices of non-perturbative shifts (shaded bands) with predictions obtained from event generators with parton showers. Taken from [180]

effects can be taken into account in first-principle calculations of jet substructure observables with a complex, non-local structure.

The tail of the jet mass distribution leads to a considerable background of jets from light quarks and gluons when identifying heavy resonance decays, as these can acquire jet masses of the order of the electroweak scale and above. This was realised early when studying the physics programme of the LHC [37, 38] and it took some time to develop more sophisticated methods to efficiently separate signal from background jets by using observables beyond the plain jet mass [39–41].

2.4.2 *Angularities and Energy Correlations*

The different colour structure of signal and background jets leads to a difference in the soft-gluon radiation pattern inside these jets. Angular and energy distributions of particles, and especially their correlation, carry information about the underlying dynamics.

N-Subjettiness

The N-subjettiness variables τ_N [209, 210] are defined similarly to the N-jettiness [188] event variables, but are calculated on jet constituents instead of full events,

$$\tau_N^{(\beta)} = \frac{1}{d_0} \sum_k p_{T,k} \min\left(\Delta R_{1,k}^{\beta}, \Delta R_{2,k}^{\beta}, \ldots, \Delta R_{N,k}^{\beta}\right) \quad \text{with} \quad d_0 = \sum_k p_{T,k} R, \tag{2.30}$$

where $\Delta R_{i,k}$ is the distance of jet constituent k to the subjet i and R is the distance parameter of the jet algorithm defining the jet. The angular exponent β is a free parameter that controls the weight given to collinear and wide-angle radiation. Usually, $\beta = 1$ is used, and assumed throughout this text if not specified explicitly. The subjet axes are obtained by minimizing τ_N, where finding the global minimum is difficult and can be computationally expensive. In practice, a one-pass minimisation procedure is usually used,[8] where seed values for the subjet axes are obtained by running the k_{T} algorithm in the exclusive mode. The value of τ_N is then minimised in using these seeds as initial values.

The value of N-subjettiness ranges between 0 and 1 and assesses the degree to which the jet constituents are localised near N axes. While a single value of τ_N has little discrimination power between one and N prong decays, as the value of τ_N can always be reduced by adding an additional axis, ratios of N-subjettiness values, τ_N/τ_M, often written as τ_{NM}, show excellent discrimination power. One important feature of N-subjettiness ratios is that these are analytically calculable [211], such that cross section measurements can be safely defined using selections based on τ_{NM}. Additionally, by comparisons with predictions from event generators, the accuracy of parton shower models can be tested and improved systematically.

The ratio τ_{21} is commonly used to identify two prong decays such as hadronic W and Z decays. Note that this observable is only Sudakov-safe [212, 213], but can be made IRC safe by either requiring a minimum jet mass or a minimum value of τ_1. It has also been pointed out that this observable has a complex singular structure [214], which leads to shoulders in the τ_{21} distribution. This implies that higher order calculations are necessary to achieve reliable theoretical predictions. The presence of singularities also leads to large non-perturbative corrections, making this observable susceptible to different tunes of MC event generators. While no analytical studies for three prong decays τ_{32} exist, it is expected that the increase to three axes will result in an even more complicated singular structure for the resolved $\tau_{32} \to 1$. A theoretically less complex variable with similar discriminative power as τ_{NM} should therefore be favoured in experimental analyses.

Energy Correlation Functions

A computational drawback of N-Subjettiness is the explicit identification of subjet axes, which also depends on the specifics of the minimisation algorithm used. Generalised energy correlation functions [215] are substructure variables independent of subjet axes. These are a generalisation of the event shape parameter C [216, 217], and the IRC-safe N-point energy correlation function (ECF) is defined as

$$\mathrm{ECF}(N,\beta) = \sum_{i_1<i_2<\ldots<i_{N_J}} \left(\prod_{a=1}^{N} p_{\mathrm{T}i_a}\right)\left(\prod_{b=1}^{N-1}\prod_{c=b+1}^{N} \Delta R_{i_b i_c}\right)^{\beta} \tag{2.31}$$

for hadron colliders, where i_a is the index of particle a inside the jet and N_J is the total number of particles in the jet. At e^+e^- colliders, where energies and angles are the natural observables, p_{T} is replaced with E and ΔR with θ in (2.31). A dimensionless variable that can be used to identify if a jet has N hard prongs is the ratio of two ECFs, $\mathrm{ECF}(N+1,\beta)/\mathrm{ECF}(N,\beta)$, since $\mathrm{ECF}(N+1,\beta)$ will be significantly smaller than

[8]This is in contrast to a multi-pass minimisation, where 100 or more minimisations are done, with different initial axes in each attempt. While this procedure is more likely to find the global minimum, the one-pass minimisation gives reasonable results even though it only finds a local minimum.

$\mathrm{ECF}(N, \beta)$ for N-prong decays. A powerful observable can then be defined similarly to N-subjettiness ratios as the dimensionless double ratio

$$C_N^{(\beta)} = \frac{\mathrm{ECF}(N+1, \beta)\mathrm{ECF}(N-1, \beta)}{(\mathrm{ECF}(N, \beta))^2}, \tag{2.32}$$

where C_N involves $(N+1)$-point correlator functions and is thus a good probe of N-prong substructure. The definition is chosen such that C_N is sensitive to the same substructure as the N-subjettiness ratio $\tau_{N,N-1}$. Theoretical advantages of C_N versus $\tau_{N,N-1}$ are insensitivity to soft and wide angle radiation and insensitivity to recoil effects [215]. It is also possible to calculate C_N analytically which allows for a validation of their modelling in event generators. On generator level, the variables C_1 and C_2 have been shown to perform well for quark/gluon (q/g) separation and the identification of heavy boson decays, while the performance of C_3 for top tagging is worse compared to other observables because the EFCs in the ratio of C_3 can be dominated by possibly different subsets of the hard emissions in the jet.

Dimensionless normalised energy correlation functions were introduced in [218] and offer a short-hand notation,

$$e_N^{(\beta)} = \frac{\mathrm{ECF}(N, \beta)}{(\mathrm{ECF}(1, \beta))^N}, \tag{2.33}$$

with unchanged properties with respect to the ECFs of (2.31).

D functions

On the basis of a power counting analysis, studying the perturbative radiation from the jet, it is possible to identify the boundary between 1- and 2-prong decays to be $e_3 \sim (e_2)^3$ [218]. This motivates the definition of

$$D_2^{(\alpha,\beta)} = \frac{e_3^{(\alpha)}}{\left(e_2^{(\beta)}\right)^{3\alpha/\beta}}, \tag{2.34}$$

which is less susceptible to soft and wide-angle radiation, and also to effects from pileup than $C_2^{(\beta)}$. Additionally, all-order factorisation theorems have been formulated for both signal and background jets for $D_2^{(\alpha,\beta)}$, allowing for systematically improvable higher-order calculations [219].

For three prong configurations like top quark decays, a variable $D_3^{(\alpha,\beta,\gamma)}$ can be constructed also from power counting arguments. In this case, splittings without energy or angular hierarchies between the subjets, strongly angular-ordered splittings and splittings with a soft subjet result in three different scalings of the relevant ECFs e_2, e_3 and e_3. Building a variable interpolating between the three different scalings leads to the fairly complex definition

$$D_3^{(\alpha,\beta,\gamma)} = \frac{e_4^{(\gamma)}\left(e_2^{(\alpha)}\right)^{3\gamma/\alpha}}{\left(e_3^{(\beta)}\right)^{3\gamma/\beta}} + x\frac{e_4^{(\gamma)}\left(e_2^{(\alpha)}\right)^{2\gamma/\beta-1}}{\left(e_3^{(\beta)}\right)^{2\gamma/\beta}} + y\frac{e_4^{(\gamma)}\left(e_2^{(\alpha)}\right)^{2\beta/\alpha-\gamma/\alpha}}{\left(e_3^{(\beta)}\right)^{2}}, \quad (2.35)$$

where x and y depend on the average jet p_{T} and the mass of the resonance, usually the top quark mass [220]. The variable $D_3^{(\alpha,\beta,\gamma)}$ performs much better for top quark identification than C_3 with a similar or better performance than τ_{32} using generator information only. If this also holds in experimental analyses has not been confirmed yet.

Ratios of Generalised Energy Correlation Functions

While the N-point ECFs capture multiple angular scales simultaneously, it is possible to generalise them to identify one angular scale at a time. The resulting generalised energy correlation functions, denoted by ${}_{v}e_N^{(\beta)}$, explicitly take into account v factors of pairwise particle-particle angles [221]. This generalisation helps to isolate redundant angular scales of particle decays and simplifies analytic calculations by facilitating a well-defined angular scaling behaviour. Dimensionless observables can be constructed from ratios of ${}_{v}e_N^{(\beta)}$, for example

$$M_2^{(\beta)} = \frac{{}_1e_\beta^{(3)}}{{}_1e_\beta^{(2)}}, \qquad N_2^{(\beta)} = \frac{{}_2e_\beta^{(3)}}{\left({}_1e_\beta^{(2)}\right)^2} \quad \text{and} \quad N_3^{(\beta)} = \frac{{}_2e_\beta^{(4)}}{\left({}_1e_\beta^{(3)}\right)^2}. \quad (2.36)$$

The M_2 and N_2 functions identify 2-prong decays, while N_3 is defined to identify 3-prong decays. It was found the ratios N_i behave similar to N-subjettiness ratios τ_i/τ_{i-1}, which is remarkable since the discrimination power is achieved in very different ways. Some of these ratios are expected to perform well only if used on groomed jets (see Sect. 2.4.3). The requirement for a grooming procedure results from regions in phase of 1-prong jets dominated by soft radiation overlapping with 2- and 3-prong signal regions. The fact that this overlap is caused by soft radiation indicates that a jet grooming procedure can help to improve the performance. Experimentally, a major challenge when using these ratios are contributions from pileup, which lead to a degradation in their performance. Another series of observables, $U_i^{(\beta)} = {}_1e_\beta^{(i+1)}$, has been designed to yield discrimination power in q/g separation. These observables probe single- and multi-particle correlations within a jet, useful to improve q/g discriminants.

Energy Flow Polynomials

Guided by the requirements for IRC safety, energy flow polynomials (EFPs) [222] have been introduced. These provide a linear basis for constructing any IRC-safe substructure observable. The EFPs are multiparticle energy correlators with specific angular structures. They can be represented as loopless multigraphs, with a set of rules to calculate their values from the corresponding vertices and edges. For a jet with M constituents, the EFP for a multigraph G with N vertices and $(k, \ell) \in G$

edges is given by [222]

$$\mathrm{EFP}_G = \sum_{i_1=1}^{M} \cdots \sum_{i_N=1}^{M} z_{i_1} \ldots z_{i_N} \prod_{(k,\ell)\in G} \Delta R_{i_k i_\ell}, \tag{2.37}$$

where the hadronic distance measure has been used and $z_i = p_{\mathrm{T,i}}/p_{\mathrm{T,jet}}$ is the transverse momentum fraction carried by particle i relative to the jet. It can be shown that many jet substructure observables can be expressed as linear combinations of EFPs, and more generally, this is true for any IRC-safe observable. For example, the EFP corresponding to the connected multigraph with two vertices, $N = 2$, and two edges, $d = 2$, is equal to $2m_{\mathrm{jet}}^2/p_{\mathrm{T,jet}}^2$. Normalised ECFs $e_N^{(\beta)}$ correspond to complete graphs with N vertices, and generalised angularities [223] can be written as linear combinations of EFPs. The EFP representation as graphs allowed for a computational simplification, resulting in a faster computation than the naive scaling with $\mathcal{O}(M^N)$, and has also led to faster methods for computing of other substructure variables [224]. The possibility to construct a complete basis from EFPs allows to build very powerful linear regression and classification models, covering the full space of jet substructure information. In many cases, linear models for jet tagging based on EFPs can compete in performance with nonlinear combinations using other substructure observables. Together with machine learning methods, EFPs can help to reveal the information learned by a trained neural network [225].

Pull Angle

An observable analysing the QCD colour connection between jets is the pull vector [226], defined as

$$\mathbf{v}_p = \sum_{i\in\mathrm{jet}} \frac{p_{\mathrm{T},i}|\mathbf{r}_i|}{p_{\mathrm{T,jet}}}\mathbf{r}_i\,, \tag{2.38}$$

where the sum runs over all jet constituents of a given jet and the vector $\mathbf{r}_i$ is the relative rapidity and azimuthal angle of the constituent to the jet axis. The pull vector points from the jet axis to the direction of dominant energy flow. The angular distance in ΔR between the pull vector of a given jet and another jet in an event can help to distinguish between jets originating from a colour singlet like W, Z or H and jets colour-connected to the beam. For example, in $H \to b\bar{b}$ decays the two b jets will form a dipole and shower towards each other, while in background processes b quarks will be colour connected to the beam remnants, with more radiation towards the beam (large $|y|$). The usefulness of the pull angle for distinguishing different colour configurations has also been proven with an analytical calculation for three-jet production in e^+e^- collisions [227]. Because the pull angle is not IRC safe, an asymmetry distribution has been proposed very recently, which can be considered an IRC safe version of the pull angle [228]. In addition, it has been pointed out that studies of variables sensitive to the colour flow should be done in the rest frame of the decaying particle in order to minimise kinematic effects from asymmetric particle decays.

Jet Shapes

Event-shape variables defined for e^+e^- collisions can be modified to include only particles clustered into a given jet. Originally, jet shapes have been defined for hadron colliders as differential $\rho(r)$ and integrated distributions $\Psi(r)$, which describe the jet energy fraction that lies within an annulus of size $r \pm \Delta r/2$ or a cone with radius r around the jet axis, respectively [60]. While the discrimination power of such jet shapes is usually limited, a combination of several variables can achieve good performance. An example are distributions of Fox-Wolfram moments [202, 203], sphericity [43] and thrust [44–46], calculated in the rest frame of the parent particle. Since the parent particle is usually unknown, a recent approach [229] defines four rest frames for W, Z, H and t hypotheses and combines the obtained boosted event shapes with an artificial neural network.

2.4.3 Jet Grooming and N-Prong Taggers

Traditionally, there has been a distinction between jet grooming techniques and N-prong finders, or taggers, where the former are designed to reduce the impact of non-perturbative effects on a jet, and the latter aim at labelling a jet to originate from the collimated hadronic decay of a given heavy particle. This distinction is somewhat misleading as grooming techniques are frequently used as taggers. By modifying the jet substructure, a better separation between signal and background is achieved with grooming techniques, resulting in an improved tagging performance when selecting jets based on substructure observables. Commonly used examples are trimming [230], as used in ATLAS, and soft drop [231], as used in CMS. The most widely used grooming methods and taggers are reviewed in the following.

Filtering

Filtering [40] was designed to resolve large jets on a finer angular scale. This is achieved by reclustering the jet constituents with a filter radius R_{sub}, much smaller than the original size of the jet R. Only the N_{filt} hardest subjets are kept, where N_{filt} should be adjusted to the particular physics case. In general, for an N-prong decay, $N_{\text{filt}} = N + 1$ subjets should be kept to capture the dominant $\mathcal{O}(\alpha_s)$ radiation from the partonic decay. Filtering has been originally introduced together with a mass drop criterion (see the mass drop tagger below) and is used in a number of applications, most notably the HEPTopTagger [232–234].

A first analytical calculation in the leading logarithmic approximation [235] for filtered jets from the $H \rightarrow b\bar{b}$ decay, showed indeed that an optimal choice is $N_{\text{filt}} = 3$ or 4 and that the parameter R_{filt} should not be chosen too large or too small, where values of R_{filt}/R between 0.3 and 0.4 lead to an acceptable agreement between the analytical fixed-order results and numerical all-order resummed results. This agreement implies not too large higher-order corrections, important for reliable predictions. Higher-order predictions in NLL accuracy and better are difficult for filtering, because of a non-trivial slicing of the phase space.

Trimming

Trimming [230] is similar to filtering. In its commonly used form, trimming reclusters all particles from a jet with distance parameter R into subjets with R_{sub}. All subjets that fulfil $p_{\mathrm{T,sub}} > f_{\mathrm{cut}}\, p_{\mathrm{T,jet}}$ are kept and merged to form the trimmed jet, where $p_{\mathrm{T,sub}}$ and $p_{\mathrm{T,jet}}$ are the transverse momenta of the subjet and the original jet, respectively. All subjets not fulfilling this requirement are discarded.[9] Experimentally, an advantage of trimming in relation to filtering is that the number of maximum subjets does not have to be specified, instead the parameter f_{cut} controls the number of subjets dynamically. Trimming became standard in ATLAS, where it is used to reduce the effects of pileup and the underlying event on anti-k_{T}, $R = 1.0$ jets. Typical parameter settings are $R_{\mathrm{sub}} = 0.2$ and $f_{\mathrm{cut}} = 0.05$.

Pruning

Pruning [236] reclusters all jet constituents, and in each step checks the angular separation $\Delta R_{ij} < R_{\mathrm{prune}}$ and for a soft splitting $\min(p_{\mathrm{T,i}}, p_{\mathrm{T,j}}) \geq z_{\mathrm{prune}}\, p_{\mathrm{T,i+j}}$. If both criteria are met, pseudojets i and j are combined, otherwise only the hardest of i and j are kept and the other is rejected. Pruning takes a different approach than filtering and trimming. While in the latter cases a jet is reclustered with $R_{\mathrm{sub}} < R$ and soft subjets are discarded, pruning rejects pseudojets during the jet clustering. Another difference is that for pruning the radius R_{prune} is adjusted dynamically through $R_{\mathrm{prune}} = m_{\mathrm{jet}}/p_{\mathrm{T,jet}}$, while R_{sub} is kept fixed for filtering and trimming. Here, m_{jet} and $p_{\mathrm{T,jet}}$ are the mass and p_{T} of the original jet. A particularity of pruning are cases where soft, wide-angle radiation leads to large values of m_{jet} and thus sets the scale for R_{prune}, while being rejected by the pruning conditions when reclustering the jet. The mass of the pruned jet is then determined by radiation at a much smaller angle and the jet shows a 1-prong structure, which is referred to as I-pruning [237]. Conversely, jets which feature at least one clustering step passing the pruning criteria show a two-prong structure, referred to as Y-pruning. For practical applications the difference between I- and Y-pruning has been shown to be mostly irrelevant, but it is important for the analytical control of calculations for pruned jets.

Mass Drop Tagger

The mass drop tagger (MDT) [40] relies on the CA clustering, where it steps backwards through the clustering history. At each step, the mass drop condition

$$\max(m_i, m_j) < \mu_{\mathrm{cut}} m_{i+j} \tag{2.39}$$

and the symmetry of the splitting,

$$\min(p_{\mathrm{T,i}}{}^2, p_{\mathrm{T,j}}{}^2)\Delta R_{ij}^2 > y_{\mathrm{cut}} m_{i+j}^2, \tag{2.40}$$

[9] In its original version, trimming was formulated using a hard scale Λ_{hard}, instead of the original jet's transverse momentum $p_{\mathrm{T,jet}}$.

are tested. If both criteria are met, the combined pseudojet is kept. Otherwise, the least massive of pseudojets i and j is discarded and the procedure continues on the other pseudojet. Typical parameter choices are $\mu_{\mathrm{cut}} = 2/3$ and y_{cut} in the range 0.09–0.15. This results in a significant mass drop, i.e. while the original jet is massive, as expected from the decay of a heavy boson, the subjets originate from massless quarks. In addition, the momentum splitting between the two subjets is symmetric, as expected from a two-body decay, whereas soft QCD radiation is rejected. In its original version, mass drop has been introduced to identify the two-prong structure of $H \to b\bar{b}$ decays and distinguish it from QCD background. The algorithm also acts as a jet groomer since it iteratively removes soft radiation.

Modified Mass Drop Tagger and Soft Drop

An improvement of the mass drop tagger has been suggested in an analytic study [237], where it was found that following the most massive branch in the declustering can lead to the algorithm following the wrong branch. This happens if a soft emission results in a large mass, and the MDT recurses into this branch even though it is soft. A modification has been introduced to avoid these configurations. In the modified Mass Drop Tagger (mMDT) the symmetry condition is replaced by $\min(p_{\mathrm{T,i}}, p_{\mathrm{T,j}}) > z_{\mathrm{cut}}(p_{\mathrm{T,i}} + p_{\mathrm{T,j}})$, and the hardest branch in p_{T} is followed in the declustering. The mass-drop criterion has been found to be sub-leading and can be dropped without a performance penalty. The mMDT has been found to greatly facilitate analytic calculations and even slightly improve the performance in relation to the MDT. The soft drop algorithm [231] is a generalisation of the mMDT, obtained by introducing an angular exponent β in the symmetry condition,

$$\min(p_{\mathrm{T,i}}, p_{\mathrm{T,j}}) > z_{\mathrm{cut}} \left(p_{\mathrm{T,i}} + p_{\mathrm{T,j}}\right) \left(\frac{\Delta R_{ij}}{R}\right)^{\beta} . \tag{2.41}$$

The angular exponent results in additional freedom to adjust the grooming strength of the algorithm. In the case $\beta = 0$, the mMDT is retained. For values $\beta > 0$ the grooming is reduced and $\beta < 0$ results in rejecting more particles in relation to the mMDT. In experimental applications the value of z_{cut} is typically chosen to be around 0.1. Note that while the soft-drop jet mass is IRC safe, the jet p_{T} after soft-drop grooming for $\beta \leq 0$ is not IRC safe, but only Sudakov safe.[10] This means that calculations with higher logarithmic accuracy will be very difficult if needed differential in groomed jet p_{T} [238]. The soft drop algorithm stops after the symmetry condition has been met and returns a maximum of two prongs. A larger number of splittings can be obtained with the recursive soft drop algorithm [239], where the number of hard prongs is a free parameter.

Splitting Scales

One of the first variables in use in experimental analyses is the k_{T} splitting scale $\sqrt{d_{ij}}$, which can be obtained by reclustering jets with the k_{T} algorithm, (2.20) with

[10]When using soft drop in tagging mode, meaning that jets failing the soft drop condition get rejected, the groomed p_{T} is IRC safe.

$k = 1$. The distance of the final clustering step of pseudojets 1 and 2 is then

$$\sqrt{d_{12}} = \min\left(p_{\mathrm{T},1}, p_{\mathrm{T},2}\right) \Delta R_{12}, \tag{2.42}$$

where the normalisation with the jet distance parameter R has been dropped [38]. In the k_{T} clustering, pseudojets with large distances d_{ij} get clustered last, and therefore the parameter $\sqrt{d_{12}}$ will typically combine the two subjets created by the decay products of a heavy two-prong resonance decay. Given a resonance mass m, the expected value of $\sqrt{d_{12}}$ is about $m/2$. Resonance decays with multiple prongs can be identified by going backwards in the clustering sequence, for example the second-to-last clustering step defines $\sqrt{d_{23}}$, which can be used to distinguish top quark decays from light flavour jets.

Johns-Hopkins/CMS Top Tagger

The first algorithm specifically designed for tagging top quarks with $p_{\mathrm{T}} > 1$ TeV is the Johns-Hopkins Top Tagger [41]. The algorithm is based on a decomposition of the primary jet into up to four subjets by reversing the CA clustering sequence. It has been adjusted by the CMS Collaboration where it is known as CMS Top Tagger (CMSTT) [240, 241]. It was used in analyses of 7 and 8 TeV data in the region of $p_{\mathrm{T}} > 400$ GeV. The algorithm has two decomposition steps, where in each step the two pseudojets are tested for the condition $\Delta R_{ij} > 0.4 - k\, p_{\mathrm{T,i+j}}$, where the minimum angular separation is allowed to decrease with increasing p_{T} with a slope of $k = 0.0004\ \mathrm{GeV}^{-1}$. If the angular separation is too small, the decomposition stops. Otherwise, the criterion $p_{\mathrm{T,i}} > \delta_p\, p_{\mathrm{T,jet}}$, is checked for each pseudojet, where δ_p is usually chosen as 0.05 and $p_{\mathrm{T,jet}}$ is the p_{T} of the initial jet. The reason for the smallness of δ_p is that $p_{\mathrm{T,jet}}$ is kept for the primary and secondary decomposition step. Depending on which decomposition step is successful, the jet is decomposed into two, three or four subjets. If the primary decomposition fails, the original jet is returned. A jet is tagged if it has three or more subjets, a mass in a window around m_t, and a minimum pairwise mass $m_{\min} = \min(m_{12}, m_{13}, m_{23})$, calculated from the three leading p_{T} subjets, greater than 50 GeV. It has been pointed out that without the ΔR_{ij} condition, or at very high p_{T} where the ΔR_{ij} condition is essentially ineffective, the CMSTT is IRC unsafe [242]. The CMSTT has achieved an average identification efficiency of 38% at 3% misidentification rate [243].

HEPTopTagger

The HEPTopTagger (HTT) [232, 233] was designed to target $t\bar{t}H$ production in the $H \to b\bar{b}$ decay channel. In $t\bar{t}H$ production the top quark p_{T} distribution peaks around 150 GeV and is steeply falling towards increasing p_{T}, where it is already an order of magnitude smaller at $p_{\mathrm{T}} \sim 400$ GeV. This results in a requirement of non-zero signal efficiency already at $p_{\mathrm{T}} \approx 200$ GeV, where the top quark decay is only moderately boosted. The HTT achieves this with a large jet distance parameter of 1.5 and a sequence of MDT declustering, filtering and re-clustering of the original CA jet. After the final filtering step, the three subjets leading in p_{T} are tested for kinematic constraints of a three-body top quark decay. An updated variant,

HTTv2 [234], introduces a variable jet radius by repeatedly reducing R in the clustering until a significant drop in the jet mass is observed. This value of R has some additional discrimination power. It is used in a Boosted Decision Tree to improve the HTT performance, where other input variables are subjet kinematics, ungroomed and groomed τ_N and Qjet volatility (described below).

Mass Jump

The vetoed jet clustering algorithm mass jump [244] is different than other veto algorithms, as it does not reject pseudojets but prevents them from participating in the next steps of the jet clustering if this would result in a large increase of the combined jet mass (a mass jump). The algorithm achieves this by a forward clustering, instead of a decomposition of an existing clustering history. At the start of the algorithm, all pseudojets are labelled active. In the clustering, only active pseudojets are considered. The condition $m_{i+j} < \mu$ is checked at each step and two pseudojets are only combined if the condition is met. Else, a mass jump criterion is introduced as $\theta m_{i+j} > \max(m_i, m_j)$ and if a mass jump is found the two pseudojets i and j are labelled as passive and not considered further in the jet clustering. It is also checked if a mass jump appears between active and passive pseudojets. If no mass jump is found, the pseudojets i and j are combined and replaced by the resulting pseudojet in the list of active pseudojets. The clustering terminates once no more active pseudojets are present, where all passive pseudojets are called jets. For $\theta = 0$ and $\mu = \infty$ the algorithm is identical to the standard sequential jet clustering without a veto. As the vetoed jets are not considered for further clustering, their effective jet radius is smaller than the parameter R, which now gives an upper bound. The mass jump algorithm has no inherent grooming, as all jets are kept in the clustering. However, it is straight forward to reject jets that were labelled passive without fulfilling the mass jump condition. The algorithm can improve the performance of the HTT, when employed instead of the MDT declustering for obtaining subjets.

HOTVR Tagger

The Heavy Object Tagger with Variable R (HOTVR) [245] is the only tagger based on the VR jet algorithm, which adapts R dynamically[11] to the p_{T} of the jet. It includes a mass jump condition [244, 246] in the clustering process, resulting in subjet finding and the rejection of soft radiation in one sequence, without the need of declustering and following grooming steps. A known shortcoming of the VR algorithm is the clustering of additional radiation into jets in QCD multijet production, resulting in a higher jet p_{T} on average and an increased rate once a p_{T} selection is applied [187]. The HOTVR algorithm approaches this issue by modifying the jet clustering procedure with a veto based on the invariant mass of the pseudojet pair, inspired by mass jump algorithm [244]. The mass jump veto prevents the recombination of two pseudojets i and j if the combined invariant mass m_{i+j} is not large enough. In case a mass jump is found and $p_{\mathrm{T,i+j}} > p_{\mathrm{T,sub}}$ the pseudojets are combined, where $p_{\mathrm{T,sub}}$ is a

[11]The text in this paragraph has been taken from [245] and has been written by the author. It has been adjusted to fit this book.

free parameter of the algorithm. The resulting pseudojet enters the next clustering step and the initial pseudojets are stored as separate subjets. In case the mass jump criterion is not fulfilled or the pseudojets are softer than $p_{\mathrm{T,sub}}$, the lighter pseudojet or the one too soft is removed from the list. This step acts as a jet grooming and stabilises the jet mass over a large range of p_{T}.

The algorithm results in groomed VR jets with an effective radius R_{eff} between R_{min} and R_{max} (see the VR algorithm in Sect. 2.3.2), containing subjets with $R < R_{\mathrm{eff}}$ with a mass jump. The number of subjets found is modified by the mass jump parameters μ, θ and $p_{\mathrm{T,sub}}$. Once the pseudojets become sufficiently heavy due to clustering, the mass jump threshold μ results in a rejection of soft and light pseudojets. For a fixed value of μ, the strength of this jet grooming depends on the parameters θ and $p_{\mathrm{T,sub}}$. For $\theta = 1$ the mass jump condition is always fulfilled and no pseudojets are rejected (equivalent to the case $\mu \to \infty$). Conversely, the case of $\theta = 0$ results in a VR jet clustering which stops as soon as a jet mass of μ is reached. The algorithm results in subjets with a maximum mass of μ. Additional jet grooming is obtained by setting $p_{\mathrm{T,sub}} > 0$. This results in subjets with a minimum p_{T} of $p_{\mathrm{T,sub}}$, effectively removing soft radiation and improving the tagging performance at small p_{T} of the heavy object. The HOTVR algorithm is IRC safe, which has also been confirmed in numerical studies [247].

The behaviour of the algorithm is visualised in Fig. 2.13 where two example $t\bar{t}$ events, generated with PYTHIA 8 [248–250] at low p_{T} (top row, Event 1) and at high p_{T} (bottom row, Event 2), are clustered with the CA algorithm (left column) and with the HOTVR algorithm (right column). The active catchment areas of the hard jets are obtained using ghost particles [185] and are illustrated by the coloured (orange/blue) areas.[12] The impact of the VR part of the algorithm is nicely illustrated by the largely different jet sizes of the two events clustered with the HOTVR algorithm (right column). The grey regions in the right panels were rejected by the mass jump criterion and are not part of the HOTVR jets. This criterion has largest impact in events at low p_{T} as exemplified in Event 1 (top, right). The HOTVR jets together with their subjets reproduce the kinematics of the top decay adequately, both at low and high p_{T}, demonstrating a better adaptation to the decay topology than CA or anti-k_{T} jets.

2.4.4 Other Methods

There are a number of jet substructure methods that can neither be classified as angularity/correlation variable nor as grooming/tagging algorithm. Some of the most widely used are described below.

[12]The exact borders of the jet areas depend slightly on the specific configuration of the ghost particles.

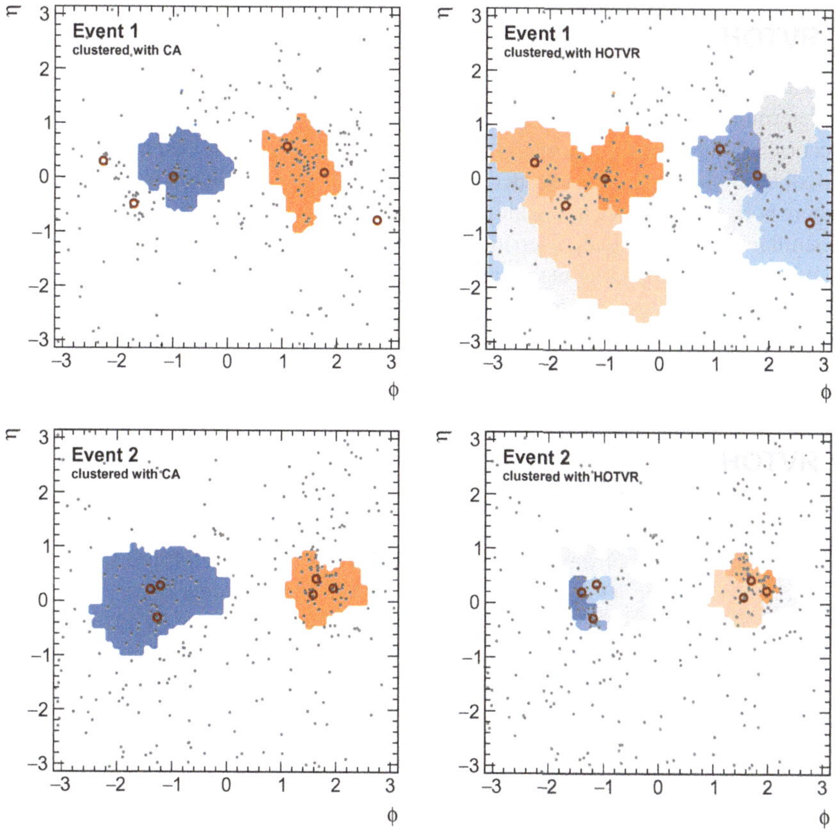

Fig. 2.13 Two simulated $t\bar{t}$ events clustered with the CA algorithm with distance parameter $R = 0.8$ (left column) and with the HOTVR algorithm (right column). The top quarks have either low p_T (top row, Event 1) or high p_T (bottom row, Event 2). The two leading jets in the events are shown as coloured areas (orange/blue). The stable particles, input for the jet finders, are drawn as grey dots. The quarks from the top quark decay are depicted by red circles and are shown for illustration purposes only. In case of the HOTVR algorithm the subjets are shaded from light to dark, corresponding to increasing p_T. The grey areas correspond to regions rejected by the mass jump criterion. Taken from [245]

Mass-decorrelating Observables

In experimental analyses, a commonly used combination for two-prong tagging is a combination of a groomed jet mass, for example the soft drop mass m_{SD}, with a selection using a ratio of N-subjettiness or (generalised) ECF values, ν_{ij}. A cut on ν_{ij} will select jets with a pronounced two-prong structure and thus sculpt the background distribution significantly. With a typical jet p_T distribution at the LHC, the Sudakov peak of the background distribution will have its position around 100 GeV, similar to signal jets. This presents a serious challenge for the background estimation in analyses, as the peak in the background distribution is difficult to model and typically introduces large uncertainties. The idea of the Designed Decorrelated Taggers (DDT) [251] is to introduce a variable transformation of the form $\nu_{ij}^{\mathrm{DDT}} = \nu_{ij} - c\log(\rho_{\mathrm{SD}})$, which results in an unaltered shape of the m_{SD} distribution after a cut on the decorrelated variable ν_{ij}^{DDT}. The scaling variable is given by $\rho_{\mathrm{SD}} = m_{\mathrm{SD}}/p_{\mathrm{T,SD}}$ and the constant c is determined from the ρ_{SD} dependence of the ν_{ij} distribution in bins of p_{T}. While the decorrelation is usually obtained from studies using MC event generators, it is also possible to obtain the decorrelated shapes from analytical calculations [252, 253], where both approaches agree within uncertainties. The performance of a DDT tagger is similar to the uncorrelated version if no selection on the jet mass is applied. In case of a jet mass selection, the tagging performance using ν_{ij}^{DDT} can be better than using the plain shape ν_{ij}, as less background events are reconstructed within the selected jet mass window (see for example [254]).

Qjets

Quantum jets [255], or Qjets, are based on the stochastic nature of parton showers. While the radiation pattern of quarks and gluons is inherently random within the physically allowed boundaries, N-prong decays adhere to kinematic constraints in addition to stochastic showering. When constructing jets, this can be used to discriminate N-prong decays from quark and gluon jets by building multiple variants of a single jet, where the jet constituents are weighted by an appropriate metric during the clustering. Then each jet in each event produces a distribution for an observable. This can lead to new discriminating variables, like the relative width of the ensemble of jet masses for a single jet, or to an improvement of the statistical stability of an observable [256]. Note that this only works in combination with a grooming step, which discards selected particles in the clustering, where pruning is used in the originally proposed version of Qjets. Due to the random modification of the distance measure d_{ij} in (2.20), different particles are discarded in each iteration, resulting in wider distributions for light quark and gluon jets than for jets with N definite hard prongs.

Shower Deconstruction

Shower Deconstruction [257, 258] is based on calculating conditional probabilities that a given set of N final state particles $\{p_N\} = (p_1, \ldots, p_N)$ originates from a signal, $P(S|\{p_N\})$, or a background hypothesis, $P(B|\{p_N\})$. The likelihood ratio

$$\chi = \frac{P(S|\{p_N\})}{P(B|\{p_N\})} \tag{2.43}$$

can then be used as a discriminator in an analysis. The signal and background probabilities are computed using an all-order QCD calculation, where all possible splittings into initial- or final-state radiation are taken into account. For each such splitting all possible shower histories are considered that could lead to the final state $\{p_N\}$. A weight is computed for each history, where each vertex receives a factor from the partonic splitting probability and a Sudakov form factor. The LO matrix elements for the decays of W, Z, H bosons and t quarks are used. By retaining the full mass dependency for the partons, leading-logarithmic accuracy is achieved. Shower deconstruction can be used for q/g separation [259], W, Z, H and t tagging. Including all jet constituents is computationally unfeasible due to the large number of possible shower histories. In practice, the jet is reclustered into small subjets with R around 0.2 and only the leading 8–10 subjets are considered for $\{p_N\}$.

The Metric Space of Collider Events

Inspired by the question of when two collider events are similar, a theoretically and experimentally robust definition of a distance between two events has been defined as the energy mover's distance (EMD) [260]. The EMD(ϵ, ϵ') is the minimum work required to rearrange an event ϵ into another ϵ' by movements of energy from a particle in one event into a particle in the other event. This approach can be applied to the substructure of jets, where the metric space between jets can be used to classify jets without requiring specially designed observables. Instead, the nearest neighbours in the metric space determine the class membership of a given jet [261]. This approach can be used for jet tagging, with comparable sensitivity to machine learning techniques [260]. In addition, this opens the possibility for unsupervised anomaly detection, where jets with the largest distance to the rest of the dataset are the most anomalous ones [262]. The EMD provides a new way to visualise the topology of jets that cluster in a given bin of an observed distribution. The jets that best describe the set of jets in a histogram bin can be obtained from the minimum sum of distances to all other jets. An example is show in Fig. 2.14, where the distribution of the mean EMD to a full dataset of jets with $p_{\mathrm{T}} = 400$ GeV is shown [262]. The most typical jets have small mean EMD and show a one-prong structure as expected from light quark and gluon jets. The most atypical jets have large mean EMD and multi-prong or diffuse structure. The anomalousness of a jet is non-trivially correlated with the

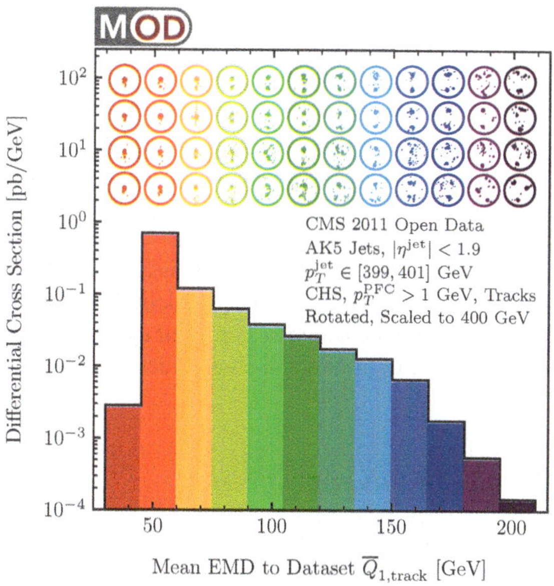

Fig. 2.14 Mean EMD to a dataset of jets with $p_{\mathrm{T}} = 400$ GeV. The four most representative jets for a given histogram bin are shown on top. Taken from [262]

jet mass, which is visually confirmed by the structure of jets best representing the set of jets in a given bin.

2.4.5 Pileup Mitigation

While jet grooming methods target the rejection of soft and wide-angle radiation and particles from the underlying event, pileup mitigation techniques are constructed to remove particles produced in additional proton-proton collisions in the same bunch crossing as the hard interaction. These also contribute to the hard jets in an event and can in principle be distinguished from particles of the leading primary interaction due to their different production vertices. While jet grooming methods also mitigate the effects of pileup, dedicated algorithms have been developed for this purpose. These rely either on the diffuse energy distribution introduced by pileup or on the identification of a pileup vertex as the source of additional particles. Applying pileup mitigation techniques results in an improved resolution of jet substructure observables and a stable performance of taggers even in very dense environments. A more detailed description of some specific pileup mitigation techniques is given in Sect. 3.3, with a focus is on experimental effects.

2.5 (Semi-)Analytical Calculations

Accurate quantitative predictions of quantities related to jets represent a serious theoretical challenge due to the complex structure of QCD, the theory of strong interactions. Besides the usual complexities of higher order calculations in perturbative QCD, additional complications arise for jet observables due to the appearance of multiple, disparate scales, reaching from the high collision energy through the electroweak scale down to hadron masses. This hierarchy of scales renders perturbative expansions unreliable at any fixed order. In order to obtain reliable predictions, all-order reorganisations of the perturbative expansion are necessary. Individual terms in this expansion can often be expressed as closed analytical expressions, but once an all-order resummation is performed, numerical methods need to be employed. Also non-perturbative effects, such as hadronisation and the underlying event, can be estimated using analytic QCD models. While these estimations will never achieve the ultimate theoretical precision, their usefulness lies in attaining an understanding of the behaviour of physical observables in presence of non-perturbative effects.

2.5.1 Perturbative Effects

Different ways of obtaining all-order results for parton level predictions exist in the literature. These can be separated in two classes, one known as direct QCD in the soft and collinear limit (see [263] for a recent comprehensive review) and Soft-Collinear Effective Theory (SCET) [264–270] (an extensive review is given in [271]). While these two approaches are formally very different, their equivalence is discussed in recent studies [272–275].

The calculations in direct QCD rely on the factorisation properties of the real and virtual amplitudes in the soft and collinear limit. While the individual real and virtual contributions can be divergent, these divergencies cancel for IRC safe [47, 159] observables due to the virtues of the Bloch-Nordsieck [276] and Kinoshita-Lee-Nauenberg [277, 278] theorems. While in the case of inclusive variables the cancellation is complete, in exclusive measurements the kinematic dependence on the observable can cause an unbalance between the real and virtual contributions. This manifests in the appearance of potentially large logarithmic corrections at any order in the perturbative series. These contributions spoil the convergence of the perturbative series and must be resummed to obtain reliable predictions. A further complication arises in the case of non-global observables [205]. Contrary to global observables, which are sensitive to emissions anywhere in the phase space, non-global ones are sensitive to emissions only in a part of the phase space. The resulting corrections emerge as non-global logarithms, which also need to be resummed to achieve next-to-leading-logarithmic (NLL) or higher accuracy. These calculations are highly non-trivial and no closed analytical solution of the resummed expression exists [235, 279–281]. In fact, when calculating observables obtained by jet substructure algorithms (e.g. jet grooming, see Sect. 2.4.3), there are several cases where the phase space for gluon emissions is sliced in a non-trivial way which leads to a further complications when aiming for an all-order resummation [237].

The importance of the resummation of global and non-global logarithms is shown in Fig. 2.15, which shows the scaled jet mass $\xi = m/p_{\mathrm{T}}$ calculated for Z+jet production at the LHC [282]. Three different approximations of the NLL result are show. The result in the small-R limit is shown in blue, where large-angle contributions from emissions other than the measured jet are suppressed. Corrections to this approximation are included as power series in the jet radius R, where effects due to the finite size of the jet are included as a resummation of global logarithms (green) and the inclusion of non-global logarithms is shown in red. Note that while the resummed

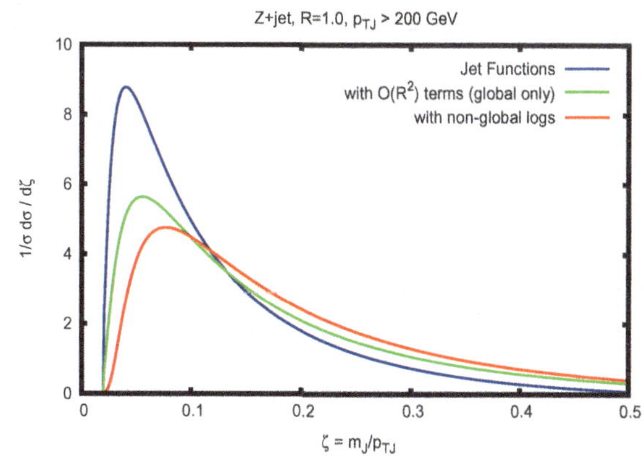

Fig. 2.15 The scaled jet mass calculated for Z+jet production in pp collisions for a jet radius $R = 1.0$. The result is shown for the small-R approximation (blue), with full resummation of the global contribution (green) and with non-global logarithms (red). Taken from [282]

result regulates the divergence of fixed-order calculations at low masses, a matching to fixed-order calculations is needed in order to obtain reliable predictions also at high values of the jet mass.

Calculations in SCET are based on a division of the multiscale problem into appropriate kinematic domains. In each of these kinematic domains an effective Lagrangian is constructed encoding the relevant dynamics of the system. In this way, the hard scattering can still be described by the full QCD Lagrangian, which is obtained by matching QCD to the effective theory. Below the hard scale the effective theory splits into several distinct collinear and soft sectors. Before and after the hard interaction takes place, the jets described by the different collinear sectors evolve independently from each other with only soft, but no hard interactions between them. The radiation between collinear jets is described by soft quark and gluon fields [283]. This simplification is the basis of factorisation theorems in SCET, which allow differential cross sections to be written as convolutions of independent pieces. These pieces are the hard functions, given by the scattering amplitudes to produce N partons, N jet functions encapsulating the evolution of partons to jets due to collinear radiation within jets, non-perturbative soft functions describing soft cross-talk between jets and beam functions, originating from initial state radiation and the usual parton distribution functions [284]. This situation is depicted schematically in Fig. 2.16, for a calculation of jet substructure. The phase space configuration in SCET is shown in Fig. 2.16a for a hard jet containing a soft subjet. The green radiation, associated with the jet functions J_n and $J_{\bar{n}}$ describes the collinear and anti-collinear radiation of the hard jet, respectively. Blue lines show the collinear soft subjet dynamics, described by the soft jet modes $J_{n_{sj}}$ and additional soft radiation at the jet boundary is described by the soft function $S_{n_{sj}\bar{n}_{sj}}$. Grey radiation denotes global soft radiation, described by $S_{n\bar{n}n_{sj}}$. In Fig. 2.16 the relevant scales for this problem, together with the related measurement functions are depicted. Increasingly differential measurements on the jet introduce multiple scales, necessitating an extension of the factorisation theorem.

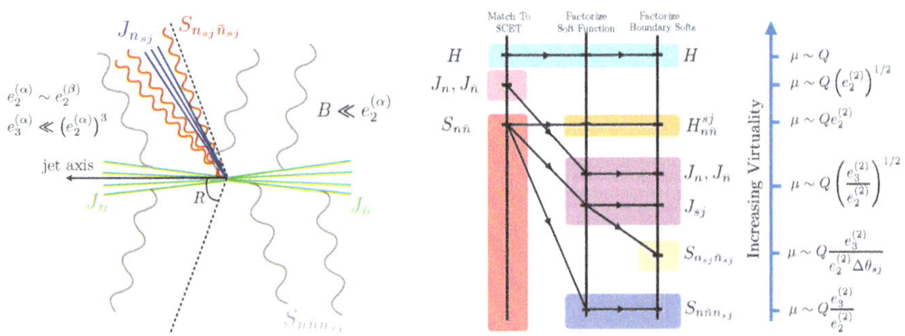

(a) (b)

Fig. 2.16 Schematic drawing of the phase space configuration describing an energetic jet with a soft subjet in SCET (**a**). Different energy regimes, relevant for describing jet dynamics for increasingly differential measurements (**b**). Taken from [285]

The virtue of SCET is that logarithmically enhanced contributions can be resummed at all orders in the coupling constant, which can be achieved using renormalisation group (RG) evolution in the effective field theory. The effective Lagrangian provides a systematic way of organising computations [271]. Factorisation theorems in SCET can also be constructed for jet substructure observables, allowing the resummation of non-global logarithms [285]. Calculations in SCET predictions have been performed for jet substructure observables at the LHC at NLL accuracy [286, 287] and even next-to-next-to-leading-logarithmic (NNLL) [195, 288] and higher accuracy [211].

2.5.2 *Non-perturbative Effects*

Non-perturbative effects are rarely taken into account in analytical calculations due to the inherent complications when approaching energy scales close to hadron masses $\mathcal{O}(\Lambda_{\text{QCD}})$. However, the impact of non-perturbative effects on jet observables as been estimated analytically using power corrections [186]. Results were obtained for inclusive jet production near the partonic threshold, considering gluon emission and separating the phase space into a perturbative and a non-perturbative region, defined by an infrared factorisation scale μ_I. Below this scale, in the non-perturbative regime, the strong coupling α_s is replaced by an effective, finite constant. This allows for the phase space integration to be performed down to vanishing scales. The contribution that would be included in perturbative calculations is subtracted, leaving a purely non-perturbative result that can be associated with the effects of hadronisation and the underlying event.

The expected modification of a jet's transverse momentum due to hadronisation is [186]

$$\langle \delta p_{\text{T}} \rangle_h = 2C_R A(\mu_I) M \left(-\frac{1}{R} + \mathcal{O}(R) \right) , \tag{2.44}$$

where C_R is the colour factor appropriate for the parton initiating the jet, i.e. $C_R = C_F$ for quark jets and $C_R = C_A$ for gluon jets. The hadronisation scale $A(\mu_I)$ is related to event shape studies, and a rough estimate gives $2C_F A(2\,\text{GeV}) \approx 0.5\,\text{GeV}$. The Mellin factor M is an algorithm-dependent quantity with $M = 1.49$ for the anti-k_{T} algorithm and $M = 1.01$ for the k_{T} algorithm [289]. The singular $1/R$ behaviour originates from the increase of momentum loss as the jet becomes narrower, which also explains the negative sign of this term in (2.44).

Corrections due to the underlying event can be calculated from emissions of dipoles not involving the outgoing jet and result in a change in p_{T} proportional to R^2, i.e. proportional to the jet area [186],

$$\langle \delta p_{\text{T}} \rangle_{\text{UE}} = \frac{\Lambda_{\text{UE}}}{2} \left(R^2 - R^4/8 + \mathcal{O}(R^6) \right) . \tag{2.45}$$

The scale Λ_{UE} receives contributions from the interactions of the proton remnants and can not be calculated, but can be estimated from experimental studies or from event generators. At the LHC, a rough estimate is $\Lambda_{\mathrm{UE}} \sim 10\,\mathrm{GeV}$. A more realistic treatment needs to take into account the fact that jets are not exactly circular, and the term R^2 in (2.45) is replaced by the active jet area [185].

Of particular importance to substructure analyses are non-perturbative corrections to the jet mass. The leading power corrections to the squared jet mass due to hadronisation are [186]

$$\langle \delta m^2 \rangle_h = 2 C_R A(\mu_I) M p_{\mathrm{T}} \left(R + \mathcal{O}(R^3) \right) , \tag{2.46}$$

and scale with p_{T} and the jet radius R. The singular $1/R$ behaviour from (2.44) is absent since gluons emitted outside of the jet do not contribute. The contribution of the underlying event to the squared jet mass is [59]

$$\langle \delta m^2 \rangle_{\mathrm{UE}} = C_R \frac{A(\mu_I)}{4} \left(p_{\mathrm{T}} R^4 + \mathcal{O}(R^6) \right) , \tag{2.47}$$

and is suppressed by three powers of R over the hadronisation correction for $R < 1$.

Contrary to direct QCD, in SCET non-perturbative effects are taken into account through the soft functions. These can encode corrections due to hadronisation and the underlying event and can be computed. Recently, progress has been made in relating quantities calculated for e^+e^- collisions to ingredients needed for jet production at hadron colliders [287, 290].

2.6 Event Generators

Multi-purpose event generators have become indispensable tools in high energy physics. Their success is due to their ability of simulating processes at all relevant stages of particle collisions, together with the availability of numerous processes, often sufficient to simulate all relevant processes for a given type of collision (e^+e^-, ep, $p\overline{p}$, pp), sometimes even including BSM effects. Event generators rely heavily on an all-order factorisation of the individual contributions simulated, which cannot be proven formally, but has been shown to be a reasonable approximation.

Event generation starts from the hard interaction obtained with LO or NLO matrix elements. Also at this stage, underlying event contributions can be simulated. This is followed by the generation of coloured partons and photons from initial (ISR) and final state radiation (FSR), known as parton showers. Each simulated physical contribution to a given process is convoluted with a parton distribution function (PDF), which encodes the number density of a parton with a given flavour and longitudinal momentum fraction x in the initial state hadron. In fact, the PDFs, multiple partonic interactions and the shower evolution are intrinsically connected and a consistent framework for these effects needs to be formulated. Once the shower evolution reaches a perturbative cut-off scale, hadronisation sets in, resulting in numerous colourless mesons and baryons. These are subsequently allowed to decay, sometimes developing decay chains. Finally, the event generation terminates with stable particles emerging from the interaction.

For jet substructure analyses, results can be obtained at formally (N)LO+LL accuracy for observables either on the level of coloured final state partons or on the level of stable particles. Often these predictions have better predictive power than their accuracy suggests, owing to the partial inclusion of formally higher order effects and tuning of (the many) free parameters to experimental data [291–295]. In general, this leads to a reliable modelling of the internal structure of jets in many different aspects (integrated and differential jet shapes, particle multiplicities, angular and energy correlations,...), which facilitates experimental analyses. This also permits event generators to be used for the validation of (semi-)analytical calculations, especially when no experimental data are available. The three general-purpose event generators used extensively at the LHC are PYTHIA [248–250, 296], HERWIG [297–300] and SHERPA [301, 302], which has been built around the multi-leg matrix element generators AMEGIC [303] and COMIX [304].

Simulated events obtained from event generators can be passed through detector simulations, either detailed ones using the GEANT toolkit [305, 306] or simplified detector simulations like PGS [307] or DELPHES [308]. With these, the effect of the granularity of detector readout units, finite efficiencies and a realistic detector response can be studied. Additionally, overlaying simulated events with minimum-bias (MB) events can be used for studies of the effects of pileup. At the end of this long chain of detailed simulation steps, simulated events on the level of the detector reconstruction are obtained and can be directly compared to recorded collision data. These simulations can be used to directly test the underlying physical principles, together with the approximations made in event generators. They can also be used to derive corrections for data in order to obtain measurements fully corrected for detector effects.

2.6.1 Parton Distribution Functions

Particle collisions involving hadrons in the initial state are described by colliding clouds of partons, with the number density of parton flavour i given by the parton distribution function $f_i(x, \mu_f^2)$. The PDFs depend on the longitudinal momentum fraction x carried by the parton relative to the hadron momentum, and the factorisation scale μ_f, usually taken to be the hard scale probed by the interaction, $\mu_f = Q$. The form of PDFs cannot be predicted in perturbative QCD, but has to be parametrised at some starting scale Q_0. The evolution of the PDFs to some scale Q, with $Q > Q_0$, is obtained by the DGLAP equations [309–312]. The free parameters of the parametrisations have to be obtained from experimental data. It is due to the freedom of choosing a parametrisation, the theoretical treatment of heavy quark mass effects, assumptions on the sea quark densities and the wealth of experimental data to choose from, that a number of different PDF sets have been proposed by several independent groups.

Global PDF fits, including experimental data from different experiments, are performed by the CTEQ collaboration, the NNPDF, and the MMHT (formerly MSTW [313] or MRST [314]) groups. Fits to HERA data only are carried out by the HERAPDF group, and the ABM collaboration provides PDF sets in the fixed-flavour-number scheme for the treatment of heavy quarks. The latest PDF sets published by these groups are the CT14 [315], NNPDF 3.0 [316], MMHT 2014 [317], HERAPDF 2.0 [318] and the ABM14 [319] sets. A detailed overview and comparisons are given in the PDF4LHC working group reports [320, 321].

While there are typically differences in the central predictions and uncertainties obtained, cross sections calculated with different PDF sets tend to agree within a few percent. The differences in PDF sets are subordinate compared to other uncertainties in jet substructure analyses, where the PDFs enter mostly in the simulation of the underlying event through initial state radiation and multiple parton interactions, and through uncertainties in the acceptance once experimental selections are introduced. Whereas the precision of recent PDF sets is sufficient in jet substructure analyses, even when considering the spread of differences obtained by different PDF sets as an additional uncertainty, precision analyses and the prediction of SM cross sections are affected by the corresponding uncertainties. Extensive studies in the framework of the PDF4LHC working group have been carried out in order to understand and quantify the differences [322–325], which lead to a much improved situation. Work is ongoing in providing combined PDF sets together with uncertainties based on Hessian reduction [326, 327] and Monte-Carlo methods [328].

2.6.2 Matrix Elements

The first step in the generation of events is the simulation of the primary hard interaction expressed for two incoming partons to produce final state X. In proton-proton collisions, due to the virtue of the factorisation theorem [329–331], the total cross section for two incoming protons, $pp \to X$, can be written as convolution with PDFs,

$$\sigma_{pp\to X} = \sum_{i,j=q,\bar{q},g} \int\!\!\int \mathrm{d}x_1 \mathrm{d}x_2 \, f_i\left(x_1, \mu_f^2\right) f_j\left(x_2, \mu_f^2\right) \hat{\sigma}_{ij\to X} \,, \tag{2.48}$$

with equivalent expressions for differential cross sections. The partonic cross section $\hat{\sigma}_{ij\to X}$ depends on the kinematics of the process, the factorisation scale μ_f, and the electromagnetic $\alpha(\mu_r)$ and strong $\alpha_s(\mu_r)$ couplings, evaluated at the renormalisation scale μ_r. A typical choice is $\mu_f^2 = \mu_r^2 = Q^2$, where the scale is either $Q^2 = p_\mathrm{T}^2$ for massless outgoing particles, or $Q^2 = p_\mathrm{T}^2 + \sum m_i^2$ when massive particles are involved.

A large number of $2 \to 1$ and $2 \to 2$ processes, as well as some $2 \to 3$ processes are implemented in the standard event generators PYTHIA and HERWIG. The event generator SHERPA allows for the simulation of tree-level $2 \to n$ matrix elements. In case of resonance production with $2 \to 1$ graphs and the resonance being either a W, Z, H boson or some BSM resonance, the decay is simulated through a Breit-Wigner shape with energy-dependent width, resulting in $2 \to 1 \to 2$ processes. A similar procedure is applied when more than one resonance is produced (e.g. $gg \to t\bar{t}$), where Breit-Wigner distributions with fixed widths are used to assign masses to the resonances.

Modern event generators often go beyond the generation of LO $2 \to 2$ processes at the matrix element level, which is accomplished through an interface to external matrix element generators. These have to be matched to the parton shower simulation to avoid double-counting of higher order contributions, as described below.

2.6.3 Parton Showers

In event generators the calculation of inclusive cross sections from (2.48) is augmented by additional particles originating from parton branchings of the form $a \to bc$. These $1 \to 2$ splittings are crucial for the description of local observables (i.e. observables sensitive to emissions in localised parts of the phase space), in particular all quantities related to jet substructure. The total inclusive cross section is unchanged by the parton shower. The simulated parton branchings result in a tree-like structure for individual events (a shower), and correspond to a resummation of the leading logarithmic (LL) corrections when considering an ensemble of all possible shower configurations. While this suggests a formal accuracy of event generators at the (N)LO+LL level, to equate a modern parton shower with LL accuracy of analytic calculations is a gross underestimation. A modern parton shower respects energy and momentum conservation through partonic recoils and includes coherence effects, which enter only at NLO in analytic calculations. The final product are predictions on a similar footing as (N)LO+NLL calculations, even though $\mathcal{O}(\alpha_s^2)$ terms are not included in the splitting kernels.[13]

The basic building blocks of parton showers are the $1 \to 2$ DGLAP splitting kernels $P_{a\to bc}$, describing the collinear splitting of parton a into partons b and c.[14] For a given branching $a \to bc$, parton b takes the fractional momentum z and parton c a fraction $1 - z$. The second variable specifying the kinematics is given by $t = -2p_a p_b$, which has dimension of a squared mass and is related to the mass or transverse momentum scale $k_\perp$ of the branching. Confinement implies a natural cut-off scale for the transverse momentum generated by the parton shower of about

[13]These would result in $1 \to 3$ splittings and corrections to the $1 \to 2$ splitting functions.

[14]In parton showers, the '$\pm$ prescriptions' and $\delta(1 - z)$ terms are absent. These ensure flavour and energy conservation in analytical calculations, which are encoded at each step of the parton shower where each step is traced in detail.

1 GeV, which results in an infrared cut-off scale t_c. This leads to an upper bound on z, given by $z < 1 - t_c/t_{\max}$, where $t_{\max}$ is the maximum parton shower scale, usually identified by the scale of the primary hard scattering. The upper bound on z results in a bound on the number of soft gluons produced, which would otherwise tend to infinity due to the soft singularity. The computation of the total number of branchings and the corresponding values of t at which these occur in individual events are the primary tasks of parton shower algorithms. The ensemble of partonic final states in many events then corresponds to the solution of an evolution equation for parton showers, similar to the DGLAP evolution. The probability for parton a to branch at a given value of t can be expressed by an integral over all allowed z values, $I(t)$. Conversely, the probability for no branching to occur between an initial value of t_0 and t exponentiates, which is known as Sudakov form factor $S_a(t_0, t)$. Thus, the actual probability for parton a to branch at scale t, is given by $I(t) \cdot S_a(t_0, t)$, if parton a was produced at scale t_0. This means that parton a can only branch at t if it has not branched already at an earlier stage $t' < t$. The Sudakov factor is the appropriate suppression factor, ensuring the conservation of total probability. Formally, the Sudakov factors account for any unresolved splittings and virtual corrections which are assumed to precisely cancel the singular real corrections when integrated over phase space [332]. The product $I(t) \cdot S_a(t_0, t)$ can be cast into an evolution equation, which is solved by any parton-shower event generator.

While this prescription is fairly general, there are a number of choices that have to be made in the exact implementation of a parton shower and the quality of parton shower predictions depends significantly on these choices. One important example is the exact choice of the evolution variable, which is one of the main differences between the most commonly used event generators. The virtuality t was used in PYTHIA versions earlier than 6.4 and SHERPA versions earlier than 1.2, whereas both generators switched to a $k_\perp$-based evolution for later versions. This change expedites the merging of parton showers with higher-order, multi-leg matrix element generators and the use of recent multiple parton interaction models. The squared energy-weighted emission angle, $E^2\theta^2$ is used in early HERWIG versions, with a generalisation implemented in recent versions. Another important choice is the splitting variable z, where the two common choices are the light-cone (LC) momentum fraction or the energy fraction taken by parton b. Coherence effects can be accounted for by employing a dipole formalism or an angular ordering (AO) of emissions. If neither option is viable, an angular ordering of emissions can be enforced through a veto. Recently, a new shower model has been developed, called DIRE [333], which is a hybrid between dipole and parton shower. The shower has been implemented in PYTHIA and SHERPA, and is the only algorithm which can be used in two independent programs, therefore easing the comparison of results and identifying differences in other parts of the programs. The choices made in common parton-shower event generators are summarised in Table 2.4.

Additional complications in the implementation of parton showers arise because of differences between final state showers with time-like virtualities and initial state showers with space-like virtualities. In final-state showers, once the shower cut off is reached, partons are put on the mass shell. Conversely, initial state showers have to be

Table 2.4 Choice of evolution/splitting variable and evolution kernels in common parton-shower programs. Table adapted from [332]

	Evolution variable	Splitting variable	Coherence	References
PYTHIA < 6.4	t	Energy fraction	Enforced	[334, 335]
PYTHIA ≥ 6.4	$k_\perp^2$	LC mom fraction	Enforced	[336]
HERWIG	$E^2\theta^2$	Energy fraction	AO	[297, 337]
HERWIG ++	$(t - m^2)/z(1 - z)$	LC mom fraction	AO/Dipole	[338, 339]
SHERPA < 1.2	t	Energy fraction	Enforced	[340]
SHERPA ≥ 1.2	$k_\perp^2$	LC mom fraction	Dipole	[341]
DIRE	dipole-$k_\perp^2$	LC mom fraction	Dipole	[333]

evolved backwards in time [334, 342], starting from the hard scale $t_{\max}$ and evolving towards decreasing values of virtuality. This is due to the unknown value of $t_{\max}$ prior to the generation of the primary hard scatter, which renders the exact handling of the splitting kinematics impossible. In initial state showers PDFs are introduced in the Sudakov form factors, reflecting the probability of parton a to come from the proton. Final state partons with time-like virtualities can also be created in initial state showers, where partons off the main branch can have positive virtualities. Some other aspects that complicate matters are quark masses, the distribution of recoil momenta, colour connections, QED branchings, and the choice of the running electromagnetic and strong couplings. Details on the treatment of these effects can be found in [248, 298, 302].

The key importance of parton showers for predicting jet kinematics and jet substructure observables has led to a number of recent developments towards increasing the accuracy of parton showers. Part of the difficulty is a quantitative estimate of the achieved accuracy for a given observable, which can be defined by considering the logarithmic divergencies in real and virtual corrections [343]. While no parton shower correct at NLL accuracy exists for the simultaneous prediction of both non-global and a wide set of global observables, promising approaches have already been formulated and are subject of much recent research [343–349].

2.6.4 *Matching Matrix Elements to Parton Showers*

Matrix element calculations rely on an expansion of the perturbative series in powers of α_s and can be performed for several (up to about 10) partons in the final state for Born-level predictions. Including higher order corrections is a challenging task due to the required cancellations of real and virtual corrections, but recent developments have led to the automated computation of one-loop amplitudes [350–355]. These resulted in an impressive list of available final states at NLO precision (which became known as the NLO revolution). Calculations at the level of NNLO are progressing, with a number of $2 \to 2$ processes already available [356–364], but until now no automatisation has been achieved due to the enormous complexity of these calculations. Due to the difficulty of calculating higher order contributions with matrix element methods, these can not be used to reliably predict the internal structures of jets. The parton shower formalism, on the other hand, has been obtained for the region of soft and collinear emissions, with numerous successive branchings leading to a high parton multiplicity in the final state. Parton showers are thus well suited to model jet substructure with perturbative methods, while the description of well separated jets is not guaranteed. The two approaches complement each other and much effort has been made to combine them without double-counting of graphs generated by the matrix elements and the parton shower, or with gaps in the phase space coverage.

Multileg tree-level calculations provide real corrections to the $2 \to n$ process of the form $2 \to n+1, 2 \to n+2, \ldots$. When merging these individual contributions, a naive combination would result in double counting, since configurations with higher parton multiplicities contribute to inclusive $2 \to n$ process with lower multiplicity. A merging with parton showers exacerbates the issue. Solutions to this problem are offered by the CKKW [365], CKKW-L [366], MLM [367] and UMEPS [368] methods. These methods introduce Sudakov form factors approximately accounting for the effect of virtual corrections, and have been extensively used in older versions of PYTHIA, HERWIG and SHERPA.

When including full higher-order information, starting with one loop diagrams leading to predictions at full NLO precision, the first parton shower emission needs to be corrected to account for changes of the Born-level kinematics. Two successful approaches used in this context are the MC@NLO [369] and POWHEG [370, 371] methods. These have become standard for the generation of simulated events at the LHC, leading to NLO predictions with varying parton multiplicities fully matched to parton showers. The fixed-order matrix element calculations at NLO are either an integral part of the event generator as in the case of HERWIG 7 [300],[15] or obtained from the external programs MADGRAPH5_AMC@NLO [372] or POWHEG BOX [373] as in the case of PYTHIA, and BLACKHAT [350] for SHERPA. Recently, a number of studies have been performed on multileg NLO merging [374–376], which can also be extended to full NNLO calculations matched to parton showers [191, 377]. In the not too distant future, these will become the new standard for event simulation at the LHC.

2.6.5 Multiple Parton Interactions

Multiple parton interactions (MPIs) are additional $2 \to 2$ scatterings that occur within the same pp interaction as the primary hard scattering. To first approximation, MPIs are scatterings of the spectator partons in the proton. Since the $2 \to 2$ parton-parton cross section diverges for $p_{\mathrm{T}} \to 0$ like $\alpha_s(p_{\mathrm{T}}^2)/p_{\mathrm{T}}^4$, the divergence needs to be

[15] HERWIG 7 also offers the possibility to use external matrix element calculations as input.

regulated. This is usually done by including a phenomenological cut off $p_{\mathrm{T},0}$ such that

$$\frac{\alpha_s(p_{\mathrm{T}}^2)}{p_{\mathrm{T}}^4} \to \frac{\alpha_s(p_{\mathrm{T},0}{}^2 + p_{\mathrm{T}}^2)}{(p_{\mathrm{T},0}{}^2 + p_{\mathrm{T}}^2)^2}, \tag{2.49}$$

which smoothly regularises the divergence [378]. The cut-off parameter $p_{\mathrm{T},0}$ can have a dependence on $\sqrt{s}$. In the process of simulating MPIs, the structure functions used for subsequent scatterings must depend on all preceding x values and flavours chosen. To achieve this, the ordinary PDFs are still used for the hardest scattering, but are split into a valence and a sea part for MPIs, where the x values of the spectator partons are rescaled to ensure energy and momentum conservation [379].

By definition, MPIs have a scale smaller than the scale of the hard scattering, $\mu_f = p_{\mathrm{T},1}$, and can be introduced as scatterings ordered in a sequence $p_{\mathrm{T},1} > p_{\mathrm{T},2} > \ldots > p_{\mathrm{T,n}}$. In order to model the competition between MPI and ISR for energy in the incoming beam, these two effects, together with FSR, are interleaved [336, 380], where ISR and FSR are modelled identical for the primary hard scattering and MPI. More complicated MPI schemes have been studied as well, like joined interactions where two partons participating in a MPI have a common ancestor [336], or rescattering, where a parton from one incoming hadron scatters against two or more partons of the other incoming hadron [381, 382]. Another important aspect of MPIs is the modelling of the dependence on an impact parameter, where the interaction rate is proportional to the overlap of the two colliding hadrons [378]. A complicated aspect of the dynamics of many parton interactions with varying scales is the way how the colours of partons are connected with each. There are different schemes implemented, also depending on the hadronisation model chosen. Generally, for substructure applications the effect of changing these choices has been found to be small, but should be tested in analyses. For a recent review on MPIs in PYTHIA, see [383].

2.6.6 Hadronisation

The parton showers from the hard scattering and from multiple parton interactions terminate at a scale t_{min}, at which the value of α_s becomes large and perturbative methods are expected to fail. At this stage, the event is populated with a number of final state partons which have to undergo a non-perturbative transition, called hadronisation, to produce the actual hadronic final state. While no rigorous approach exists to describe non-perturbative hadronic phenomena, two independent models, namely the string and cluster models, have been very successful in describing a wealth of data. These models are inspired by QCD and incorporate features based on phenomenological observations.

The hadronisation in the PYTHIA event generator is based on the Lund string model [384–386]. The model is based on the expectation from confinement that the potential between two colour charges increases linearly with distance, $V(r) = \kappa r$, for distances larger than about 1 fm, with the string tension $\kappa \approx 1\,\mathrm{GeV/fm}$. This form of the potential has also been confirmed by lattice QCD simulations [387]. This potential can be viewed as the one-dimensional colour flux tube stretched between a colour-anticolour state. Originally formulated for $q\overline{q}$ states, the formalism has been extended to multi-gluon states [388]. The string produces a linear confinement potential, which breaks up through new $q\overline{q}$ pairs produced in the intense colour field, resulting in regions with no colour field between singlet states. If the invariant mass of the individual systems is large, further breaks may occur and after $n-1$ breaks the system fragmented into n primary hadrons. A pictorial representation is shown in Fig. 2.17. In order to produce new $q\overline{q}$ pairs with invariant mass or relative transverse momentum, the q and $\overline{q}$ must be produced some distance apart such that the field energy between them can be transformed into a transverse mass $m_\perp$. This implies a quantum mechanical tunnelling process to reach the classically allowed region, with probability proportional to $\exp(-\pi m_\perp^2/\kappa)$. The tunnelling gives rise to a flavour-independent Gaussian p_T spectrum of the $q\overline{q}$ pair, which translates into p_T of the produced hadron. The tunnelling implies a suppression of heavy quark production of $u:d:s:c \approx 1:1:0.3:10^{-11}$. Thus, charm and heavier quarks are not produced in the hadronisation [390]. The formation of baryons is more complicated to model due to the three valence quarks needed. In the simplest approach, a system of diquark-pairs is produced instead of a $q\overline{q}$ pair [391] and leads to the string breaking and the formation of baryons. In a more complex approach (the "popcorn model") [392, 393], baryons are constructed from the successive production of several $q\overline{q}$ pairs, which leads to less strong correlations in momentum between the baryon and the anti-baryon pair.

The cluster hadronisation model is based on the idea of pre-confinement [394], which means that at any shower cut-off scale t_min colour singlet combinations (clusters) of partons can be built, featuring a universal mass distribution. If $t_\mathrm{min} \gg \Lambda_\mathrm{QCD}$, this distribution is perturbatively calculable and is independent on the scale of the hard process [395]. Clusters are formed by connecting colour lines with anti-colour lines in the colour plane of the shower. Adjacent lines form colour singlet states and imply closeness in phase space, leading to the suppression of large cluster masses.

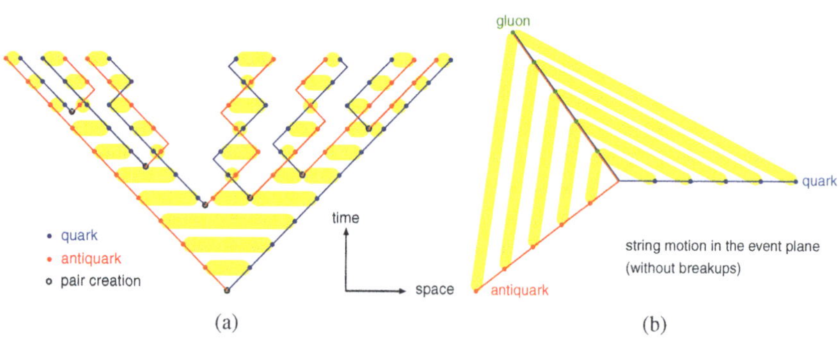

Fig. 2.17 Schematic drawing of a string breakup in a $q\overline{q}$ event (**a**). Horizontal bars represent the string colour connections and diagonal lines the quark propagation. String configuration in a $q\overline{q}g$ event (**b**). Taken from [389]

In order to obtain clusters with mesonic quantum numbers, non-perturbative gluon splittings $g \to q\overline{q}$ must be enforced at the scale t_{min}. The decay of clusters to hadrons is modelled by treating clusters as excited hadrons, where all decay channels allowed by flavour constraints and kinematics need to be included. Most clusters undergo a sequential chain of two-body decays. The limited cluster mass spectrum leads to limited transverse momenta of the produced hadrons and a natural suppression of heavy flavours. Heavier clusters with masses above 3–4 GeV appear in events with little parton showering and are typically forced to fission into lighter clusters before the decay into hadrons is started. Parameters can be introduced to steer the momentum distribution of light quarks from gluon splittings, the flavour distribution, and the gluon splitting to light diquark/anti-diquark pairs for baryon production. The two implementations of cluster hadronisation in use in HERWIG and SHERPA are based on [396] and [397], respectively.

A number of unstable hadrons are produced in the hadronisation phase of event generation. The decay of these hadrons to particles stable on the time scale of collider experiments is simulated using experimental data wherever possible (most notable the decay tables from the Particle Data Group (PDG) [398]), and theoretically well motivated choices in cases where no data is available. The level of sophistication for the simulation of decays is quite high nowadays, including matrix elements for certain decay modes and spin correlations. The simulation of particle decays is closely related to the hadronisation process, such that the free parameters of both models need to be adjusted simultaneously when comparisons to data are made.

In the context of jet substructure, hadronisation and particle decays lead to a dispersion of the energy flow and angular distances within jets. Ultimately, this results in a broadening of distributions in substructure observables, comparable to the effect of finite resolutions and efficiencies from particle detectors, albeit hadronisation typically results in smaller modifications. The difference between the effects from hadronisation and multiple parton interactions is due to the characteristic radiation pattern. While hadronisation is related to the dynamics of the partonic final state and thus leads to a broadening of distributions, multiple parton interactions result in uncorrelated radiation with respect to the hard interaction, leading to a dilution of jet substructure observables.

2.6.7 Tuning

It is apparent from the discussion above that event generators are complicated algorithms with many free parameters. These parameters are usually adjusted such that a variety of measurements are well described (referred to as tuning). While final state showers are constrained by data from e^+e^- collisions (mostly by using measurements from the LEP and SLD collaborations), the parameters affecting initial state showers can only be adjusted through data from hadronic collisions (ep, $p\overline{p}$, pp). Additional complications arise because the parameters of the parton shower are intricately connected with parameters from the modelling of multi-parton interactions and hadronisation. A full tune can only be successful in describing all aspects of collider data if all free parameters are constrained simultaneously. A parametrisation tool called Professor [399] has been developed for this purpose, facilitating the fast calculation of the event generator response to parameter variations. Various tunes [291–295] are available for the three common event generators PYTHIA, HERWIG and SHERPA, and are updated regularly when new measurements become available or major modifications are made to the programs. Only once all relevant measurements are described satisfactorily, an event generator can be used to make reliable predictions for unknown regions of phase space. However, even then caution is advised and some understanding is needed of the approximations made in event generators and their limitations for successful use. This is especially true when event generators are used to predict backgrounds in analyses searching for new physics, where the interesting regions have usually not been explored so far. One recent example comes from a study using jet substructure to separate quark- from gluon-initiated jets [400]. This study observes large differences in the quark/gluon discrimination power under variations of generator settings, with some trends in opposite directions when comparing different generators. A possible conclusion is that, contrary to common belief, e^+e^- data does not constrain all aspects of final state showers sufficiently. Gluon radiation patterns are largely unconstrained and new global tuning strategies including more LHC measurements may be needed in the future.

Chapter 3
Jet Substructure at the LHC

Abstract Before applying jet substructure methods to experimental data, the detector response to jets, to regions of high energy density within jets and even to individual particles has to be studied. This chapter introduces the ATLAS and CMS detectors in the light of jet substructure applications. The different approaches by the two collaborations for reconstructing and calibrating jets are described and an overview is given of the methods used to calibrate the jet mass and substructure observables. The mitigation of effects from pileup, i.e. contamination from radiation originating from other proton-proton collisions than the primary one, and the underlying event is discussed. The different approaches used in the experiments for the identification of the origin of jets (jet tagging) are summarised.

3.1 ATLAS and CMS Detectors[1]

The ATLAS [401] and CMS [402] detectors are designed to observe leptons, photons, and hadrons resulting from LHC *pp* and heavy ion collisions. The physics of the hard interaction takes place at the point of collision (the primary vertex) within the beam pipe. Beyond the beam pipe,[2] at 4.4 cm (3.3 cm) in CMS (ATLAS), the first cylindrical layer of detectors encountered are silicon pixels and strips for identification of charged particles. CMS provides a 3.8 T magnetic field via a solenoid positioned outside the silicon tracking detector, the Electromagnetic Calorimeter (ECAL) and most of the Hadronic Calorimeter (HCAL). ATLAS has an additional tracking layer composed of straw drift tubes (Transition Radiation Tracking or TRT), with a 2 T magnetic field encompassing the silicon and TRT detectors, while the ECAL and HCAL are situated outside the solenoidal magnet. The calorimeters are surrounded

[1]The text in this section has been taken from [26] and has been written by the author. It has been adjusted to fit this book.

[2]The LHC collaborations are continuously working to improve the detectors; the numbers given here are for the detectors that operated in 2015–2017. Before and after this time, the exact values are not the same as reported here.

R. Kogler, *Advances in Jet Substructure at the LHC*, Springer Tracts in Modern Physics 284, https://doi.org/10.1007/978-3-030-72858-8_3

Table 3.1 ATLAS and CMS detectors in the barrel regions. The granularity is in pseudorapidity and azimuth ($\eta \times \phi$) and d_0 is the transverse impact parameter resolution with respect to the beam-line. The tracker momentum resolution is from muons while the d_0 resolution is from generic charged particles (mostly pions) in $t\bar{t}$ events. The ECAL energy resolution is presented for electrons. The granularity for the ATLAS calorimeters are for the middle layers only, which collect the largest amount of energy. For the ATLAS EM calorimeter, the innermost layer has $\Delta\eta = 0.0031$ for γ/π^0 separation. Taken from [26]

	ATLAS	CMS
Tracking		
$1/p_T$ resolution	$0.05\% \times p_T/\mathrm{GeV} \oplus 1\%$ [403]	$0.02\% \times p_T/\mathrm{GeV} \oplus 0.8\%$ [404]
d_0 resolution (μm)	20 [405]	20 [404]
ECAL		
E resolution	$10\%/\sqrt{E} \oplus 0.2\%$ [401]	$3\%/\sqrt{E} \oplus 12\%/E \oplus 0.3\%$ [402]
Granularity	0.025×0.025	0.017×0.017
HCAL		
E resolution	$50\%/\sqrt{E} \oplus 5\%$ [401]	$100\%/\sqrt{E} \oplus 5\%$ [406]
Gtranularity	0.1×0.1	0.087×0.087

by muon spectrometers which build the outermost part of the ATLAS and CMS detectors. Both detectors are nearly hermetic and can therefore measure the missing transverse momentum.

The energy and momentum ranges and resolutions for the barrel regions[3] of ATLAS and CMS are shown in Table 3.1 along with the measurement granularity, which limits the angular resolution. The better energy resolution of the CMS ECAL is due to the use of lead tungstate ($PbWO_4$) crystals, as opposed to the Liquid Argon (LAr) used by ATLAS. The differences in the ATLAS and CMS calorimeter designs are a result of the different ranking of priorities decided by the two collaborations; ATLAS chose a radiation-hard technology with sufficient resolution in a fine sampling LAr calorimeter, while CMS prioritised the excellent resolution of a total absorption crystal calorimeter (the focus was Higgs mass reconstruction), and accepted the accompanying limitations in radiation-hardness associated with this technology. The CMS ECAL crystal response varies under irradiation, which is partially recovered in a few hours at room temperature.

The ATLAS ECAL is segmented into two and three longitudinal layers for $|\eta| > 2.5$ and $|\eta| < 2.5$, respectively. The granularity of the ATLAS ECAL in Table 3.1 refers to its second layer (as most of the electromagnetic energy is deposited there); the first layer has a finer granularity in η. The multiple layers allow for a finer granularity than the cell size in any of the individual layers, being advantageous over

[3]For example, the ATLAS ECAL barrel covers the pseudorapidity range $|\eta| < 1.475$, the end-caps cover $1.375 < |\eta| < 3.2$ and the forward ECAL layer extends the coverage up to $|\eta| < 4.9$. The CMS ECAL barrel covers $|\eta| < 1.48$, the end-caps extend the coverage up to $|\eta| < 3$.

a laterally segmented calorimeter, and additionally provide pointing information. The difference between ATLAS and CMS for the HCAL resolution is particularly large at higher energies: a 1 TeV jet has $\frac{\sigma(E)}{E} \sim 2\%$ in ATLAS, in contrast to $\frac{\sigma(E)}{E} \sim 5\%$ in CMS. This is one reason why CMS fully adapted a particle flow technique since the beginning of the LHC (see Sect. 3.2).

3.2 Jet Reconstruction and Calibration[4]

The ATLAS and CMS experiments have dedicated algorithms to reconstruct particle kinematics from calorimeter and tracker information designed to minimise the fake rate, maximise the efficiency, and minimise the bias and resolution of the particle candidate parameters. As there is no algorithm that can simultaneously optimise all of these objectives, the various approaches trade off optimality under one metric for improvements under another. ATLAS and CMS have also developed different algorithms that cater to the experiment's hardware as well as the collaboration's goals for the tradeoffs. By default, CMS combines tracker and calorimeter information into unified particle flow (PF) objects as inputs to jet reconstruction [74, 75, 407]. ATLAS has traditionally used calorimeter-only information for jet reconstruction, with tracking information used to augment/enhance the performance. While ATLAS is current migrating to a variation of particle flow [408], most of this review will focus on calorimeter-only jets as they are still the most widely used setup. ATLAS benefits less than CMS from particle flow because of its weaker magnetic field, the longitudinally segmented calorimeter and the better jet energy resolution obtained with the calorimeter.

ATLAS and CMS combine calorimeter cells using topological clusters [407, 409]. These clusters are three dimensional in ATLAS as a result of the longitudinal segmentation. Cluster seeds are started from highly significant energy (high cell signal compared to average electronic and pileup noise) deposits which are combined (or split) based on the distribution of the significance of energy in nearby cells. Calorimeter-cell clusters in CMS are obtained using a Gaussian-mixture model, which results in one or more calorimeter clusters within each topological cluster. HCAL clusters can be split according to the number and energy distribution of associated ECAL clusters. Cluster splitting is critical to achieve a better estimate of the spatial energy distribution as input to jet substructure algorithms [410, 411].

The topological clusters are calibrated using simulations to account for the non-compensating calorimeter response to hadrons, signal losses due to energy deposited in inactive detector material and signal losses on cluster boundaries caused by the topological clustering algorithms. In ATLAS, the calibration scheme relies on a classification of clusters as hadronic or electromagnetic in origin based on the energy and position of the cluster, the longitudinal depth (λ_{clus}) and normalised signal

[4]The text in this section has been taken from [26] and has been written by the author together with B. Nachman. It has been adjusted to fit this book.

energy density; hadronic showers tend to occur deeper in the calorimeter and be less dense [409]. Charged and neutral pions are used to derive this classification and calibration, called the Local Cell Weighting (LCW). In CMS, dedicated ECAL (based on photons) and HCAL (based on neutral kaons) calibrations are combined to account for energy and $|\eta|$-dependent non-linearities in the hadron calorimeter response [407]. Both ATLAS and CMS validate the performance of these calibrations with single particle studies in data [407, 412].

Different strategies are used by ATLAS and CMS to reconstruct tracks from their inner detectors. ATLAS focuses first on maintaining a high efficiency with a rather inclusive first pass through inner detector hits. A second step known as ambiguity solving reduces the fake rate. In contrast, CMS uses a sequential approach with multiple passes through the remaining inner detector hits. With each pass, the efficiency increases while maintaining a low fake rate. Both procedures are effective at identifying about 90% of charged pions above 1 GeV with a percent-level (or smaller) fake rate. Lower momentum particles can be reconstructed, at the cost of a higher fake rate and lower efficiency. Due to its weaker magnetic field, ATLAS is able to reach low track momentum of 100 MeV for physics analysis [413], although most jet substructure measurements and searches use a threshold of 500 MeV. In contrast, the momentum resolution in CMS is excellent up to higher momenta than in ATLAS. The TRT can be used to improve the momentum resolution of high p_T tracks [414], but the weaker magnetic field despite a comparable inner detector radius is a fundamental limitation.

Both experiments have implemented dedicated strategies for track reconstruction in high density environments such as the core of high p_T jets. In such environments, pixel and strip clusters can merge resulting in a loss in tracking efficiency and degraded resolution. ATLAS has implemented a stacked neural network (NN) approach to examine pixel clusters to identify multi-particle clusters, estimate the position of the particles passing through the clusters, and also predict the residual resolution of the position estimates [415–419]. CMS has introduced a dedicated tracking step in which a cluster splitting procedure attempts to split merged clusters exploiting the information of the jet direction, predicting the expected cluster shape and charge.

For particle flow in CMS, tracks and calibrated clusters are combined taking the tracking and calorimeter resolutions into account. First, a link is created between tracks in the central tracker and calorimeter clusters. Links are also created between clusters in the ECAL and HCAL, if the cluster position in the ECAL is within the cluster envelope in the less granular HCAL. Tracks with a p_T uncertainty in excess of the calorimetric energy resolution expected for charged hadrons are masked, which allows the rate of misreconstructed tracks at large p_T to be reduced.

Charged hadrons are created from ECAL and HCAL clusters, linked to tracks. If the calibrated calorimetric energy is compatible with the corresponding track momenta under the charged-pion hypothesis, no neutral particles are created. Otherwise, the excess energy is interpreted to originate from photons and neutral hadrons for deposits in the ECAL and HCAL, respectively. The remaining ECAL and HCAL

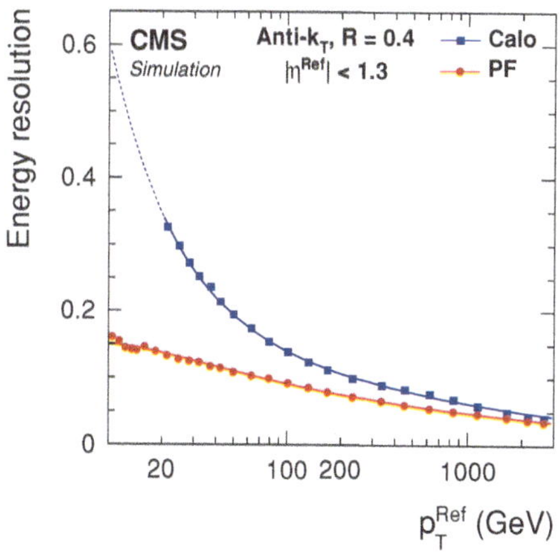

Fig. 3.1 Jet energy resolution for particle flow (red, lower line) and calorimeter-only (blue, upper line) jets in the barrel region in CMS simulation, with no pileup, as a function of the p_T of the reference jet. Taken from [407]

clusters not linked to any track give rise to photons and neutral hadrons. The PF algorithm in ATLAS is similar to the one used by CMS and is described in more detail in [408].

The combination of tracking and calorimetric measurements results in an optimal input for jet substructure measurements, making use of the superior angular resolution from the tracking detector and calibrated calorimeter clusters. Once the calibrated PF objects are clustered into jets, their relative momenta and angular distances are kept constant, and only the total energy response of jets is corrected with factorised jet energy calibrations.

The PF algorithm improves the energy resolution as shown in Fig. 3.1. A similar performance gain is observed in ATLAS [408], but the weaker magnetic field means that the point where calorimetry and tracking are comparable is lower (about 100 GeV).

The ratio of the measured energy E_{reco} to the deposited energy E_{true} is the jet energy response which depends on the energy, pseudorapidity and other features of the jet. Due to the properties of tracking detectors and calorimeters, the average response is not unity. For example, calorimeter jets in ATLAS with $E_{\text{true}} = 30\,\text{GeV}$ may have responses below 0.3, while jets of higher energies may have responses above 0.8. For this reason, the jet energy scale (JES) is calculated in bins of the particle-level jet energy E_{true} and η_{det} as the mean of a Gaussian fit to the response distribution and a numerical inversion procedure is used to derive calibration factors in bins of the reconstructed jet energy from E_{true} [420–423].

In ATLAS, the calibration of the JES is undertaken in several stages, starting from jets either at the electromagnetic (EM) or LCW (built from calibrated inputs) scale. Using calibrated inputs bring the JES to within 10% of unity for $E = 30\,\text{GeV}$ and $|\eta| < 0.3$ [420]. The Global Sequential Calibration [421, 424, 425] was introduced in 2015 and reduces the sensitivity to differences in the responses of quark versus gluon-initiated jets. This additional calibration results in a significant jet p_T resolution improvement of up to 35% depending on the p_T and η of the jet [425]. The JES

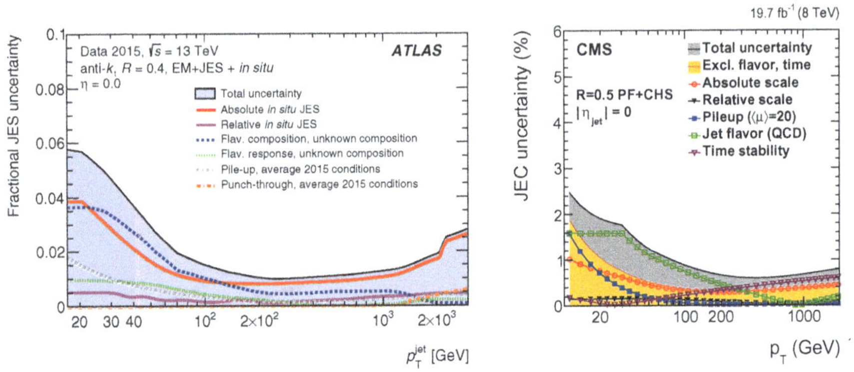

Fig. 3.2 ATLAS (left) and CMS (right) jet energy scale uncertainty. Taken from [70, 421]

uncertainty varies between 1–6% in the central region with $\eta = 0$ as shown in Fig. 3.2 (left).

In CMS, jets are clustered from calibrated PF objects, thus the uncalibrated JES is within 6% of the expected value of 1 for central jets with $\eta < 0.7$ and $p_T > 30$ GeV [70]. To account for deviations from unity, factorised JES calibrations are applied in multiple stages including pile-up corrections, simulation-based response corrections and small residual corrections for tracking inefficiencies and threshold effects, derived in-situ from γ+jet, Z+jet and dijet samples [422]. These factorised JES corrections are not used when jet substructure observables are constructed, but dedicated corrections are derived as described below. Figure 3.2 (right) shows the calibrated JES uncertainty obtained in CMS, which is below 1% for jets with $p_T > 100$ GeV in the central region with $\eta = 0$. Even for jet p_T as low as 10 GeV the uncertainty is below 3%, owing to the excellent performance of the particle flow reconstruction. A detailed discussion of the different approaches for deriving jet energy scale uncertainties in ATLAS and CMS can be found in [426].

Of special importance to the application of jet substructure techniques is the calibration of the jet mass. While the reconstruction of jet energies mainly relies on the capability of a detector to measure the total energy of all particles deposited in the detector, the measurement of jet mass requires detection of the deposited energy with a granularity finer than the size of a jet. The mass of a jet can only be estimated if the energy is deposited in at least two detector elements, as it depends on both the energy and opening angle between the jet constituents. For jet substructure techniques that rely on the rejection of soft particles, it is also important to be able to reconstruct particles with low p_T separately from harder particles in a jet.

The jet mass response distribution is constructed from the calibrated, reconstructed jet mass divided by the particle-level jet mass. The jet mass scale (JMS) is defined as the mean of this response distribution,the jet mass resolution (JMR) as one standard deviation.[5]

ATLAS has developed a data-driven approach using forward-folding to extract the JMS and JMR. In this approach, the jet mass distribution at the particle level is modified by a response function such that the JMS and JMR are scaled by appropriate scale parameters, which can be functions of jet mass and p_{T}. The values of these parameters for which the folded distribution best matches the data are extracted from a two-dimensional fit [427, 428]. With this approach, the JMS and JMR for hadronically decaying boosted W bosons with $p_{\mathrm{T}} \gtrsim 200$ GeV are determined with 2–3% and 20% systematic uncertainties, respectively [429]. The results from this measurement are combined with the so-called R_{trk} method, which constrains the mass scale by comparing the calorimeter jet mass to the mass calculated from track jets. The R_{trk} method extends the jet mass calibration up to $p_{\mathrm{T}} = 3000$ GeV [429]. It can be generalised to other variables and is used in ATLAS to constrain the p_{T} scale of large-R jets as well as to derive systematic uncertainties on jet substructure variables. ATLAS has also studied combining calorimeter and tracking information to maintain a stable jet mass reconstruction for highly boosted particles, where the calorimeter granularity is not sufficient any longer. The track-assisted mass [427] benefits from the excellent angular resolution of the tracking detector and shows a better JMR at $p_{\mathrm{T}} \gtrsim 1000$ GeV, while the calorimeter mass performs better below this value. The combined mass is a weighted average of these two reconstruction methods, based on the expected resolution.

A similar approach of correcting topoclusters based on tracking information has been studied as well, the method based on Track-CaloClusters (TCC) improves the resolution of jet substructure variables at high p_{T} [410]. The TCC method is complementary to the PF algorithm, but is designed to work in a very different energy regime. While the PF algorithm is especially useful to improve the jet reconstruction at low p_{T}, the TCC approach is based on a weighting scheme where the track momenta are used to spatially redistribute the energy measured in the calorimeter. By doing this, the TCC improves the angular resolution of the calorimeter measurement, without changing the total energy measured in the calorimeter. This approach improves the resolution of jet substructure observables at high p_{T}, although the JMR is slightly worse than for the combined mass below $p_{\mathrm{T}} < 1.5$ TeV. Another approach studied in ATLAS is a track-assisted reconstruction of substructure observables [430]. This approach is a generalisation of the track-assisted mass [427], taking into account local fluctuations inside the jet. There are different flavours of the track-assisted jet reconstruction, but they all have in common that in a first step tracks are assigned to calibrated small-R jets.[6] In a second step, the p_{T} of each track is scaled such that the total p_{T} of the track-jet is equal to the p_{T} of the calibrated calorimeter jet. In this way,

[5] ATLAS often uses half of the 68% interquartile range of the response distribution, which is robust against non-Gaussian tails and is equal to one standard deviation for a Gaussian response distribution.

[6] These can be subjets of large-R jets or all small-R jets found in the event, which are re-clustered into large-R jets in a later stage of the algorithm [431, 432]. This track-assignment is done via ghost association [433], but in some cases un-matched tracks are added using a ΔR criterion.

the missing momentum from neutral particles is accounted for in small-R track-jets, which are then clustered to form large-R jets. In this way, local fluctuations in the neutral components can be accounted for in track-jets, which improves the experimental resolution of several substructure observables. A more unified approach, combining the ATLAS particle-flow algorithm with the TCC method has recently been studied for large-R jets [434]. The resulting list of unified flow objects (UFOs) can be further processed by pileup mitigation techniques (see below), and is then used as input to jet clustering algorithms. The resulting jets show better performance in terms of substructure reconstruction than previous approaches. This approach will improve the efficiency and precision of future jet substructure measurements and searches for new physics.

In CMS, the jet mass is by default reconstructed as a combination of track and calorimeter measurements via the virtues of the PF algorithm. Thus the strategy for calibrating the jet mass in CMS differs from the one in ATLAS. In CMS, the individual PF objects are input to the jet reconstruction, and are locally calibrated to account for the detector's single particle response. After correcting the individual inputs, the jet four-vector is corrected using JES corrections. Small residual differences in the jet mass between data and simulation are corrected using dedicated samples. The residual jet energy corrections are not applied when reconstructing jet masses. Therefore, dedicated corrections are derived from simulation and data. The jet mass response is measured using W jets in a data sample enriched in $t\bar{t}$ production [435, 436]. After a dedicated selection, large-R jets in this sample show a peak at the W mass in the jet mass distribution (see Sect. 4.1.2). The excellent performance of the PF algorithm results in a JMR of about 10%. The absolute response and the resolution are well described by the simulation, and are within 1–2% for the JMS and about 10% for the JMR. Since these measurements are performed in samples of W jets with $p_{\mathrm{T}} \approx 200\,\mathrm{GeV}$, additional systematic uncertainties apply at higher p_{T} [437]. A detailed study of the various contributions to the JMS has also been performed for fully merged top-jets in the context of an unfolded top-jet mass measurement [438].

The relative JMR in CMS is shown in Fig. 3.3 as a function of the ungroomed jet mass m_{u} for anti-k_{T} $R = 0.8$ jets. The JMR is obtained from a sample of jets

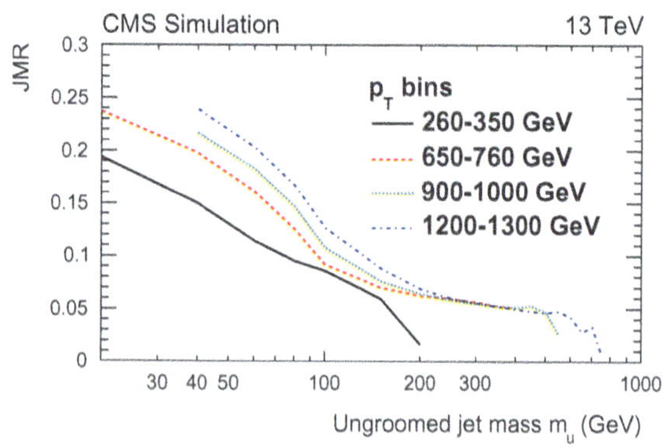

Fig. 3.3 The CMS JMR as a function of the ungroomed jet mass m_{u} in different generated p_{T} bins. Taken from [439]

initiated by quarks and gluons. The resolution improves with increasing m_u and is around 9–13% for the most probable value of $m_u \approx 100$–150 GeV. For a given value of $m_u < 200$ GeV, the resolution worsens with increasing jet p_T due to a higher degree of collimation. Remarkably, the resolution obtained in CMS is comparable to the combined mass in ATLAS, even though quark/gluon jets are compared with W/Z jets and very different technologies are used to reconstruct the jet mass.

3.3 Pileup Mitigation

Pileup originates from simultaneous pp collisions that occur in addition to a hard scattering, triggering the readout and reconstruction of an event. The interaction vertex of the hard scattering is referred to as leading vertex (LV), and is usually taken to be the vertex with the highest scalar sum of p_T, calculated from reconstructed objects associated to it. Pileup interactions are uncorrelated to the primary interaction (unlike MPIs) and typically consist of an admixture of inelastic, elastic and diffractive pp processes. Vertices from pileup interactions are distributed along the longitudinal direction in the sensitive detector region. As the detector response is not instantaneous, pileup events from both the same (in-time) and neighbouring (out-of-time) bunch crossings can contribute. This section focusses on the mitigation of in-time pileup, though out-of-time pileup is also mitigated by dedicated reconstruction algorithms developed by the ATLAS and CMS Collaborations. During the LHC data taking between 2010 and 2012 the mean number of pileup interactions reached $\langle \mu \rangle = 21$, and μ values up to 60 were attained in certain runs of 2017 as shown in Fig. 3.4 (left). For the data taking in 2022 even higher values are expected, culminating at $\langle \mu \rangle = 140-200$ for the high-luminosity LHC (HL-LHC). On average, pileup leaves approximately 0.5 GeV of energy in the detector per unit area in (η, ϕ), per pileup vertex. The effects of this are present in all aspects of LHC physics, from detector design and software performance to the final sensitivity of measurements and searches.

3.3.1 *Mitigation Methods*

Properties of pileup interactions can be exploited to remove energy contributions from pileup to individual jets. Pileup can be approximated as a spatially uniform deposition of energy. Using this approximation, the so-called area subtraction [185] corrects the jet four-momentum through $p_T^{\text{corr}} = p_T^{\text{orig}} - \rho A$, where A is the jet area. The estimator ρ for the contribution from pileup per unit area is obtained from reconstructed jets in the event, associated to pileup. An example of ρ is shown in Fig. 3.4, with a slope of approximately 0.5 GeV per pileup vertex. There are many subtleties in defining both ρ and A, which are discussed in e.g. [185, 441, 442]. An

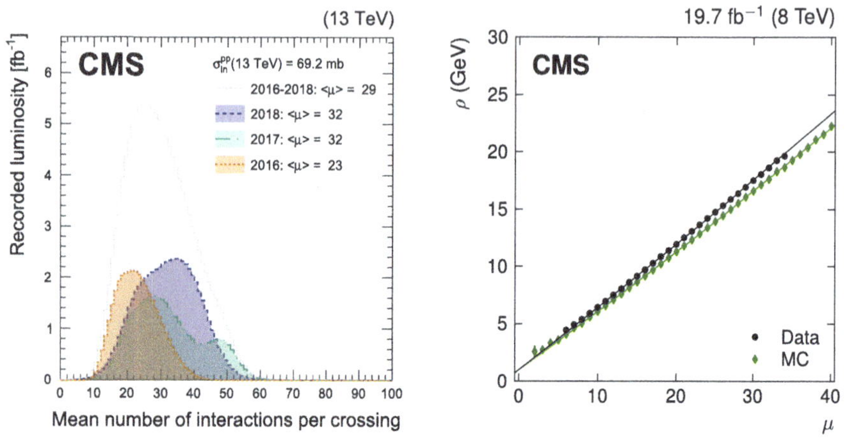

Fig. 3.4 Recorded luminosity as a function of the mean number of interactions per bunch crossing, taken from [440] (left). Average pileup contribution to the jet p_T as a function of the number of pileup interactions per bunch crossing for data and simulation in CMS, taken from [70] (right)

extension to this method is shape subtraction [443], where the assumption is dropped that pileup leads to a spatially uniform deposition of energy. In shape subtraction, randomly distributed ghost particles are used to calculate the sensitivity of a jet shape to pileup. This is then used to correct the jet shape for non-uniformities in the spatial distribution of pileup particles.

Identification of pileup jets, formed predominantly by particles originating from one or many pileup vertices, is a technique for removing contributions from pileup to the whole event, instead of correcting individual jets. Once a jet has been identified to originate from pileup, it can be removed. This leads to improvements in the resolution of event variables, such as the missing transverse momentum or the total scalar sum of jet p_T. Pileup jets can be identified using charged particles pointing to the LV [442, 444]. This can be supplemented by information from jet shapes, using the fact that pileup radiation is softer and uncorrelated to the radiation from the LV. Topoclustering [409], used by the ATLAS Collaboration, is deployed at the formation of clusters in the calorimeter requiring radiation to have a certain topological profile. In the forward region, where no tracking information is available, topological correlations and jet shapes can be used to identify pileup [440, 445]. A combination of variables results in efficiencies to correctly identify quark jets from the LV of 80–99% for misidentification rates from pileup jets of 1–10% in the central part of the detector, depending on the chosen working point [440, 442].

It is also possible to classify individual particles as originating from pileup interactions. Charged particle tracks can be classified based on their longitudinal position z along the beam direction. The charged hadron subtraction (CHS) [441] method removes all charged particles associated to pileup vertices in the track fit. This removes a large part of the charged pileup radiation from the event, including calorimeter signals that are linked to tracks through the PF algorithm. This method is usually combined with the area subtraction. First, the the pileup contribution from charged particles is removed. In a second step, the remaining contributions from neutral particles are removed with the area subtraction method.

While the above methods have been successfully deployed in ATLAS and CMS, they each have some deficiencies for the reconstruction of jet substructure observables. Ideally, one would hope to remove pileup at the most granular level possible, i.e. at the level of reconstructed particles or calorimeter clusters, in order to be as generic as possible. For example, while area subtraction is very effective for correcting the jet p_T, it is not capable to mitigate the pileup dependence of jet substructure observables as it only removes pileup contributions on average. In fact, jet substructure variables are among the most difficult observables to correct for pileup because they are so reliant on radiation profiles. A number of hybrid methods have been proposed operating at the event constituent level, such as jet cleansing [446], jets without jets [447], SoftKiller [448], Constituent Subtraction [448–450] and PUMML [451].

An example of a method extensively used in CMS is the pileup per particle identification (PUPPI) algorithm [450]. This algorithm uses information related to charged particle tracks, local particle densities and event pileup properties to determine if a particle originates from pileup. A discriminator value α_i is calculated event-by-event for each particle i using $p_{T,i}$ and its angular distance to nearby particles. To translate this value into a probability for a particle to originate from the LV, charged particles assigned to pileup vertices are used to calculate the expected distribution in α per event. The value of α_i of each neutral particle is compared to the expected value for pileup particles, assuming that charged and neutral particles result in the same distribution in α. The difference between α_i and the expected mean value is used to calculate a weight for the particle four-vector, with large weights for parton shower-like radiation and small weights for pileup-like radiation. Almost all pileup particles have values within a few standard deviations of the median of the α distribution and are assigned small weights. Values that deviate significantly from the mean value are indicative of a hard scatter, and these particles are assigned large weights. This weighting method allows for experimental information, such as tracking, vertexing and timing information, to be included.

3.3.2 *Performance Studies*

Pileup removal algorithms are commissioned for use in ATLAS and CMS via detailed studies of jet observables in terms of the resolution, absolute scale and pileup dependence. Other observables include lepton identification efficiencies, missing transverse momentum resolution and the performance of jet substructure taggers. As an example, pileup mitigation techniques can improve the lepton identification performance, where the energy in a small cone around the lepton is summed, to reduce the susceptibility to pileup.

For observables like jet p_T, dependencies on the number of reconstructed vertices are observed with area subtraction methods for the pileup levels currently observed

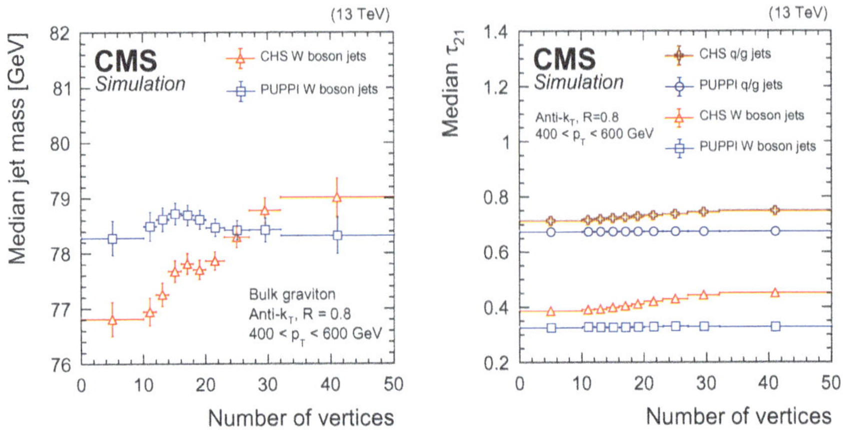

Fig. 3.5 Comparison of the performance of CHS and PUPPI for the median value of the soft drop jet mass (left) and the N-subjettiness ratio τ_{21} (right) as a function of the number of vertices. Taken from [440]

at the LHC, $\langle\mu\rangle \sim 25$. To correct for these effects, an additional residual correction is applied [70, 421]. Improvements are also obtained by combining area subtraction methods with particle-based methods, e.g. CHS.

For jet substructure observables, particle- or constituent-level pileup mitigation strategies have been shown to improve the reconstruction performance and stability. An example is given in Fig. 3.5, showing the median values of the soft drop jet mass and N-subjettiness ratio τ_{21} distributions, as a function of the number of vertices. The reconstruction using PF with PUPPI improves the stability significantly when compared to CHS, which is also observed in other characteristic observables [440]. Note that a purely area based pileup removal technique would show an even larger slope than CHS. The effect of different pileup removal techniques on the groomed jet mass also depends on the choice of the grooming algorithm as discussed in [441, 452]. The excellent performance of PF with PUPPI has also been recently reported by the ATLAS Collaboration in a comprehensive study of grooming and pileup mitigation techniques [453]. The improved performance observed in simulation has also been verified in collision data [436]. Preliminary studies of advanced hybrid techniques at the high pileup levels anticipated for the HL-LHC suggest that they are even effective in the $\langle\mu\rangle = 140$–200 range [454, 455].

3.4 Grooming Methods

Jet grooming can be used for two purposes: the mitigation of pileup effects on jets and the removal of soft and wide-angle radiation. Grooming techniques are usually utilised on large-R jets with R in the range of 0.8–1.5, where the effects from pileup and the UE on jet substructure are much larger than on small-R jets.

The ATLAS experiment has adopted trimming for anti-k_T, $R = 1.0$ jets with $f_{cut} = 0.05$ and $R_{sub} = 0.3$ in analyses of 8 TeV data, and $R_{sub} = 0.2$ for analyses of 13 TeV data. Extensive studies of different algorithms showed that this choice gives the best results in terms of pileup stability, jet mass resolution and jet tagging performance [452]. The reclustering of small-R jets [431] has the experimental advantage of using fully calibrated inputs when clustering jets with a larger distance parameter, and has already been used in several ATLAS searches for Supersymmetry [456–458]. This method allows for flexibility of optimizing the jet distance parameter depending on the considered phase-space of the analysis without a need to re-derive jet energy corrections [432]. Note that this method has similarities to trimming with R_{sub} set to the distance parameter of the small-R jets. The difference is in the value of $p_{T,sub}$, which is fixed to the minimum p_T of the small-R jets for reclustering, but dynamically adjusted to the p_T of the large-R jet for trimming. The groomed jets obtained with reclustering are different from trimmed jets also because reclustering considers all small-R jets in an event, while trimming uses only the ones obtained from the constituents of large-R jets. In practice these differences lead to small effects only, resulting in a very similar performance of reclustering and trimming. A recent ATLAS study [453] on grooming algorithms includes PF reconstruction and a number of hybrid pileup removal techniques. It is shown that gains in tagging performance are possible by using soft drop instead of trimming, when used with PF and a combination of Constituent Subtraction and SoftKiller.

The CMS experiment has employed PF with CHS already early in the data taking, which resulted in a small impact from pileup on jet substructure during the 8 TeV data taking. Contributions from the neutral pileup component have been mitigated using the pruning algorithm for anti-k_T, $R = 0.8$ jets with $z_{prune} = 0.1$ and $R_{prune} = 0.5$ [435]. In analyses of 13 TeV data PUPPI has become the standard pileup mitigation technique, reinstating the role of grooming algorithms to their original purpose of removing soft and wide-angle radiation. For this purpose, soft drop with $z_{cut} = 0.1$ and $\beta = 0$ is used in jet substructure analyses. The combination of PUPPI and soft drop shows a similar performance as pruning in substructure analyses [436, 459]. The soft drop algorithm has been chosen over pruning due to a better theoretical control of substructure observables.

The role of grooming algorithms in the performance of W, Z, H and t taggers is discussed in detail in the respective sections below.

3.5 Jet Substructure Tagging

Particle identification is an experimental challenge that is traditionally met using charged-particle detectors, pre-shower detectors, high-granularity calorimeters and muon chambers. Particle identification of electrons, muons, tau leptons, photons and jets originating from b quark fragmentation processes played an important role in the design considerations for the ATLAS and CMS detectors and is used extensively in physics analyses. Jet substructure techniques used for the identification of the particle origin of jets are a much more recent development, though. While the detectors have not been specifically designed for measurements of jet substructure, ATLAS and CMS have the specifications necessary to efficiently identify the origin of jets using substructure observables. In what follows, the term 'tagger' indicates the use of one or more of these observables, sometimes together with a grooming algorithm, for jet identification. The performance of taggers is classified in terms of their efficiencies and misidentification rates. Experimental use of jet substructure taggers hinges upon detailed measurements of these numbers and the verification of a similar performance in data as in simulation. In addition, care has to be taken to exactly define how these numbers are evaluated, as differences in the definition of efficiencies and misidentification rates can lead to very different conclusions. Systematic uncertainties play an important role in these studies, as performance gains obtained by the use of more advanced taggers can be negated by larger uncertainties relative to a less complex approach.

3.5.1 Quark/Gluon Discrimination

Experiments at the LEP, Tevatron and HERA colliders have already used jet substructure to distinguish quark-initiated (quark) versus gluon-initiated (gluon) jets. In these early studies, the dynamics of quark and gluon scattering were probed. At the LHC, quark versus gluon jet (q/g) tagging also serves the purpose of separating signal from background, where often signal jets are quark jets (for example in production processes through vector-boson scattering) and gluon jets are background. The probability for a $g \rightarrow gg$ splitting is enhanced by a factor of $C_A/C_F = 9/4$ over the probability $q \rightarrow qg$ at the same opening angle and energy fraction [460]. As a result, gluon jets tend to have more constituents and a broader radiation pattern than quark jets, which can be used to discriminate between them. There are also more subtle differences due to quark and gluon electric charges and spins, which can be exploited when constructing substructure variables for q/g tagging.

A complication arising when studying q/g tagging is the ambiguous definition of a quark and gluon jet. Quarks and gluons carry colour charges, but jets reconstructed in the detector result from sprays of colour-singlet particles such that these can not be connected unambiguously. A number of definitions have been suggested, ranging from matching jets to outgoing partons at the level of the hard scattering, to parsing the entire parton shower and jet clustering history [461, 462], to using operational definitions at the level of observable distributions [463, 464]. In practice, differences in these definitions have a small impact on the experimental performance of q/g taggers as long as used consistently throughout an analysis. However, different choices may render results from experimental studies incommensurable. Note also that quark and gluon jet radiation depends on the production mechanism [465], such that the calibration and application of q/g taggers must be treated with additional care compared to other taggers.

The sensitivity to partonic radiation and non-perturbative effects makes q/g tagging particularly sensitive to modelling choices in simulation. Typically, PYTHIA tends to describe quark jets better than HERWIG, whereas the opposite is observed for gluons [436]. For this reason, PYTHIA tends to overestimate the q/g tagging performance with respect to data, while HERWIG underestimates it [466]. The modelling of observables for q/g tagging is an active field of research, see for example [467] for a recent study. A reduction of uncertainties related to the modelling of quark and gluon jets can be achieved with dedicated measurements, sensitive to the radiation pattern in jets. A very promising example is the measurement of the Lund jet plane [468, 469], which has very recently been performed on dijet events by ATLAS [470]. Since distinct regions in the Lund jet plane are dominated by contributions from different QCD processes, the measurement will be useful for tuning non-perturbative models and parameters of event generators (see Sect. 4.1.4).

Numerous jet substructure observables have been studied for their q/g discrimination power. Some of the most useful ones are particle multiplicity, thrust [44–46], broadening [471, 472], girth [473], integrated jet shapes [60] or the fragmentation function p_{T}^{D} [444, 474]. The most powerful single variable has been found to be particle multiplicity which is not IRC safe, but can be modified using an iterative version of soft drop grooming. The resulting soft drop multiplicity is IRC safe and can be calculated perturbatively. It shows similar discrimination power as particle multiplicity [475]. In general, counting observables are sensitive to multiple emissions at LL accuracy, resulting in better discrimination power than IRC-safe observables which are dominated by a single emission at LL accuracy, giving rise to Casimir scaling of the q/g discrimination power [223]. While particle multiplicity performs best as single variable tagger, a gain in the q/g tagging performance can be obtained by combining it with a Casimir scaling observable. Only marginal gains are obtained by including more observables [476, 477].

Particle multiplicities are the most important input to the q/g taggers developed by the ATLAS [466, 478–480] and CMS [436, 481, 482] Collaborations. While ATLAS uses the number of tracks as an approximation for the number of jet constituents and the jet width [480], CMS utilises the number of particle-flow constituents, p_{T}^{D} and the angular opening of the minor jet axis σ_2, computed from the p_{T}^2-weighted average of the angular distance between the jet constituents [436]. The performance of the CMS q/g tagger in terms of quark-jet efficiency versus gluon-jet rejection is shown in Fig. 3.6. The gain by including p_{T}^{D} and σ_2 is an improvement of about 2% in gluon-jet rejection at a quark-jet efficiency of 60%. The performance of q/g tagging degrades considerably in the region outside of the tracker acceptance, where only the granularity of the calorimeters can be used to infer the particle multiplicity. In addition, this region has a higher susceptibility to pileup radiation than the central region.

Since the distributions of variables used for q/g tagging depend on the jet kinematics and the pileup activity in the event, likelihood discriminators are built differential in jet η, p_{T} and the value ρ of the event. In order to mitigate the effect of modelling

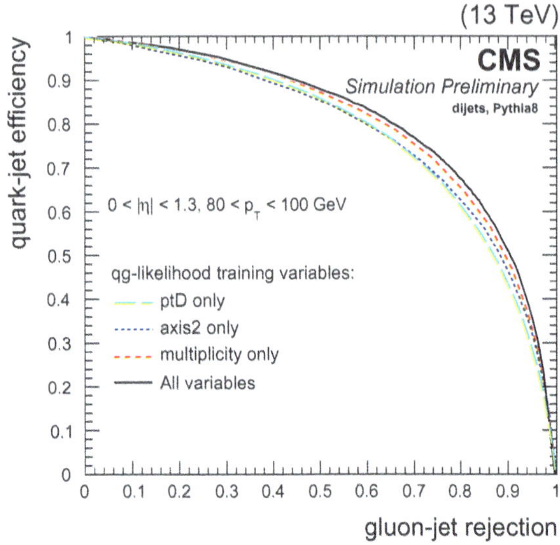

Fig. 3.6 The CMS q/g tagging performance in simulation for three individual observables and their combined likelihood for central jets with $|\eta| < 1.3$ and $80 < p_T < 100\,\text{GeV}$. Taken from [436]

uncertainties on the q/g tagging efficiency, samples of different but known quark and gluon fractions are used to calibrate simulated events by correcting the corresponding distributions. ATLAS and CMS use Z/γ+jets and dijet samples, which are enriched in quark and gluon jets, respectively. A template-based fit to the two q/g likelihood distributions is used to determine weights for quark and gluon jets, correcting the simulation for effects from mismodelling [436, 466]. Alternatively, the rapidity dependence of the q/g fraction in dijet events can be used to extract the track multiplicity separately for quarks and gluons [479]. The template fit results in typical uncertainties of 2–5% for a wide range in p_T, estimated on the same samples as used for calibration. However, residual differences between data and simulation are observed once the resulting calibrated tagger is applied to another final state, which leads to additional uncertainties [478].

3.5.2 *Vector Boson Tagging*[7]

The hadronic two-prong decays of weak vector bosons V have a distinct radiation pattern compared to individual high-p_T quarks or gluons. In particular, boosted bosons tend to have two distinct subjets with relatively equal momentum sharing (see Sect. 2.2.7). In contrast, most generic quark and gluon jets will have one prong and if they have two, the second one tends to be soft. Furthermore, the mass of quark and gluon jets scales with their p_T and is lower than the electroweak boson masses

[7] The text in this subsection has been taken from [26] and has been written by the author. It has been adjusted to fit this book.

for low jet p_T and higher for very high p_T jets (see Sect. 2.4.1). Good separation power between W and Z bosons is desirable in a number of analyses, most notably searches for diboson resonances (see Sect. 5.1).

In experimental studies, simulated samples containing W jets rather than Z jets are primarily used, as W jets are abundant in data and can be selected efficiently thanks to the large quantity of $t\bar{t}$ events produced at the LHC. ATLAS and CMS performed a broad range of studies, systematically identifying the influence of pileup mitigation and grooming techniques on jet substructure observables used for V tagging [411, 435, 441, 452, 483, 484]. The discrimination power for a number of jet substructure variables has been studied, including N-subjettiness [209, 210], Qjet volatility [255], ratios of energy correlation functions $C_2^{(\beta)}$ [215], $D_2^{(\beta)}$ [218, 219] and $N_2^{(\beta)}$ [221], angularities and planar flow [485], splitting scales [38, 486], the jet and subjet quark/gluon likelihood, and the jet pull angle [226]. In addition to a good background rejection at a given signal efficiency, the design of a V tagging algorithm also has to take into account the accurate detector response to jet substructure observables.

Both ATLAS and CMS developed simple taggers that rely on the combination of the jet mass with one other variable that improves the discriminating power between the signal and background. The standard ATLAS V tagger for analyses of 13 TeV data was chosen to be the trimmed jet mass and $D_2^{(\beta=1)}$ [483], while CMS decided to use the soft drop jet mass and the N-subjettiness ratio $\tau_{21} = \tau_2/\tau_1$. Despite the different choices of tagging observables and detector design, ATLAS and CMS reach a very similar background rejection at a given tagging efficiency. An active field of developments is the usage of multivariate techniques for boosted V identification which have shown to be able to significantly improve the background rejection [435, 535].

In the ATLAS studies the variable $C_2^{(\beta=1)}$ in combination with the trimmed jet mass has been shown to be as good a discriminator as τ_{21}[8] as shown in Fig. 3.7. This is in contradiction to the study by CMS, where C_2 is one of the weaker observables; however, a direct comparison is difficult, since in ATLAS groomed substructure variables are used, calculated for trimmed jets, while in CMS ungroomed variables are used. Also, the particulars of particle reconstruction have a large impact on the performance of individual observables. While a study of the performance of D_2 at CMS is still pending, the soft drop N_2 observable was found to give similar performance to τ_{21} in CMS [489]. The ATLAS measurements of signal efficiencies versus background rejection power in Fig. 3.7 lie on the predicted performance curves, giving confidence in performance studies using simulated events.

The CMS Collaboration studied the q/g likelihood discriminator for its potential in V tagging applications [411], finding that a combination of the groomed jet mass and the q/g likelihood achieved a similar discrimination power as the groomed jet mass and τ_{21}. When adding the q/g likelihood to a V tagger utilizing pruned jet mass and τ_{21}, the misidentification rate was reduced slightly from 2.6 to 2.3% at a

[8]A different axis definition for the subjet axes is used in ATLAS when calculating τ_N, known as the-winner-takes-all axis [488], which is consistently found to perform slightly better than the standard subjet axis definition in tagging bosons.

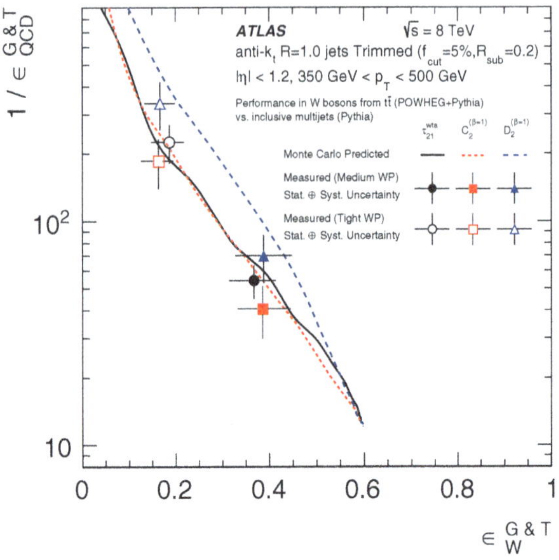

Fig. 3.7 Signal efficiency versus background rejection power compared with measurements from ATLAS for $350 < p_{\mathrm{T}} < 500\,\mathrm{GeV}$. Taken from [483]

constant signal efficiency of 50%. A similar reduction of the misidentification rate was observed when adding $C_2^{(\beta=2)}$, showing that C_2 carries additional information with respect to the groomed jet mass and τ_{21}. However, the q/g likelihood and C_2 exhibit a considerable pileup dependence, resulting in a degradation of their discrimination power with increasing activity. This pileup dependence is expected to be reduced when using PUPPI in place of particle flow with CHS.

The most important systematic uncertainty in V-tagging is the detector response to jet substructure observables. In ATLAS, this has been studied by comparing calorimeter-jets with track-jets [483, 484], whereas CMS relies on PF for deriving efficiencies due to a jet substructure selection [435, 436]. In both experiments the distributions in data lie between the ones derived with PYTHIA and HERWIG, leading to large modelling uncertainties. Improving the modelling of jet properties and thereby reducing the differences between different event generators is a major task, but crucial for future precision studies using jet substructure.

An important aspect of V tagging is the derivation of background rates from multijet production in real collision data when performing measurements. A commonly used method is the extrapolation from one or more control regions, which are defined orthogonally to the signal region. Usually, these control regions are obtained by inverting the jet mass window selection, see e.g. [490–495]. Transfer functions are derived from simulation, extrapolating the rates and shapes from the control to the signal regions. Even though these transfer functions are ratios of distributions, which results in a reduction of the impact of modelling uncertainties, a residual dependence on the simulation can not be eliminated. However, additional uncertainties in the high-p_{T} tails of the transfer functions can be removed by ensuring a constant behaviour as a function of p_{T}. The requirement is thus a flat signal or background efficiency (depending on the needs of the analysis). In order to achieve a flat signal efficiency, ATLAS developed a p_{T}-dependent selection on the value of $D_2^{(\beta=1)}$, as this distribution shows a strong dependence on p_{T} [484]. No p_{T}-dependent selection is made on the trimmed jet mass, as the calibrated jet mass is used to define the V tagging working point. While the jet mass resolution increases with p_{T}, a constant window of $\pm 15\,\mathrm{GeV}$ around the mean reconstructed W or Z boson mass is used. This results in a p_{T}-dependent signal and background efficiency, which can also be countered with the p_{T}-dependent cut on $D_2^{(\beta=1)}$. This leads to a constant signal efficiency, while the background efficiency shows a residual p_{T} dependence [484].

Another possibility has been explored by CMS. Instead of introducing p_{T}-dependent selection criteria, a linear transformation of the ratio τ_{21} has been studied [436], given by $\tau_{21}^{\mathrm{DDT}} = \tau_{21} - M \cdot \log(m^2/p_{\mathrm{T}}/1\,\mathrm{GeV})$ [251], where M is a constant determined from simulation. The replacement of τ_{21} with DDT version τ_{21}^{DDT} does not affect the overall performance of the tagger, but results in an approximately flat misidentification rate as a function of p_{T}, as shown in Fig. 3.8 (left). The effect of the DDT method on the V tagging efficiency is shown in Fig. 3.8 (right). The efficiency increases as a function of p_{T} with a slope somewhat smaller than the slope for the decreasing efficiency obtained with plain τ_{21}. The development of decorrelated jet substructure taggers is an active field with new techniques e.g. described in [252, 496, 497].

A less-studied possibility to lift the p_{T}-dependence of substructure observables is the application of variable-R jets [187]. By shifting the p_{T}-dependence to the jet-clustering level with a distance parameter proportional to p_{T}^{-1}, a stable position of the jet mass and jet substructure variables with respect to changes in p_{T} can be achieved [498]. This can lead to a stable tagging performance without the necessity

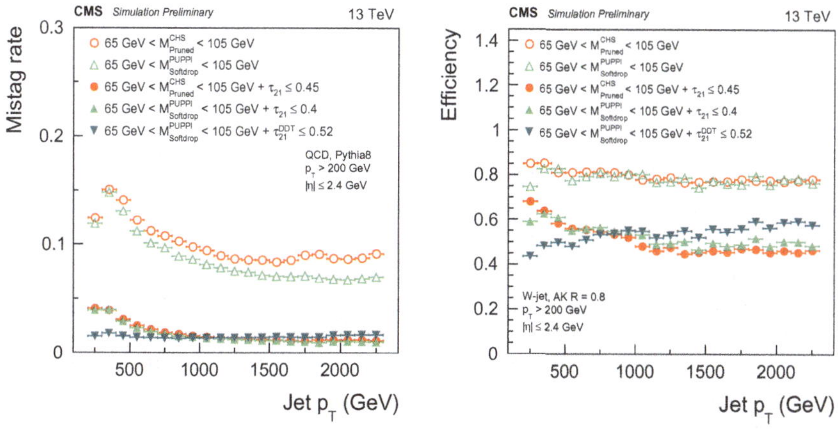

Fig. 3.8 Misidentification rate and efficiency of various identification techniques for boosted W tagging. Taken from [436]

of p_T-dependent optimisation steps, but further experimental studies are needed to commission this strategy for use in analyses.

For some analyses the requirement of $p_T \gtrsim 200$ GeV is too restrictive, and hadronically decaying V bosons with lower p_T need to be selected. This poses a particular challenge due to the abundance of light flavour jets at the LHC and their indistinguishability from jets from resolved W/Z decays. An attempt was made by CMS to discriminate 'resolved' (non-merged) hadronic W decays from multijet background using the q/g likelihood, the sum of the jet charges of the dijet pair and the jet pull angle. Combining these variables into a boosted decision tree, a misidentification rate of about 25% is achieved for a signal efficiency of 50% [411]. While this is a first success, the performance is about an order of magnitude worse than V tagging for fully merged decays, showing the power of substructure techniques in this field.

In addition to developing tools for distinguishing boosted hadronically decaying W and Z bosons from generic quark and gluon jets, ATLAS has also built a tagger to further classify a boson jet as either originating from a W boson or a Z boson [499]. While theoretically clean due to the colour singlet nature of the W and Z bosons, this task is particularly challenging because the jet mass resolution is comparable to the difference $m_Z - m_W$. In order to improve the sensitivity of the tagger, jet charge and b tagging information are combined with the jet mass. The jet mass distribution depends on the type of W or Z decay due to semi-leptonic B and D decays, so a full likelihood tagger is constructed by summing over the conditional likelihoods for each flavour type. To maximise the discrimination power from b tagging, multiple efficiency working points are used simultaneously in the tagger. A W^+ rejection near 8 (corresponding to a misidentification rate of 12.5%) is achieved at a Z boson efficiency of 50%. At this moderate Z boson efficiency, all of the inputs offer useful discrimination information. At low efficiencies, below the $b\bar{b}$ branching ratio for Z bosons, b tagging dominates over the jet mass and jet charge.

3.5.3 Higgs Boson Tagging

The Higgs boson decays with highest probability to $b\bar{b}$ and $WW^* \to 4$ quarks with branching fractions of 58.1 and 9.8%, respectively (see Sect. 2.2.4). Both are hadronic decays, which are difficult to identify. In fact, it has taken six years after the discovery of the Higgs boson to measure the $H \to b\bar{b}$ decay with a significance of more five standard deviations [35, 36]. The decay $H \to WW^* \to 4$ quarks is still undiscovered, while the leptonic $H \to WW^*$ decay channels have been important for the discovery and classification of the Higgs boson properties [1, 2, 500–503].

The identification of $H \to b\bar{b}$ relies on the distinct signature of jets originating from the fragmentation of b quarks (b tagging). The fragmentation leads to the presence of B hadrons; whose decay results in a secondary vertex within a jet due to the long lifetime of about 1.5 ps. The b tagging is crucial in many analyses at the LHC and has been developed for small-R jets from isolated b production. ATLAS and CMS use multivariate techniques with various input parameters related to the secondary vertex or charged particle tracks originating from the B hadron decay. For 13 TeV analyses, CMS uses the CSVv2 algorithm [504] and ATLAS uses the MV2c10 algorithm [505, 506]. Typically, efficiencies of around 70% with misidentification rates of 0.3–1% for light quark and gluon jets and 10–20% for charm jets are achieved with these algorithms [507, 508].

The challenge of identifying boosted $H \to b\bar{b}$ decays lies in the busy hadronic environment, where secondary vertices are difficult to identify and properties of the B decay vertices calculated relative to the jet axis have less discrimination power than for isolated b production. To meet this challenge, various algorithms have been developed. For large-R jets, algorithms in use by ATLAS and CMS are subjet b tagging and double-b tagging. Once the b content of the large-R jet has been identified, adding other substructure variables to the $H \to b\bar{b}$ tagger can give additional improvements. An advantage of the a tagger using only b tagging is that other jet substructure observables like the jet mass are unaffected by an $H \to b\bar{b}$ selection and can be used as discriminating variables in H analyses or searches.

Subjet b tagging [509–514] is a technique where the standard b tagging algorithm is applied to each of the subjets of the large-R jet. In ATLAS, the subjets are track jets with a radius of 0.2, matched to the large-R jet using the ghost-association technique [433]. In CMS, either subjets from the N-subjettiness minimisation or from the soft drop algorithm are used. At $p_T = 800$ GeV and higher, the subjets start to overlap, causing the subjet b tagging techniques to break down due to double-counting of tracks and secondary vertices. In this regime, b tagging on the large-R jet gives better performance when requiring the presence of two B hadrons. The ATLAS Collaboration has also performed studies to improve the efficiency at high p_T, using variable-radius track jets, exclusive k_T subjets and calorimeter subjets reconstructed in the centre-of-mass frame of the Higgs jet candidate [515]. For highly boosted Higgs bosons, these reconstruction techniques promise to improve the performance compared to the usage of fixed-radius track jets.

Double-b tagging in ATLAS means that the two leading p_T track-jets must pass the same b tagging requirement [514]. In CMS, the double-b tagger [504, 516] is a dedicated algorithm combining several discriminating variables using a boosted decision tree. The double-b tagger has been designed for jets with a pruned mass of $50 < m_{\text{jet}} < 200$ GeV and $p_T > 300$ GeV. The algorithm exploits not only the presence of two B hadrons inside the large-R jet, but also the correlation between the directions of the momenta of the two B hadrons and the energy flows of the two subjets. It employs the same variables as used in the CSVv2 algorithm. It is constructed such that its performance is largely uncorrelated to the jet mass and p_T. A recent improvement of double-b tagging in ATLAS uses VR track-jets [515] for subjet b tagging. The reason for this development is a decreasing $H \to b\bar{b}$ identification efficiency with increasing large-R jet p_T. For larger H boson boosts, track-jets with $R = 0.2$ start to merge, making it difficult to identify two B hadron decays within the H jet. The shrinking size of VR track-jets with increasing p_T helps to reduce the inefficiency at high p_T. These are obtained by clustering tracks with the VR algorithm with the parameters $\rho = 30$ GeV, $R_{\text{min}} = 0.02$ and $R_{\text{max}} = 0.4$. The VR track-jets are ghost-associated to large-R jets for double-b tagging. For large-R

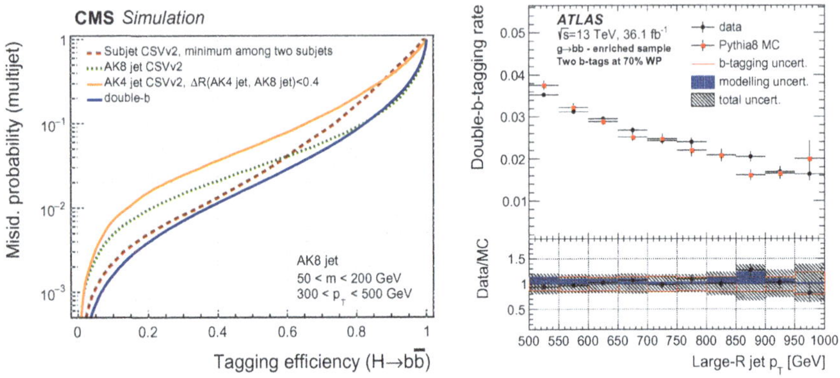

Fig. 3.9 The H tagging efficiency for $H \to b\bar{b}$ versus the multijet rejection for light quark and gluon jets for jets with $300 < p_T < 500\,\text{GeV}$, taken from [504] (left). The double-b tagging efficiency for large-R jets in a sample enriched with $g \to b\bar{b}$ decays in ATLAS, taken from [514] (right)

jet $p_T = 1$ TeV, the selection efficiency of $H \to b\bar{b}$ jets can be improved by 30% and more for the same background rejection [515]. Another possibility is obtained by clustering subjets in the rest frame of the large-R jet [517]. Since the $H \to b\bar{b}$ decay is back-to-back in the H centre-of-mass frame, there is a natural angular separation in this topology. Two subjets are obtained with an exclusive k_T clustering. The tracks associated to the subjets in the laboratory rest frame are input to the MV2c10 b tagging algorithm for subjet-b tagging. The gain in $H \to b\bar{b}$ tagging performance is comparable to VR track-jets [515], and the algorithm should be chosen based on the best sensitivity in the design of an analysis.

The efficiency for $H \to b\bar{b}$ tagging versus the misidentification rate for light-quark and gluon jets is shown in Fig. 3.9 (left). The performance of four different algorithms is shown: subjet b tagging with a requirement of two b-tagged subjets, large-R jet and small-R jet b tagging with a single b tag, and the double-b tagger. At a given efficiency, the misidentification rate for small-R jets is always worse than for large-R jets. The reason is that in the p_T range between $300 < p_T < 500\,\text{GeV}$ not all tracks from the $H \to b\bar{b}$ are reconstructed in a jet with $R = 0.4$, resulting in a worse identification efficiency than for jets with $R = 0.8$. At p_T of 1000 GeV and higher, the performance for these two jet distance parameters becomes comparable. The subjet b tagging shows a better performance for H tagging efficiencies below 60% compared to large-R jet tagging; above this value the large-R tagging is better. The double-b tagger shows the overall best performance, where the performance improvement compared to the other approaches is even larger at high p_T. Due to the kinematic information included in addition to b-tagging observables, the double-b tagger also improves the background rejection for jets from $g \to b\bar{b}$ splittings from 60 to 50 at 80% signal efficiency and from 20 to 10 at 35% signal efficiency, when compared to subjet b tagging.

The commissioning of $H \to b\bar{b}$ tagging algorithms is challenging since efficiency measurements are not possible on jets from boosted $H \to b\bar{b}$ and $Z \to b\bar{b}$ production due to the smallness of the production cross sections. Instead, jets originating from $g \to b\bar{b}$ splittings are being used. ATLAS and CMS select jets with one or two muons associated associated to them, where the muons originate from semi-leptonic B hadron decays. While jets associated with two muons are more similar to the topology of jets from $H \to b\bar{b}$ decays, a selection of jets with one muon results in a larger dataset for the performance measurement. It should be noted that the jet mass depends on the number of muons and a large fraction of the signal will not contain two muons.

A measurement of the double-b tagging efficiency in ATLAS is shown in Fig. 3.9 (right). One of the two ghost-associated track subjets is required to contain a muon track. About 6% of the jets originate from $g \to b\bar{b}$ splittings and another 6% from $g \to c\bar{c}$ before applying the ATLAS double-b tagger. The flavour fractions in simulation have been adjusted to the ones in data by a template fit to the distribution of the mean signed impact parameter significance, calculated from the three leading-p_{T} tracks. The flavour fraction corrections are found to vary between 0.7 and 1.7 depending on the jet p_{T}, with a statistical uncertainty below 10%. The rate of double-b tagged jets is measured to be 3.5% at $p_{\mathrm{T}} = 500$ GeV and 1.6% at $p_{\mathrm{T}} = 1000$ GeV, consistent with the expectation of a 70% efficiency for $H \to b\bar{b}$ tagging. The decrease in efficiency with increasing p_{T} is a result of the increasing collimation and eventual merging of the two track jets, as well as the degradation of the tracking performance inside high p_{T} jets. The uncertainty in the measurement is dominated by uncertainties in b tagging variables and modelling of the $g \to b\bar{b}$ signal. The background rejection of the double-b tagger for jets from boosted top quarks is about an order of magnitude smaller than the one for light quark and gluon jets [514, 516], caused by the presence of a b quark from the $t \to bW$ decay. This misidentification rate has been measured in a sample enriched in high-p_{T} $t\bar{t}$ production [504]. The misidentification rate in data and simulation agree within the uncertainties, which are of the order of 10%, dominated by the statistical precision of the data.

In contrast to $H \to b\bar{b}$, the decay $H \to WW^* \to 4q$ has not been studied experimentally so far, except for a search for VH resonances in all-hadronic final states by CMS [518]. In this analysis, $H \to WW^* \to 4q$ jets are identified using a selection in the pruned jet mass and the N-subjettiness ratio $\tau_{42} = \tau_4/\tau_2$. Since the distribution in τ_{42} is p_{T} dependent, the signal efficiencies have been observed to decrease from approximately 30 to 20% with increasing p_{T} from 500 to 1000 GeV, for a selection of $\tau_{42} < 0.55$. A direct measurement of the signal efficiency proves difficult because of the inability to enrich a sample with jets from four prong decays without applying a jet substructure selection. Instead, the efficiency correction of a W tagger based on τ_{21} measured in a $t\bar{t}$ sample has been extrapolated using the difference between PYTHIA and HERWIG in modelling the $H \to WW^* \to 4q$ decay as systematic uncertainty. Further studies are clearly needed for the applicability of this decay channel in future analyses, where advanced machine learning concepts offer the opportunity to make this channel complementary to $H \rightarrow b\bar{b}$. In addition to H analyses, tagging $WW \rightarrow 4q$ jets may also become important in searches for new physics, for example in boosted diboson resonances [519].

3.5.4 *Top Tagging*[9]

The three-prong decays of highly boosted top quarks in the fully hadronic decay channel offer richer phenomenology for their identification than the two-prong decays of W, Z and H bosons. This has been exploited in a number of algorithms, which usually aim at an optimal performance[10] in a particular kinematic regime. Flavour tagging also plays a key role for top tagging, which offers its own challenges because the jet from the b quark may not be isolated from the radiation resulting from the associated W boson decay. Due to the heavier mass of the top quark compared with the electroweak bosons, top tagging must also operate in a moderate boost regime where the decay products may not all be contained inside a single jet with $R \lesssim 1.0$.

The techniques for tagging boosted top quarks have evolved as fairly complex methods in comparison to the V and H taggers. Some commonly used techniques include the CMSTT and HTT. While the CMSTT is only studied in CMS [240, 241, 243], the HTT is studied by ATLAS and CMS on 7 and 8 TeV data [243, 433, 520]. Efficiencies of the HTT of 10% with misidentification rates of 0.5% for jets with $200 < p_T < 250$ GeV are observed. The efficiency increases with increasing jet p_T, where a plateau is reached for $p_T > 400$ GeV, with efficiencies of approximately 40 at 3% misidentification rate, very similar to the performance achieved with the CMSTT. One of the best performing taggers for top quark tagging is shower deconstruction, with efficiencies from 30 to 45% at misidentification rates from 1 to 4% for jets from $p_T = 350$ to 700 GeV [520]. Recently, the shower deconstruction algorithm was optimised for top quarks with $p_T > 800$ GeV in context of the W' to tb hadronic search [521] by using exclusive k_T subjets.

In addition to the dedicated techniques described above, simpler algorithms using grooming and substructure similar to V tagging methods have been investigated by ATLAS. A performance study at 7 TeV [433] investigated a variety of performance metrics relating to the usage of groomed jets. Different grooming algorithms were investigated for their resilience to pileup and mass resolution. It was concluded that trimmed anti-k_T jets with $R = 1.0$ and trimming parameters of $R_{sub} = 0.3$ and $f_{cut} = 0.05$ were the best option. This jet definition became standard in ATLAS for W, Z, H and top quark tagging in 7 and 8 TeV analyses. The ATLAS choice of $R = 1.0$ jets compared to CMS with $R = 0.8$ jets results in an earlier rise of the tagging efficiency with increasing jet p_T. A later ATLAS study [520] inves-

[9]The text in this subsection has been taken from [26] and has been written by the author. It has been adjusted to fit this book.

[10]The text in this subsection has been taken from [26] and has been written by the author. It has been adjusted to fit this book.

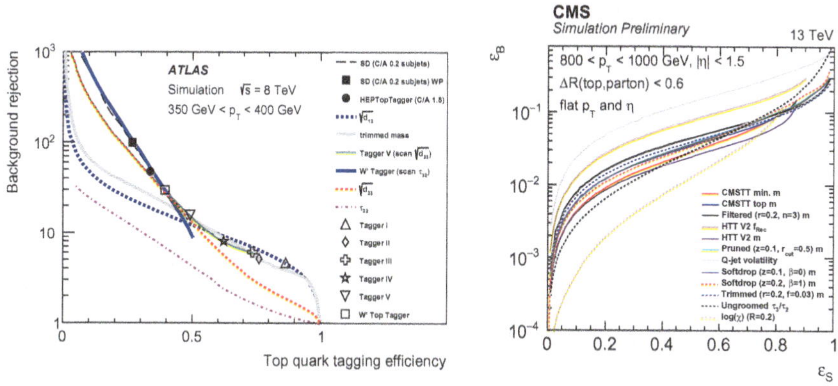

Fig. 3.10 Top quark tagging efficiency versus background rejection for various substructure variables and combinations in ATLAS, taken from [520] (left). Background versus signal efficiency for the single variables studied in the optimisation of top tagging for 13 TeV data in CMS, taken from [459] (right)

tigated the various methods available for top tagging. The so-called Tagger V has $m_{\text{jet}} > 100\,\text{GeV}$, $\sqrt{d_{12}} > 40\,\text{GeV}$ and $\sqrt{d_{23}} > 20\,\text{GeV}$, where $\sqrt{d_{ij}}$ is the k_T-splitting scale [38]. The efficiency versus rejection is shown for various taggers in Fig. 3.10 (left). The difference between Taggers III and V is the additional requirement on $\sqrt{d_{23}}$ in Tagger V. At efficiencies smaller than 45%, the W' tagger, based on $\sqrt{d_{12}}$ and the N-subjettiness ratios τ_{21} and τ_{32}, has better background rejection than Taggers III and V. ATLAS also tested the HTT and shower deconstruction tagger [522], which have been found to have good background rejection (larger than 50) for efficiency values smaller than about 35%. However, similar as for the CMS experiment, the background efficiencies of the two taggers show a significant rise with increasing p_T.

CMS has focused on enhancing the performance of the CMSTT and HTT by identifying observables which carry discriminatory power, but have only small or moderate correlations with the observables used in the main algorithm. Typically, correlation coefficients of about 0.3 or less are required for noticeable improvement when augmenting an algorithm with additional variables. Examples for discriminating variables which fulfil this are N-subjettiness ratios, energy correlation functions and their ratios, and b tagging. A study by CMS showed that at 20% signal efficiency, the background rejection of the CMSTT can be improved by a factor of 5 when adding information from τ_{32} and subjet b tagging information [243]. At higher efficiencies, the improvements become smaller. For the HTT, improvements of similar size are observed for $p_T > 200\,\text{GeV}$, becoming less significant at higher p_T.

The large difference in performance of the single variable τ_{32} between ATLAS and CMS (Fig. 3.10) is due to jet grooming. Although the CMS study shows only the receiver operating characteristic (ROC) curves for $800 < p_T < 1000\,\text{GeV}$, the overall picture does not change when studying top quarks in the region of $p_T \approx$

400 GeV. Instead, in ATLAS τ_{32} is calculated from trimmed jets, which results in less discrimination power when used as sole tagging variable compared to ungroomed τ_{32}. However, groomed τ_{32} can still lead to considerable improvements when combined with other variables [523].

ATLAS and CMS have studied various top taggers also for their stability with respect to the number of pileup interactions [459, 524]. Single variables and their combinations are studied and compared with shower deconstruction, CMSTT, HTT, and an improved version thereof with shrinking cone size (HTTv2) [234].

Figure 3.10 (right) shows a comparison based on simulation of the single variable performance in CMS, where signal jets are generated through a heavy resonance decaying to $t\bar{t}$ and background jets are taken from QCD multijet production. Note that for this study reconstructed jets are matched to a generated parton, and the distance between the top quark and its decay products must be less than 0.6 or 0.8 for a reconstructed $R = 0.8$ and 1.5 jet, to ensure that the top quark decay products are fully merged and reconstructed in a single jet. The best single variable in terms of efficiency versus background rejection is the discriminator $\log\chi$, calculated with shower deconstruction. The second best variables are the N-subjettiness ratio τ_{32} at low efficiency and the jet mass calculated with the HTTv2 at high efficiency values. The individual groomed jet masses show similar performance, and CMS moved to using the soft drop mass due to its beneficial theoretical properties [24]. The default for CMS analyses of 13 TeV data was chosen to be the soft drop jet mass combined with τ_{32} for top tagging at high p_{T}. Generally, at high boost, the combination of a groomed mass with τ_{32} leads to a large gain in background rejection.

The CMS study [459] also investigated combining single variables with more complex taggers. Combining shower deconstruction with the soft drop mass, τ_{32}, and subjet b tagging can lead to improvements. However, the efficiency and misidentification rate for this combination were found not to be stable as a function of jet p_{T} (the combined algorithms were studied using working points corresponding to a background efficiency of 30%). At low boosts, the dedicated HTTv2 shows the best performance. In this kinematic region, using groomed τ_{32}, obtained by using the set of particles from the soft drop jet instead of the original jet, helps to improve the performance.

ATLAS has commissioned a top tagger with little complexity for use by physics analyses of 13 TeV data. The rationale behind this approach was the potential benefit of having an efficient top tagger with well-understood efficiency and associated systematic uncertainties for use in early analyses. The supported top tagger makes use of anti-k_{T} $R = 1.0$ trimmed jets, but with a parameter of $R_{\mathrm{sub}} = 0.2$ instead of 0.3. Candidate top jets are required to satisfy a calibrated mass window requirement and a p_{T}-dependent, one-sided cut on τ_{32} [524]. The variable τ_{32} has been chosen since it shows the best background rejection in combination with a small correlation with m_{jet}, a reduced p_{T}-dependence, and good performance across a large range in p_{T}. A comparison of this two-variable tagger, and more generally a tagger based on the jet mass and an angular variable like τ_{32}, and more complex taggers shows that the background rejection can be improved by about 50% when using the HTT or shower deconstruction [525].

A common problem of top tagging algorithms is the rise of the misidentification rate with increasing p_T, which is due to the Sudakov peak in the jet mass distribution shifting to higher values for quark- and gluon-initiated background jets. For some taggers, for example the CMSTT, this shift also results in a decrease of the efficiency once a very high p_T threshold is crossed (larger than 1 TeV) [241]. A possible solution is offered by the VR algorithm. ATLAS studied the performance of the VR algorithm for top tagging and reported a stabilisation of the position of the jet mass peak for a large range of p_T [498]. The VR jets are shown to improve the performance of the jet mass, $\sqrt{d_{12}}$ and τ_{32} for top tagging, when compared to trimmed jets. In CMS, the HOTVR algorithm [245] has been shown to result in an efficiency increasing from 25 to 40% for jet p_T from 300 to 2000 GeV with constant background rejection of 2% [526]. Further improvements can be obtained by combining this algorithm with other substructure variables and subjet b tagging.

Most top taggers target either the region of low to intermediate boosts, or the highly boosted regime. However, in typical searches for new physics at the LHC non-vanishing efficiency for the full kinematic reach is crucial. Several attempts of combining different reconstruction and identification algorithms have been made. A search for resonances decaying to $t\bar{t}$ by ATLAS uses a cascading selection from boosted to resolved [527], where the resolved topology is reconstructed and identified using a χ^2-sorting algorithm. To efficiently identify top quarks over a broad p_T range in the search for top squark pair production, reclustered variable-R jets are used with $R = 0.4$ jets as inputs to the jet reclustering algorithm [457, 528].

A search for supersymmetry in CMS [529] uses three distinct topologies: fully-merged top quark decays with soft drop and τ_{32} top tagging (Monojet), merged W boson decays (Dijet) and resolved decays (Trijet). The efficiency of the three categories is shown in Fig. 3.11 (left), where the turn-on of the combined efficiency starts at values as low as $p_T \approx 100$ GeV. The resolved trijet category is identified using three anti-k_T jets with $R = 0.4$. The large combinatorial background is suppressed through a multivariate technique, which achieves a misidentification rate of

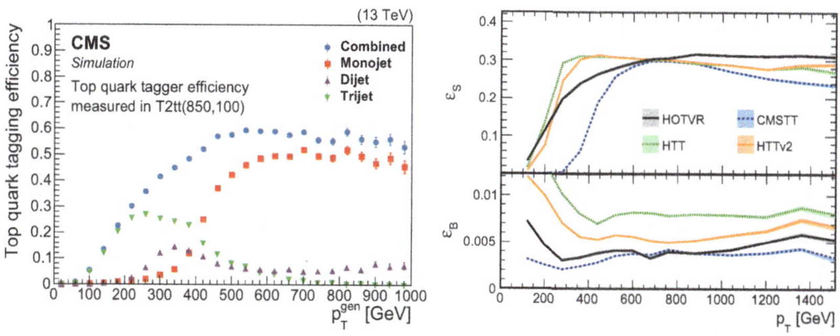

Fig. 3.11 Top tagging efficiency of three different top tagging methods and the combined efficiency, as a function of the generated top quark p_T; taken from [529] (left). Comparison of top tagging efficiency (top) and misidentification rate (bottom) for four dedicated top tagging algorithms (right)

approximately 20%. There exist other approaches to cover the transition from low to high Lorentz boosts, using a single algorithm. In the HTTv2 algorithm, the jet size is reduced until an optimal size R_{opt} is found, defined by the fractional jet mass contained in the smaller jet. This results in better performance at high p_T, while keeping a low misidentification rate at low p_T. Similar performance is obtained with the HOTVR algorithm, although with less algorithmic complexity. A comparison of the efficiencies and misidentification rates for the CMSTT, HTT, HTTv2 and HOTVR as a function of p_T is shown in Fig. 3.11 (right). This comparison has been obtained with the parameter settings described in [245]. The tagger working points have been chosen to obtain an efficiency of 30% at $p_T = 700$ GeV. The turn-on of the HOTVR algorithm is not as sharp as for the HTT or HTTv2 taggers, but the misidentification rate is smaller in this region. Overall, the HOTVR tagger achieves a misidentification rate comparable to the one of the CMSTT over the full range in p_T, while achieving similar efficiencies as the HTT or HTTv2 taggers at low p_T. The high efficiency of the HOTVR tagger starting from $p_T > 300$ GeV, together with a flat background efficiency, will make this tagger an important tool in analyses targeting intermediate to high top quark boosts.

An important step towards the commissioning of top taggers within an experiment are measurements of the efficiency and misidentification rate in real collision data. Generally, high-purity samples of top jets in data are obtained using a tight signal selection (a lepton, well-separated from a high-p_T large-R jet, and an additional b-tagged jet) to ensure that events contain a fully-merged top quark decay in a single large-R jet. This can never be fully achieved, as no requirements on the substructure of the large-R jet can be imposed without biasing the efficiency measurement. This results in an efficiency measurement that will be based on a sample containing partially-merged and non-merged top quark decays. These can be subtracted from the efficiency measurement by using simulated events, as done in a measurement by ATLAS [520]. This has the drawback of relying on a specific simulation and the ambiguous definition of a fully merged top quark decay. By not correcting for non-merged top quark decays, efficiency values are obtained smaller than the ones suggested by ROC curve studies, see for example [459]. Instead of subtracting the t backgrounds, recent measurements perform a simultaneous extraction of the efficiencies for fully and partially merged categories [525, 526, 530, 531]. This is done by separating events into a passing and a failing region, where pass and fail are determined by a selection on τ_{32}. The obtained distributions in the (groomed) jet mass are used for a template fit of the fully merged, semi-merged and un-merged categories to data, as shown in Fig. 3.12. The jet mass is used since the three contributions have different shapes in this distribution, allowing for a precise determination of the top tagging efficiencies. An additional advantage is that systematic uncertainties can be included consistently. The resulting corrections are close to unity with typical uncertainties between 5 and 10%.

Measurements of misidentification rates can be carried out by selecting a dijet sample with two high-p_T jets, which are dominated by gluon and light-flavour jets. In order to test different flavour compositions, Z/γ+jets selections are used as well. Due to the high p_T threshold of unprescaled jet triggers, measurements on dijet

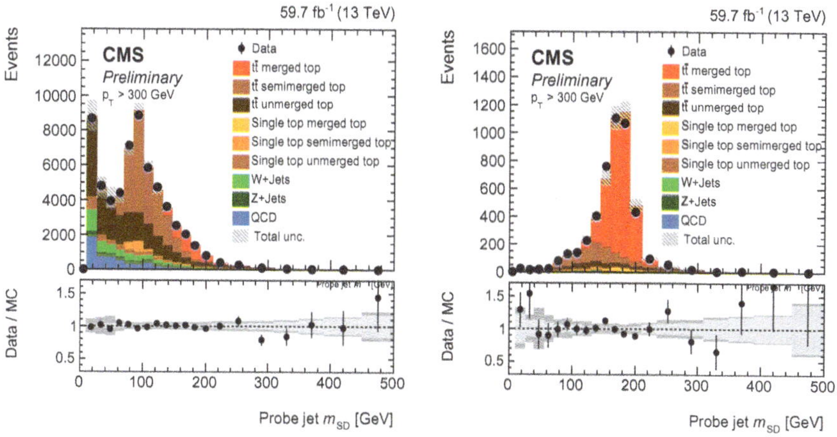

Fig. 3.12 Softdrop mass distributions for anti-k_T jets with PUPPI and p_T > 300 GeV in CMS data recorded in 2018. Shown are the top tagging fail (left) and pass (right) regions, where pass is defined by τ_{32} < 0.46. The simulation is shown after the template fit to data of the merged, semi-merged and un-merged contributions. In the ratio at the bottom, the inner area displays the statistical uncertainty of the simulation and the outer area shows the total uncertainty. Taken from [531]

events start from p_T > 400 GeV or higher, whereas it is possible to probe p_T > 200 GeV in γ+jets samples thanks to lower thresholds in photon triggers [525, 526]. Another approach is to use a non-isolated electron trigger, where the electron fails offline identification criteria. This yields events mainly from light-flavour multijet production, where a jet is misidentified as an electron at the trigger level. While the top-tag misidentification rate can be measured starting from smaller values of p_T with this strategy, a non-negligible amount of $t\bar{t}$ contamination has to be subtracted after requiring a top-tagged jet [520].

3.5.5 *Machine Learning Taggers*

Soon after the first studies were conducted on jet substructure at the LHC, it was realised that multivariate analyses can help to identify variables of importance for tagging. Due to the wealth of substructure observables based on different approaches, it is far from obvious which ones carry additional information relative to other observables. In addition, the information carried by an observable changes when calculated on groomed jets, and usually gets reduced by detector effects. Studies by ATLAS and CMS have used boosted decision trees (BDTs) [532, 533] and multilayer perceptron (MLP) neural networks [534, 535] to gain information on the importance of variables for substructure taggers [411, 435, 459, 535]. Typically, these BDTs and MLPs take jet substructure variables as input and perform a classification into signal and background, where the output distribution is obtained through an optimisation of a loss function. The free parameters of the algorithms are determined with training sets, which is known as supervised learning. It has been found that in many cases, adding more than the two or three variables with highest significance to the tagging algorithm does not result in a significant performance gain. In these studies, also the linear correlations of substructure variables have been examined, indicating the mutual information carried by observables. No new observables linearly uncorrelated to the output distribution could be identified, which would have indicated a potential gain in performance when adding these to the new tagger.

New developments in machine learning (ML) allow for more complex architectures of artificial neural networks (ANNs), with several hidden layers and a large number of neurons in each layer—so called deep neural networks (DNNs). These networks are capable of learning complex, non-linear correlations when trained on a sufficiently large sample of simulated data [536]. While previous approaches like the ones mentioned above have combined shallow ANNs with high-level observables, DNNs allow to obtain the classification power directly from the raw data using low-level inputs such as the four-vectors of all reconstructed jet constituents or just energy deposits in the detector. A further advantage of DNNs is the possibility to have multi-class classification with one output per class, instead of only discriminating between background and signal. There are several representations conceivable which can be used to pass the full low-level information to DNNs. Jet images are pixelated images, where the pixel intensity represents the momentum of all particles that deposited energy in a particular angular region [537–540]. Additional information, such as the identified particle type, can be encoded by additional image layers, similar to colour images [477]. Jet images can be analysed with convoluted neural networks (CNNs), which have a smaller number of parameters to be determined in the training than recurrent neural networks (RNNs). These RNNs can be used to analyse sequential information, such as the four-vectors of all jet constituents ordered in p_{T} [541]. This approach has already been been used for jet flavour tagging (b or c tagging), with charged particle tracks and secondary vertices as inputs [504, 542–544]. A generalisation of this approach, where jets are represented as graphs, has been studied in the context of pileup mitigation [545]. Lastly, advances in ML tools on point clouds have allowed the study of jets as unordered sets with the possibility to encode additional information about particles beyond their kinematic properties [225, 546]. More information on these ML tools in the context of jet substructure can be found in a recent review [24].

Recently, advanced machine learning methods have been applied to jet substructure taggers by ATLAS [525] and CMS [526]. In a first step, ATLAS has used 10–13 high-level observables to train a BDT and a DNN for W and t tagging. It has been found that the BDT and DNN performances are nearly identical, indicating that the advanced DNN can not surpass the BDT performance through an algorithmic improvement if the inputs are the same.[11] A comparison of the ROC curves of these two taggers (BDT top and DNN top) is shown in Fig. 3.13 (left). The ML taggers

[11] A similar conclusion has been obtained in a comprehensive comparison of DNN-based top taggers, where only small performance differences were observed due to differences in the algorithms [547].

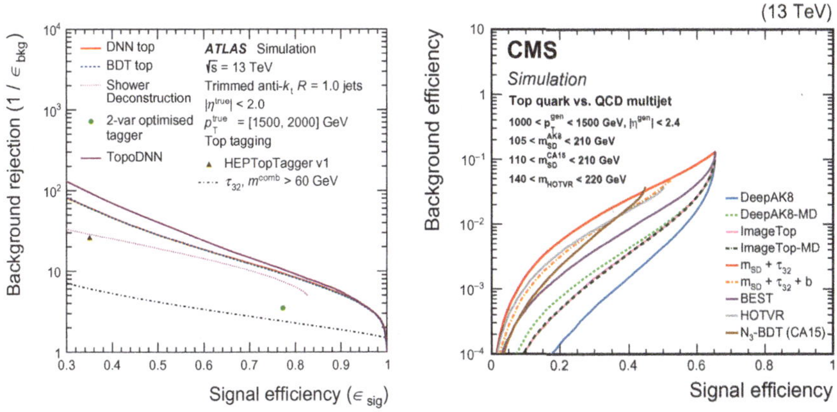

Fig. 3.13 Performance comparison of top taggers in ATLAS (left) and CMS (right) at high p_T. Taken from [525] (left) and [526] (right)

perform better than the HTT, a two-variable optimised tagger or the shower deconstruction algorithm, but do not achieve the optimal performance. A DNN based on four-vectors of a fixed-number of topoclusters (TopoDNN) shows the best performance. The algorithm consists of a fully connected neural network with four hidden layers, following closely the approach outlined in [548]. The gain in background rejection at low signal efficiency with respect to the DNN top tagger indicates that a DNN based on low-level inputs can learn correlations hidden to algorithms based on high-level observables. CMS has studied multi-class taggers and the use of jet images to further improve the performance of ML taggers. Figure 3.13 (right) shows the ROC curves of nine different top taggers, out of which six are based on ML techniques. The best performance is achieved with the DeepAK8 algorithm, which exploits the particle-level information to obtain a multiclass classifier for the identification of the five main categories, $W/Z/H/t$/other. In addition to particle four-vectors, information such as the energy deposit, the charge, the displacement and quality of the tracks, etc. are included for each particle. A secondary vertex list helps the network to extract features related to the heavy-flavour content of the jet. The particle and secondary vertex lists are processed with two CNNs. The outputs of these are combined using a fully connected network to perform the jet classification [526]. In the case of H and t tagging, the secondary vertex information helps to improve the performance significantly due to the heavy-flavour content of the jet. The other ML algorithms studied by CMS are a top tagger based on jet images (ImageTop), a multiclass classifier based on event shapes in the rest frames of W, Z, H and t [549, 550], called the boosted event shape tagger (BEST) [229], and a BDT based on eleven ECF ratios (N_3-BDT). Their performance lies between the cut-based approaches and the DeepAK8 tagger. It should be noted that the N_3-BDT has been designed for low p_T, such that it shows the best performance for $300 < p_T < 500$ GeV. Interestingly, in this region the HOTVR algorithm has a comparable performance to the ML taggers, even though no b tagging information is included. Further variants of the DeepAK8 and ImageTop taggers are mass-decorrelated (MD) taggers. In the case of DeepAK8, an adversarial training is used to achieve the decorrelation, whereas for ImageTop the decorrelated version is obtained by training on samples with similar shaped mass distributions. The MD taggers have the advantage of an unchanged jet mass distribution when selecting jets based on the algorithm's output. This can be crucial in analyses where the non-resonant background contribution is predicted from side-band regions. While the MD procedure results in a decrease in performance for the DeepAK8 algorithm, the performance of ImageTop is very similar between the nominal and MD versions. Efficiency and misidentification rate measurements have been carried out in data [525, 526], verifying the ML tagger performances from studies in simulation. The derived correction factors can be used to adjust the efficiencies in simulation. The corrections are typically smaller than 10% with uncertainties of the same size. These measurements allow for the use of these ML taggers in future analyses, promising a large gain in sensitivity relative to traditional approaches.

Chapter 4
Standard Model Measurements

Abstract Jet substructure measurements at the LHC can broadly be divided into two classes: measurements of jet substructure observables and measurements using jet substructure. The first class enables detailed studies of the dynamics of jet production and the energy flow within jets. These measurements can be used to test the modelling of jet substructure observables, verify analytical calculations and extract fundamental parameters of the theory. The second class are measurements made possible by jet substructure techniques. These include studies of top quark or electroweak boson production at very high transverse momenta. In most cases, the channels analysed would not be accessible without the use of jet substructure techniques. The high momenta probed in these measurements allow to study the validity of the SM up to the highest scales reachable at the LHC. These measurements can also be used to place important constraints on BSM theories, which predict deviations from the SM expectations in the high-energy tails of the measured distributions.

4.1 Measurements of Jet Substructure Observables

4.1.1 Jet Mass of Light Flavour and Gluon Jets

The most important observable in jet substructure studies is the jet mass, which is indispensable in measurements at high p_T and in searches for BSM effects. Its reconstruction and modelling is complicated, subject to the influence of many sources of experimental and theoretical uncertainties see Sects. 2.4.1 and 3.2). Measurements of the jet mass are crucial for our understanding of the detector response and the theoretical description of jet substructure.

The first data recorded at the LHC in the year 2010 with $\sqrt{s} = 7$ TeV enabled a measurement of the jet mass distribution in dijet events by ATLAS [551]. The inclusive jet cross section as a function of the jet mass has been measured for anti-k_T jets with $R = 1.0$ for different regions in p_T, where two regions in p_T are shown in Fig. 4.1. From today's point of view this measurement is special as events are selected to have only a single primary vertex, such that the measurement is free

R. Kogler, *Advances in Jet Substructure at the LHC*, Springer Tracts in Modern Physics 284, https://doi.org/10.1007/978-3-030-72858-8_4

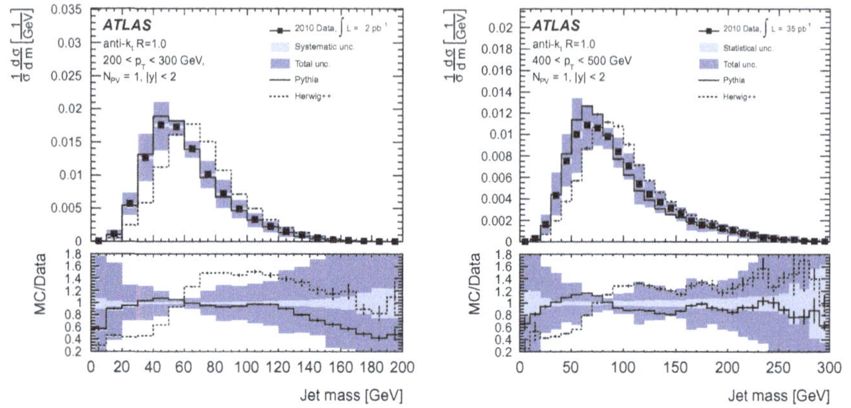

Fig. 4.1 Normalised inclusive jet cross sections for dijet production at $\sqrt{s} = 7$ TeV as a function of the jet mass for two different regions of jet p_T, as measured by the ATLAS Collaboration. The data are compared to predictions from PYTHIA and HERWIG++. Taken from Ref. [551]

from pileup effects. The data are described by the leading-order event generators with parton showers, giving confidence in the modelling of the jet mass and jet substructure variables. However, the measurement shows a feature that is observed in many substructure measurements: the data lie between the predictions from PYTHIA and HERWIG due to the differences in light flavour quark and gluon jets in these two event generators (see Sect. 3.5.1). This difference often leads to non-negligible uncertainties when correcting data for detector effects, as the detector response will vary for these two simulations. Jet grooming can reduce this difference. In the ATLAS measurement, it is shown that filtering leads to a better description of the data and a reduced difference between PYTHIA and HERWIG.

The first measurement of the jet mass distribution on trimmed and pruned jets has been performed using 5 fb^{-1} of 7 TeV data by CMS [552]. The measurement selects events from dijet and W/Z+jet production, where the latter sample has a larger fraction of quark jets than gluon jets. It is observed that the jet mass in W/Z+jet production is better described, suggesting that quark jets are better modelled than gluon jets. Indeed, quark jets are better constrained by LEP data than gluon jets when adjusting the free parameters of event generators [467]. Differences between data and simulation are larger at small mass values, where soft radiation is more important. Grooming methods are found to reduce these differences, where best agreement is found for pruned jets. In this study, the parameters of the grooming algorithms have been chosen such that pruning removes larger fractions of soft contributions to the jets than the other grooming algorithms, suggesting that the removal of soft radiation leads to the improved modelling.

Measurements of the soft drop (scaled) jet mass distribution in bins of p_T using 13 TeV data have been performed by ATLAS [553] and CMS [439]. Both collaborations have used anti-k_T $R = 0.8$ jets in a dijet sample, where the soft drop parameters were chosen to be $z_{\text{cut}} = 0.1$ and $\beta = 0$ ($\beta = 1$ and 2 have also been considered in the ATLAS measurement). The usage of soft drop grooming allows for a comparison of the data with semi-analytical resummation calculations at next-to-leading order with next-to-leading-logarithm (NLO+NLL) accuracy [238] and leading order with next-to-next-to-leading-logarithm (LO+NNLL) accuracy [288]. While ATLAS and CMS use different detector technologies, reconstruction algorithms, calibration methods, unfolding techniques and simulations, the results are compatible. In both measurements, uncertainties related to the jet and cluster energy response dominate the experimental uncertainties with a relative size of about 5%. Uncertainties due to the QCD modelling are typically between 5–10%, but can be as large as 20% at low mass where non-perturbative effects are important. The ATLAS measurement [553] in the scaled jet mass $\rho = \log(m_{\text{SD}}^2/p_{\text{T}}^2)$ is shown in Fig. 4.2, where m_{SD} is the soft drop jet mass and p_{T} is the transverse momentum of the ungroomed jet. Good agreement between data and the predictions is observed for $-3.7 < \rho < -1.7$, where the resummation is expected to yield accurate results. At high values of ρ perturbative radiation is important, such that the NLO+NLL calculations describe the data better than the LO+NNLL calculations. As β increases, the fraction of radiation removed by the soft drop procedure decreases and the impact of non-perturbative effects increases. For $\rho < -3.7$ the LO+NNLL calculation does not describe the data for $\beta = 1$ since it does not include non-perturbative corrections. However, for $\beta = 0$ non-perturbative effects are reduced by the soft drop grooming and the NLO+NLL and LO+NNLL calculations describe the data equally well. Recently, a calculation based on SCET has been presented, which reliably predicts not only the shape but also the absolute normalisation of the data [554]. CMS has also measured jet mass distributions for ungroomed jets and compared the results to the soft drop measurement. The precision of the measurement is reported to improve for soft drop, where especially modelling and pileup uncertainties get reduced because of the removal of soft radiation in jets [439]. Overall, the event generators Pythia, Herwig and Sherpa describe the data well in all measurement regions.

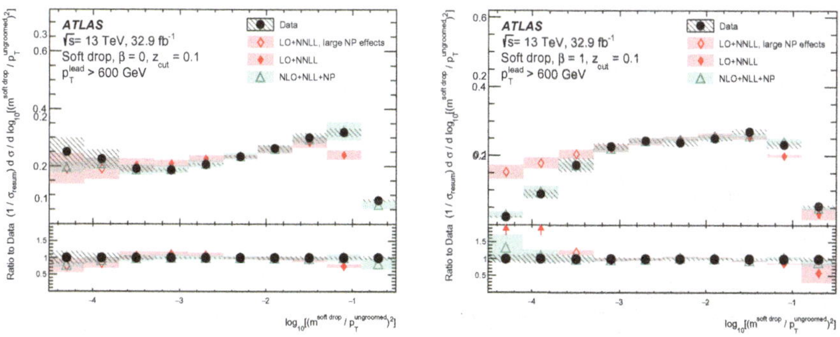

Fig. 4.2 Normalised distributions in ρ for soft drop light quark and gluon jets with $\beta = 0$ (left) and $\beta = 1$ (right), as measured by ATLAS. Taken from Ref. [553]

4.1.2 Jet Mass of W and Z Jets

From the beginning of jet substructure measurements at the LHC, the two prong decay of boosted W bosons has served as a means of calibrating the jet mass scale and resolution [427–429, 435, 436]. Usually, W jets are selected in a sample enriched with high-p_T $t\bar{t}$ production, for example with a selection on the angular distance ΔR between the large-R W jet and the closest b-tagged small-R jet, where ΔR is required to be less than 2.0 [241]. The abundance of high-p_T $t\bar{t}$ events, their identification through high energy charged leptons and b jets, and the well known mass of the W boson have resulted in precise calibrations of the jet mass scale and resolution. An example is shown in Fig. 4.3 (left), which shows a high purity of W jets in $t\bar{t}$ and tW events. The largest background originates from unmerged $t\bar{t}$ events, where a light quark or gluon jet is misidentified as a W jet candidate. The peak position in data is observed at a value of 80.8 ± 0.4 GeV and in simulation of 82.2 ± 0.3 GeV, where the uncertainties are of statistical nature [254]. The jet mass resolution is found to be about 10%, which agrees with the simulation within uncertainties. While W jets are used frequently as standard candles, no measurement of the W jet mass unfolded at the particle level exists. A sensitivity study suggests that such a measurement in the all-jets final state could lead to a precision in m_W of 30 MeV for the HL-LHC with 3000 fb^{-1} [555].

ATLAS has recently reported on a measurement of the trimmed and soft drop jet mass distributions for boosted Z bosons, decaying to $b\bar{b}$ [556]. The measurement is performed in $Z\gamma$ events, with the advantages of a clean trigger signature from the photon and an opportunity to directly estimate the backgrounds from data at the cost of a very small predicted cross section of 9–15 fb in the fiducial region. The study gives access to the jet mass of $Z \to b\bar{b}$ jets, important for assessing the modelling

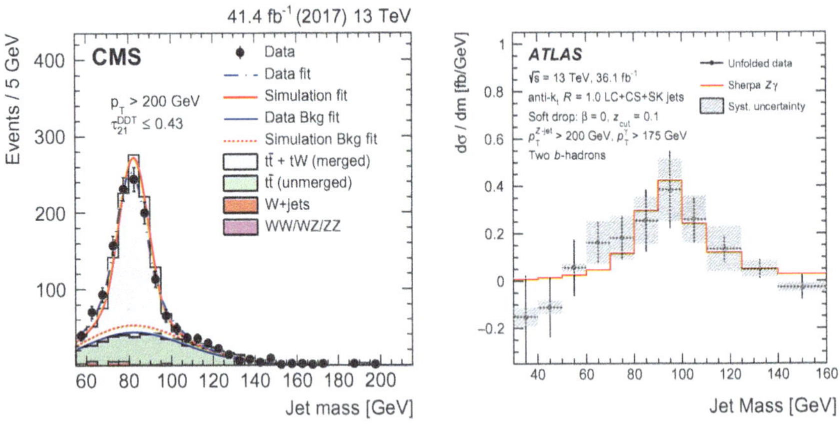

Fig. 4.3 Reconstructed jet mass distribution in $t\bar{t}$ candidate events enriched with boosted W jets with $p_T > 200$ GeV and $\tau 21^{\mathrm{DDT}} < 0.43$ (left); taken from Ref. [254]. Soft drop jet mass of $Z \to b\bar{b}$ jets in γZ production, unfolded at the particle level (right); taken from Ref. [556]

and systematic uncertainties of high p_T VH and H+jet production in the $H \to b\bar{b}$ channel. After a high p_T photon selection, Z jets are identified using two b tags from associated track jets. The dominant background originates from γ+jets production with $g \to b\bar{b}$ splittings. After the signal region selection, this background yields about a factor of 20 more events than signal events. It can be estimated in data from control regions with either one b tagged jet, or a less restrictive photon selection. It should be noted that for this background estimation to work, simulated contributions from $t\bar{t}$+γ and W+γ production need to be subtracted from data, as the shape of the mass distribution differs from that of γ+jets. The resulting distribution in the soft drop jet mass of γ+jet events, where the jet has two b tags, shows a peak from $Z \to b\bar{b}$ on a falling distribution. After subtracting the background, the observed significance over the background-only hypothesis is 2.7 standard deviations for $Z \to b\bar{b}$, with an expected significance of 2.7. The unfolded soft drop jet mass distribution at the particle level is shown in Fig. 4.3 (right). It peaks at around 95 GeV and is described well by the SHERPA event generator. The trimmed jet mass distribution peaks at lower values, around 85 GeV, and shows a higher significance of 3.9 standard deviations, because of a narrower jet mass distribution at the particle level.

4.1.3 Jet Mass of Top Quark Jets[1]

The jet mass of top quark jets, with the full top quark decay merged into a single large-R jet, has been measured in ℓ+jets $t\bar{t}$ events, where ℓ stands for an electron or muon, recorded by CMS at 8 [438] and 13 TeV [557]. The 8 TeV measurement is the first jet mass distribution unfolded at the particle level probing three prong decays. Large-R jets are reconstructed with the CA algorithm using a distance parameter of $R = 1.2$. The larger value of R in this measurement compared to the default $R = 0.8$ applied for top tagging applications in CMS is due to an optimisation of statistical precision versus the width of the jet mass distribution at the particle level and the JMR. The number of fully merged top quarks grows with increasing R, but so does the width of the jet mass distribution and the susceptibility to pileup and the underlying event. The leading jet p_T is required to be above 400 GeV to ensure the hadronic top quark decay to be fully captured within the large-R jet. No substructure selection is applied on the high-p_T large-R jet in order not to bias the jet mass measurement. The shape of the particle-level differential $t\bar{t}$ cross section as a function of the leading jet mass is well described by the simulations. The simulations predict a larger cross section than observed in the measurement, consistent with the $t\bar{t}$ cross section measurements by ATLAS and CMS at high p_T (see Sect. 4.2.3). The experimental systematic uncertainties are dominated by the uncertainties on the jet mass and energy scale, but are smaller than the uncertainties due to the signal modelling, coming from the choice of the top quark mass, the parton showering

[1]The text in this subsection is based on Refs. [26, 557], written by the author. It has been adjusted to fit this book.

and the choice of the factorisation and renormalisation scales. The normalised mass distribution from boosted top quarks can be used to extract the top quark mass. The normalised distribution is used since only the shape can be reliably calculated, and it has the additional benefit that systematic uncertainties partially cancel. The top quark mass is measured to be $m_t = 170.8 \pm 6.0\,\text{(stat)} \pm 2.8\,\text{(sys)} \pm 4.6\,\text{(model)} \pm 4.0\,\text{(theo)}$ GeV in agreement with top quark mass measurement in resolved $t\bar{t}$ events (see e.g. Refs. [558–561]), albeit with a much larger uncertainty. This constitutes a proof-of-principle, presenting the possibility to extract a fundamental SM parameter from a jet mass distribution. This is of particular interest, as ambiguities arise in the interpretation of traditional m_t measurements [562, 563], which can be circumvented by measurements and analytical calculations in the highly-boosted regime [564, 565].

A recent CMS measurement using 13 TeV data [557] has reported a precision improved by a factor of three relative to the 8 TeV measurement. The measurement is based on XCone jets, obtained by a two-step jet clustering [180]. First, the exclusive XCone algorithm is applied with a distance parameter of $R = 1.2$ and the specification of returning two jets, corresponding to the two boosted top quarks in the event. Using the constituents of these two large-R jets as input, XCone is run again with the distance parameter $R = 0.4$ and the parameter of the number of subjets in each jet $N = 3$. This procedure results in exactly two large-radius XCone jets with three XCone subjets each. Jet energy corrections [70] derived for anti-k_{T} jets are applied to the XCone subjets. An additional correction applied to the XCone-subjet momenta is obtained from simulation to account for differences between the XCone-subjet momenta and the momenta of anti-k_{T} jets. This correction is parametrised as a function of XCone subjet p_{T} and $|\eta|$, and has an average size of 2%, with an average uncertainty of 0.3%. The four-momenta of the three XCone subjets are combined to form the final XCone jet, where the jet mass is the invariant mass of all PF candidates clustered into the three XCone subjets. The resulting distribution in m_{jet} at the particle level has a width half as large as for CA jets with $R = 1.2$, as used in the 8 TeV measurement. The improvement is due to the two-step XCone jet clustering procedure, which acts as a grooming algorithm, similar to trimming, on the large jet. The advantage of XCone over other grooming algorithms in this measurement is its dynamical interpolation between the resolved and boosted regime, i.e., between three well-separated subjets and three subjets close in ΔR, which would not be resolved by other reconstruction methods. The XCone-jet reconstruction results also in a large improvement of the experimental resolution in m_{jet}, due to the accurate subjet calibration. With XCone a resolution of 6% is achieved, compared to a resolution of approximately 14% for CA jets with $R = 1.2$. Figure 4.4 (left) shows the reconstructed XCone jet mass for jets with $p_{\mathrm{T}} > 400$ GeV. The distribution shows a pronounced and narrow peak close to the value of m_t. The fraction of fully merged $t \to Wb \to q\overline{q}'b$ decays in the region of the top quark peak with $140 < m_{\text{jet}} < 200$ GeV is approximately 75%. The peak position is stable as a function of the number of pileup vertices, thanks to the inherent grooming of the two-step XCone jet clustering and the area-based jet energy corrections. The stability of the peak position is also verified using m_{jet} from fully merged W decays, as calculated from the two XCone subjets with the smallest pairwise mass. The simulation,

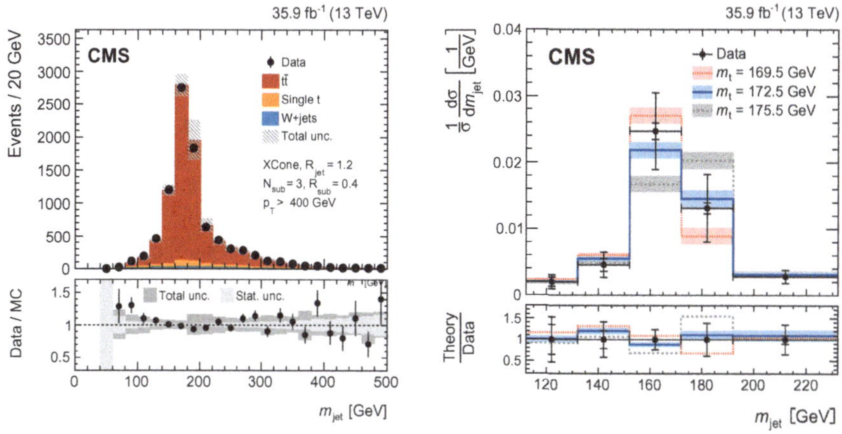

Fig. 4.4 Reconstructed jet mass distribution in $t\bar{t}$ production after selecting events enriched in fully merged top quark decays (left). Normalised particle-level $t\bar{t}$ differential cross section as a function of the XCone-jet mass (right). Taken from Ref. [557]

in this case an NLO calculation with POWHEG [370, 371, 373, 566–568] interfaced with PYTHIA for parton shower and hadronisation, describes the data very well. Also MADGRAPH5_aMC@NLO [369, 372] is able to describe the shape of the m_{jet} distribution, giving confidence in the different techniques for matching matrix elements to parton showers and the string hadronisation model for energy flow observables.

Figure 4.4 (right) shows the measured normalised differential cross section as a function of m_{jet}. The normalised differential cross section benefits from a partial cancellation of systematic uncertainties. The largest sources of systematic uncertainties originate from the jet energy scale and the modelling of final state radiation. The normalised differential cross section shows high sensitivity to the value of m_t, which is measured to be $m_t = 172.6 \pm 0.4\,(\text{stat}) \pm 1.6\,(\text{exp}) \pm 1.5\,(\text{model}) \pm 1.0\,(\text{theo})$ GeV. The improvement in precision by a factor of 3.6 relative to the measurement at 8 TeV [438] is attributed primarily to the novel jet reconstruction using XCone. The improvement by a factor of two in both, the m_{jet} width at the particle level and experimental resolution, together with more integrated luminosity and an increased value of $\sqrt{s}$, provides a reduction by a factor of about 14 in the statistical uncertainty. The systematic uncertainties are also reduced through the XCone-jet reconstruction, which enables a more precise calibration of the XCone-subjet energies and a better stability against contributions from pileup and the UE. Uncertainties from modelling are reduced through the use of additional sideband regions with higher granularity in the unfolding. The jet reconstruction with the XCone algorithm results in the accuracy necessary for precision measurements at large top quark momenta, which will become increasingly important in the future.

4.1.4 Jet Substructure in Light Flavour and Gluon Jets

Beyond measurements of the jet mass, more complex jet substructure observables can be used to study the energy flow inside jets, localised spots of high energy density, angular particle-particle correlations or particle multiplicities. These measurements help to improve our understanding of jet formation, as observables have different sensitivities to contributions from perturbative and non-perturbative effects. These measurements can also be used to study the approximations made in parton shower and hadronisation models. Once unfolded at the particle level, distributions of jet substructure observables can be used to constrain the free parameters in event generators, or for the determination of fundamental parameters of the SM, like the value of the strong coupling [569, 570].

The first measurement of jet substructure at the LHC has been a measurement of differential and integrated jet shapes from ATLAS using 3 pb^{-1} of 7 TeV data [571]. While this measurement is limited in statistical precision, it shows the power of jet substructure observables for the determination of the quark and gluon fractions in jets. Unfolded measurements of observables related to tagging studies have also been performed by ATLAS and CMS on 7 TeV data recorded in 2010 [551, 572, 573]. These measurements have been carried out on inclusive jet data, dominated by dijet production with a large fraction of gluon jets at low p_{T}. Distributions of k_{T} splitting scales, N-subjettiness, jet width, transverse jet size, eccentricity, planar flow and angularity have been measured. The event generators PYTHIA and HERWIG are in good agreement with the data, which is a striking achievement considering that the simulations available in 2011 have been tuned mostly to LEP, SPS and Tevatron data [291, 574], with only charged particle multiplicities entering the tunes [575–577]. These measurements gave confidence in the use of jet substructure at the LHC and have preceded further developments of substructure techniques.

Measurements of charged particle multiplicities n_{ch} are important for several aspects of physics at the LHC. Multiplicity distributions are affected by the details of particle production, from quark/gluon prodution to perturbative effects in the parton shower, to hadronisation and the decay chains of short-lived hadrons. The underlying event and pileup interactions also have a large impact on multiplicity distributions. Charged particle multiplicities constitute an important input to a number of substructure taggers, most notably q/g discrimination (see Sect. 3.5.1). ATLAS and CMS have measured n_{ch} distributions in jets using 7 TeV data [573, 578]. ATLAS has also performed measurements using 8 [579] and 13 TeV [580] data. The challenge of these measurements lies in a precise understanding of the track reconstruction efficiencies over a large range of p_{T} and $|\eta|$. Also, the n_{ch} distribution depends on the q/g content of the sample, on jet p_{T} and η, and on the track p_{T}. Dominant experimental effects include the track reconstruction efficiency, the rate of fake and secondary tracks, the track momentum scale, and density effects from pixel and strip cluster merging for high p_{T} jets [415–418]. Nevertheless, precise measurements with uncertainties at the few percent level have been performed, enabled by detailed studies of these effects. The measured average n_{ch} as a function of jet p_{T} is shown in Fig. 4.5 (left) using

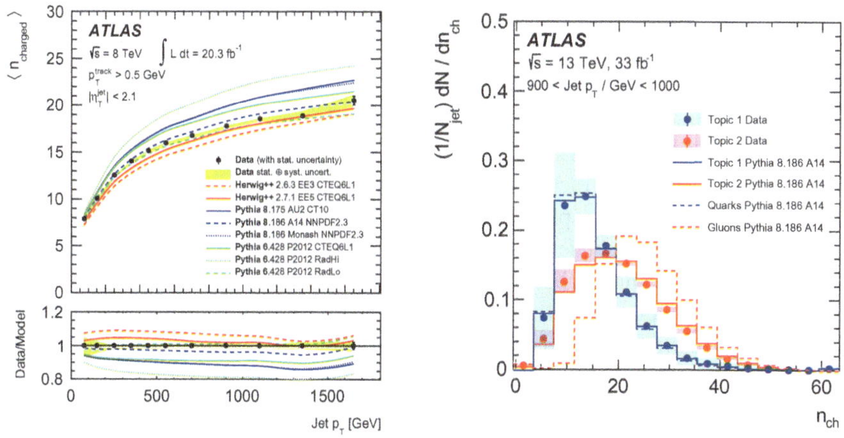

Fig. 4.5 Average charged particle multiplicity as a function of jet p_T for track $p_T > 0.5\,\text{GeV}$; taken from Ref. [579] (left). Jet topics extracted from charged particle multiplicity for jets with $900 < p_T < 1000\,\text{GeV}$, taken from Ref. [580] (right)

8 TeV data. The predictions from PYTHIA and HERWIG with different tunes scatter around the data. The sensitivity of the data to subtleties of the final state modelling can be seen from the variations of the PYTHIA 6.428 Perugia 2012 tune [291], labelled RadHi and RadLo. There, the α_s value that regulates final state radiation is changed by factors of one half and two with respect to the nominal Perugia 2012 tune. The n_{ch} distribution can be used to extract information on the quark or gluon origin of jets [579, 580]. For a fixed jet p_T, jets with higher $|\eta|$ are more often quark-initiated due to valence quarks scattering off low-x gluons. For a fixed $|\eta|$, the quark fraction increases with jet p_T due to the relative increase in valence-quark scattering relative to gluon–gluon scattering. Differential measurements of n_{ch} as a function of jet p_T in different $|\eta|$ regions can therefore be used to extract information on the underlying q/g fractions. Such an extraction is shown in Fig. 4.5 (right), where distributions of jet topics [463, 464] have been obtained. These topics are not purely quark and gluon jets, but combinations of them. Topic 1 resembles quark jets rather closely, while Topic 2 is a mixture of quark and gluon jets. It approaches gluon jets more closely at high p_T. The advantage of topic modelling is that theoretical ambiguities due to the definition of quark and gluon jets are reduced. The Topic 1 data are well described by PYTHIA, whereas Topic 2 data have a lower average value of n_{ch} than PYTHIA. The jet p_T dependence for quark and gluon jets can be calculated analytically [581, 582]. The analytic calculations agree with the predictions from PYTHIA, but a different slope for gluon jets is observed in data. The average value of n_{ch} rises less steeply with jet p_T in data than in the predictions [580]. A similar effect is also seen for the jet topics, when compared to the predictions from PYTHIA. More studies and higher statistical precision at high p_T will be needed to understand this discrepancy.

Jet fragmentation functions are closely related to the measurement of n_{ch}. They are defined as the probability that a particle carries a longitudinal momentum fraction z of a parton initiating the jet. Jet fragmentation functions can be probed by measurements of the relative momentum of charged particle tracks inside jets [583], providing valuable input for the modelling of fragmentation in event generators. A recent measurement by ATLAS [580] uses $\zeta = p_{\mathrm{T}}^{\mathrm{particle}}/p_{\mathrm{T}}^{\mathrm{jet}}$ instead of z, which is a better proxy for the starting scale of the parton shower evolution at hadron colliders. Measurements of partial fractions of the fragmentation functions probe how much of the jet energy is carried by particles of a given p_{T} fraction. The fraction of particles carrying 10% or less of the jet p_{T} changes very little across a jet p_{T} of 300–2500 GeV and asymptotically approaches 96%. A strong p_{T} dependence is introduced when ζ is lowered. The fraction of jet particles with $\zeta < 1\%$ increases logarithmically with jet p_{T}, while the fraction of particles with $\zeta < 0.1\%$ increases more rapidly [580]. Additional information is carried by the moments of the ζ distribution, which are well described by PYTHIA and HERWIG, while SHERPA displays some discrepancies with the data. A less-well probed aspect of jet fragmentation is the fragmentation of $g \to b\bar{b}$, important for H tagging and analyses probing final states with b quarks. While studies on well separated b quark pairs exist, data with $\sqrt{s} = 13$ TeV allow for the first time a measurement of jet fragmentation in collimated $g \to b\bar{b}$ production at high p_{T} [584]. In this analysis, quark and gluon flavour fractions have been fit to the data to remove contributions from processes other than $g \to b\bar{b}$. The fitted fractions significantly disagree with the PYTHIA predictions, suggesting that further studies can lead to an improvement in the modelling of heavy flavour production at high p_{T}. Comparisons of unfolded distributions at the particle level with simulations suggest that these data have the power to constrain the modelling of gluon polarisation.

Another variable related to n_{ch} is the jet charge, defined as the p_{T}-weighted sum of the electric charges of the jet constituents. Jet charge is sensitive to the charge of the parton initiating the jet. At the LHC, the mean of the jet charge distribution increases with increasing p_{T} for the more forward of the leading p_{T} jets, due to the increasing fraction of jets from valence up-type quarks. The jet charge is sensitive to non-perturbative effects and can be used to test and improve the modelling of final state radiation. Unfolded measurements of jet charge distributions have been reported by ATLAS [585] and CMS [586] using 8 TeV data.

Observables less sensitive to non-perturbative effects are k_{T} splitting scales. These probe the energy scale of the last combination step in the clustering process. An unfolded measurement in W+jets production with 7 TeV data [587] reveals that higher order perturbative corrections are important to model the hard region with $\sqrt{d_k} > 20$ GeV, while resummation and non-perturbative effects are important in softer regions. Another observable characterising the parton splitting is $z_g = p_{\mathrm{T},2}/(p_{\mathrm{T},1} + p_{\mathrm{T},2})$ [213], calculated on the branch of a jet fulfilling the soft drop condition in Eq. (2.41). This variable is closely related to the QCD splitting function, which is singular in the collinear limit and thus not experimentally accessible. However, the cross section as a function of z_g approximates the QCD splitting function in the high-energy limit and can thus be used to probe this universal function. The distribution in z_g can be predicted from first principles because of its Sudakov safety, making z_g a potent probe of the underlying $1 \to 2$ splittings in QCD. The distribution in z_g has first been measured using CMS open data[2] on an inclusive jet sample [589, 590]. The theoretical prediction at modified leading-logarithmic accuracy describes the data well, even though the data have not been corrected for detector effects and the all-particle prediction is compared to z_g calculated from reconstructed tracks only. Other quantities, like the opening angle $\theta_g = \Delta R_{12}/R$ or angularities $e_\beta^{(g)}$ are not described by the analytical calculations [590], which can be attributed to larger detector effects. In a later measurement, ATLAS has reported unfolded distributions of ρ, z_g and θ_g (called r_g in this publication) [591]. The analytical calculations, including non-global logarithms to achieve full NLL accuracy [592], describe the unfolded distributions in θ_g very well. The θ_g distribution for soft drop grooming with $\beta = 0$ is very similar for quark and gluon jets, as expected, since θ_g is independent of α_s to leading order. For $\beta > 0$, different shapes are observed, where the gluon jet distribution tends towards a larger splitting. The splitting function z_g has also been measured by CMS in pp and Pb-Pb collisions [593]. The results from pp and peripheral Pb-Pb collisions agree within 15%. The z_g distribution in central Pb-Pb collisions is steeper, indicating that the parton splitting process is modified by the hot medium created in heavy ion collisions.

The full radiation pattern of a jet can be conceived as emissions in a plane in $\ln 1/z$ and $\ln 1/\theta$, first described by Lund diagrams [468], where the ungroomed versions of z_g and θ_g are used. The Lund jet plane [469], also called primary Lund plane, is obtained by reversing the clustering history of a jet and following the harder branch at each clustering step. Each emission corresponds to one entry in the Lund jet plane, such that any jet can be represented by a number of entries in the plane of $\ln 1/z$ and $\ln 1/\theta$. Different regions in this plane correspond to different physical effects as shown in Fig. 4.6. While only primary emissions are followed in the construction of the Lund jet plane and secondary emissions are discarded, its measurement visualises the salient features of jet fragmentation and radiation patterns inside jets. The soft drop splitting variable z_g is identical to the value of z at the first splitting that fulfils the soft drop condition (2.41), such that the variable pair (z_g, θ_g) is also included in the Lund jet plane. A first measurement has been presented very recently by ATLAS using a dijet selection on the full available 13 TeV data [470], and is shown in Fig. 4.6 (right). As expected, hard-collinear (bottom right) and soft-wide angle (top left) radiation have the highest probability. The average density of emissions is approximately uniform for hard-wide angle radiation (bottom left). The measurement can entangle regions with high sensitivity to the choice of parton shower and hadronisation modelling, and will provide important input in future developments of these processes.

Instead of probing the substructure of a mixture of quark and gluon jets, it is possible to probe primarily quark jets in identified $t\bar{t}$ events. ATLAS has exploited this idea in a measurement of differential and integrated jet shapes in dileptonic and lepton+jets final states of $t\bar{t}$ production at 7 TeV [594]. The two final states enable a comparison of jet shapes between light quark and b jets. At low p_T, it is observed that

---

[2] In 2014, CMS has publicly released the 7 TeV pp collision data recorded in 2010 through the CERN open data portal [588]. By now, also data from the years 2011 and 2012 have been released.

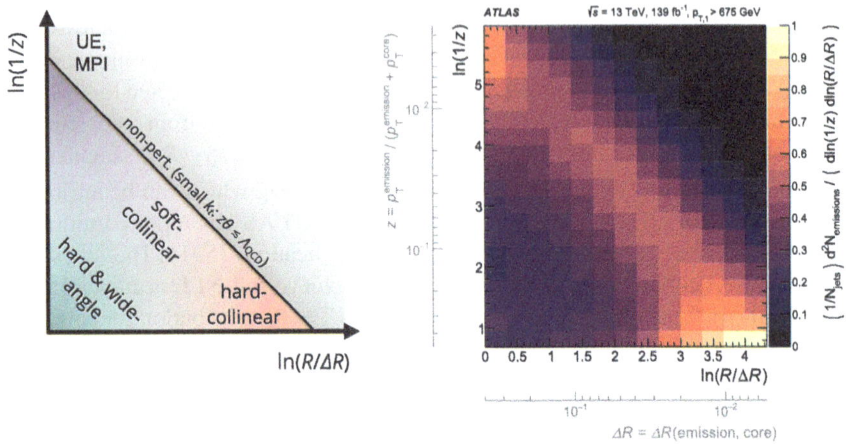

Fig. 4.6 Schematic representation of the Lund jet plane (left). Unfolded measurement at the particle level of the Lund jet plane using jets with $p_{\mathrm{T}} > 675\,\mathrm{GeV}$ (right). Taken from Ref. [470]

b jets have a wider and more diffuse energy distribution than light quark jets. With increasing p_{T} the jet shapes become more similar and at $p_{\mathrm{T}} > 100\,\mathrm{GeV}$ the light quark and b jet shapes are nearly indistinguishable. It should be noted that jet shapes measured on jets from t and W decays are generally narrower than those obtained in QCD jet production. The differences are due to differences in the colour flows in the different final states. Nevertheless, the conclusion that b jets are broader also holds for jets in QCD production, as shown in a recent CMS measurement [595].

The presence of light quark, gluon and b jets in the lepton+jets final state of $t\bar{t}$ production allows for a coherent measurement of jet substructure observables on different flavour jets. Such a measurement has been performed by CMS using 13 TeV data [596], where b tagging and a W mass constraint have been used to obtain very pure samples of b jets (44%), light quark jets (46%) and gluon jets (10%). A number of unfolded substructure distributions has been measured at the particle level, including n_{ch}, p_{T}^{D}, generalised angularities [223], eccentricity [597], z_g, θ_g, soft drop multiplicity n_{SD} [475], τ_{ij}, $C_N^{(\beta)}$, $M_2^{(\beta)}$, $N_2^{(\beta)}$ and $N_3^{(\beta)}$. Overall, state-of-the-art event generators model the data well, but some discrepancies are observed. For a number of distributions, the best description is obtained by the DIRE and SHERPA generators, which have the highest formal accuracy in simulating the parton shower. The largest differences between data and simulation are observed in the related quantities n_{ch} and n_{SD}, where on average smaller multiplicities are observed than predicted. While this trend is observed for all jet flavours, it is strongest for b jets, highlighting the need for improvements in the simulation of heavy flavour jets. The conclusions from the ATLAS jet shape analysis also hold for this measurement, b jets are broader than light quark jets and resemble gluon jets for variables sensitive to particle multiplicities and local energy densities like n_{ch}, n_{SD}, p_{T}^{D} and z_g. For variables sensitive to angular correlations like θ_g, jet width and Les Houches angularity, differences between the distributions in all three jet flavours are observed. These observations can be generalised using a single observable like the energy correlation ratios $C_1^{(\beta)}$. By changing the angular exponent β, the importance of energy and angular correlations can be adjusted. For $\beta = 0$, the distributions in $C_1^{(0)}$ for gluon and b jets are similar, while light quark jets behave differently. For $0.2 \leq \beta \leq 1$, the distributions in $C_1^{(\beta)}$ are different for all three jet flavours and for $\beta \geq 2$, b and light quark jets are similar, but gluon jets are different. Similar observations are made for the $M_2^{(\beta)}$, $N_2^{(\beta)}$ and $N_3^{(\beta)}$ distributions. An interesting observation is the high sensitivity in the θ_g distribution to the value of α_s used in the final state shower. This has been exploited to determine $\alpha_s(M_Z^2)$ for the first time from the substructure of b jets. The value obtained is $\alpha_s(M_Z^2) = 0.115^{+0.016}_{-0.013}$, where the dominating uncertainty of $^{+0.014}_{-0.012}$ originates from scale variations in the final state shower. While the data have the precision for a very accurate determination of $\alpha_s(M_Z^2)$ with an experimental uncertainty of ± 0.001, higher order corrections are needed to improve this determination.

The $t\bar{t}$ final state uniquely allows studies of the colour flow in events where the colour configuration is known. In hadronic decays of the W boson, the $q\bar{q}$ pair is in a colour singlet state. A measurement of the pull angle θ_p between the two jets initiated by the $q\bar{q}$ system is sensitive to the colour configuration. A first measurement by ATLAS using 8 TeV data has shown that the colour-connected light quark jets result in a falling distribution with a peak at $\theta_p \approx 0$ [598]. This measurement also resulted in an exclusion of an alternative colour-octet configuration of the $q\bar{q}$ pair by more than three standard deviations. Despite being IRC unsafe, the pull angle can be calculated analytically from first principles using an all-order resummation [599]. The resummed calculation and the prediction from PYTHIA are in agreement, but are slightly more peaked than the observed data. The presence of two b jets from the $t\bar{t}$ decay can be used to study the pull vector for two different colour configurations in the same final state. An ATLAS measurement using 13 TeV data has reported unfolded distributions of the pull angle between the light quark jets from the W decay and the two b jets from the $t\bar{t}$ decay [600]. In contrast to the colour-connected light quark jets, the θ_p distribution for the two b jets is approximately flat. This behaviour is qualitatively reproduced by the simulations tested, but quantitative differences are observed. Understanding these differences is difficult because of the IRC unsafety of the pull angle, resulting in large uncertainties in the theoretical calculations. It has been found that the projections of the pull vector along and perpendicular to the line joining the two centres of the jets of interest are IRC safe, allowing for all-order calculations and more conclusive theoretical insights [228].

4.1.5 *Jet Substructure in W and Top Jets*

Substructure observables of W and t jets play a key role in the development and commissioning of jet substructure taggers. Their experimental reconstruction and resolution, performance and simulation is usually studied at the level of reconstructed objects using the full detector simulation [243, 435, 436, 459, 483, 522, 525, 525, 526, 530, 531]. Substructure measurements unfolded at the particle level are rare for boosted W and t jets. So far, only a single measurement has been performed by ATLAS using 13 TeV data [601]. In this measurement a number of jet substructure distributions has been measured and compared to simulations for light quark and gluon jets, W jets and t jets. Trimmed large-R jets are used for a selection of events, before the original jet constituents are groomed with the soft drop algorithm. There are three distinct selections, all using anti-k_{T} $R = 1.0$ jets. Inclusive jets from a dijet sample are selected with $p_{\mathrm{T}} > 450$ GeV and $|\eta| < 2.5$. Events from $t\bar{t}$ production in the lepton+jets final state are used to select W and t jets with $|\eta| < 1.5$ and $p_{\mathrm{T}} > 200$ and 350 GeV, respectively. The soft drop jet mass is required to be $60 < m_{\mathrm{jet}} < 100$ GeV for W jets and $m_{\mathrm{jet}} > 140$ GeV for t jets. In addition, a condition of angular separation to an identified small-R b jet is required for W jets. In the selection of t jets, the b jet has to overlap with the t jet. After this selection, the kinematic phase space is sufficiently different for the inclusive, W and t jets such that a direct comparison of substructure distributions between these is not meaningful. However, comparisons with simulation are very useful to study the modelling of substructure observables. The observables studied are subjet multiplicity, Les Houches angularity, ECF(2, 1), ECF(3, 1), $e_2^{(1)}$, $e_3^{(1)}$, $C_2^{(1,1)}$, $D_2^{(1,1)}$, τ_{21} and τ_{32}, where the N-subjettiness ratios have been calculated with the winner-takes-all (WTA) axis [488]. The largest uncertainties in the unfolded distributions originate from modelling differences between different simulations used for the correction of detector effects. These uncertainties are typically around 15% for W and t jets, highlighting the need for a better understanding of differences in the simulations. The largest experimental uncertainties can be attributed to uncertainties in the cluster energy, derived with the R_{trk} method (see Sect. 3.2). These amount to about 8% for W and t jets and are smaller for inclusive jets. Overall, the simulation describes the data within the uncertainties, but some trends are visible. In Fig. 4.7 the measured distributions of τ_{21} for W jets and τ_{32} for t jets are shown. These are important observables for W and t tagging, used in several tagger studies and analyses. A potential mismodelling of these distributions results in efficiency differences between data and simulation for signal jets. It should be noted that soft drop groomed N-subjettiness ratios have been measured, whereas ungroomed versions are usually used for tagging. However, the data provide valuable input for improving the simulation of W and t jets, which have not been considered so far in the tuning of event generators. The combination of POWHEG+PYTHIA describes the distribution in τ_{21} adequately, but discrepancies are observed for τ_{32}. A similar disagreement between data and simulation is observed in CMS [531]. The τ_{32} distribution has been shown to be sensitive to the tuning of MPI and the underlying event [602]. In fact, N-subjettiness ratios can help to constrain uncertainties in the simulation of the final state radiation. Initial studies suggest that the value of α_s used in final state shower is preferred to be smaller for t jets than the value used in the CMS default simulation with POWHEG+PYTHIA [603], similar to the observation for light quark jets [596]. However, a complete analysis of

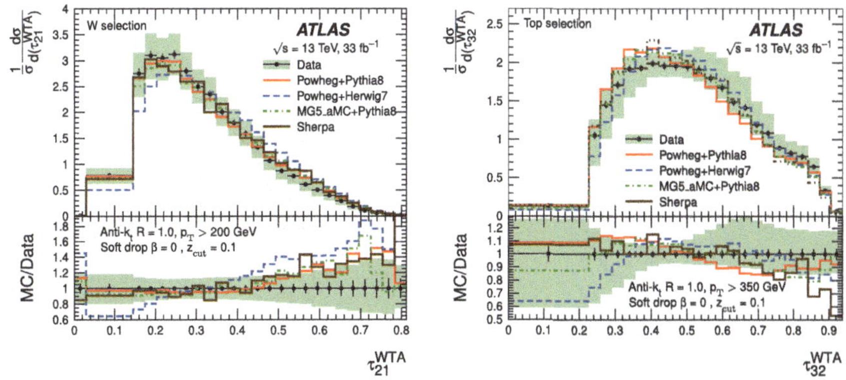

Fig. 4.7 Measured distributions in τ_{21} for W jets (left) and τ_{32} for t jets (right) unfolded at the particle level. Taken from Ref. [601]

jet substructure measurements, together with data sensitive to the underlying event, MPI and hadronisation, is needed in the context of tuning the free parameters of final state radiation in event generators for a firm conclusion.

4.2 Measurements Using Jet Substructure

Measurements of jet substructure observables help to study fundamental properties of QCD and can be used to extract SM parameters. An accurate understanding of the detector response and reliable modelling of signal and background processes facilitate measurements of production cross sections of heavy SM particles. Jet substructure techniques make final states accessible for which traditional methods fail. Reasons can be an overwhelming amount of background, as in the case of inclusive production of $H \rightarrow b\bar{b}$, or highly collimated particle decays resulting in final states with less jets than expected for resolved decays. Substructure taggers enable SM measurements in hadronic final states with high p_{T} and small angular separation. These measurements probe the production of heavy SM particles at the highest scales, testing the validity of the SM in regions unexplored so far. Based on the final state, SM measurements with jet substructure can be divided into EW bosons, H boson or top quark measurements.

4.2.1 Electroweak Boson Production

The first measurement using substructure tagging of W and Z bosons has been performed by ATLAS using 7 TeV data [604]. The analysis targets W/Z+jets pro-

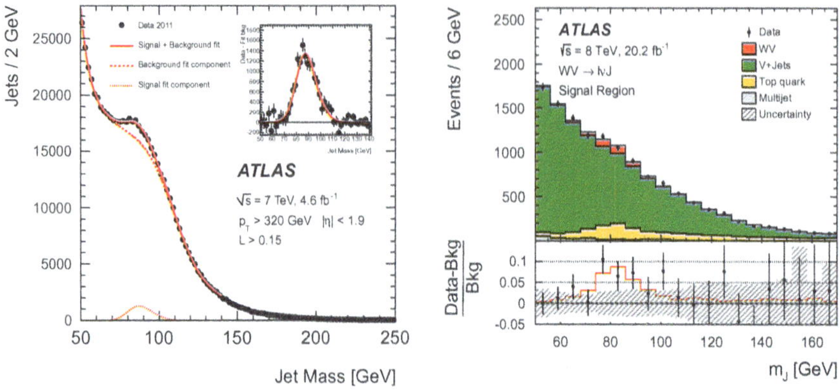

Fig. 4.8 Jet mass distribution in the measurement of the W/Z +jets production for anti-k_T $R = 0.6$ jets with $p_T > 320\,\text{GeV}$. The inset shows the background subtracted distribution, taken from Ref. [604] (left). Jet mass distribution for trimmed anti-k_T $R = 1.0$ jets in the WZ measurement, taken from Ref. [605] (right)

duction at high boson p_T in the all-jets final state. A dijet selection is made using anti-k_T $R = 0.6$ jets with $p_T > 320\,\text{GeV}$. The ungroomed jet mass is required to be $m_{\text{jet}} > 50\,\text{GeV}$. The signal contribution is enhanced by a likelihood function L, built from thrust minor, sphericity and aplanarity. The jet mass distribution is shown in Fig. 4.8 (left) after a selection of $L > 0.15$ and subtraction of the expected background from $t\bar{t}$ production. A binned likelihood fit to this distribution results in a measured signal cross section of $\sigma_{W+Z} = 8.5 \pm 1.7\,\text{pb}$ in agreement with the SM expectation. While this measurement constitutes a successful application of jet substructure tagging, there is a lesson to be learned from it. The shoulder in the background distribution at the signal mass results from two effects. The ungroomed jet mass results in a maximum of the Sudakov peak around the signal mass (see Sect. 2.4.1), and the selection on the likelihood discriminator results in a sculpting of the mass distribution. The first effect can be mitigated by jet grooming, resulting in a less pronounced, but still visible shoulder due to the requirement of $L > 0.15$ [604]. The second effect needs a dedicated decorrelation of the likelihood discriminator. The importance of this lies in the difficulty to model the background without a prediction of its shape. Changes in the parametric form of the background function have not been considered in this analysis, but can lead to sizeable effects. Instead, when using a discriminator not affecting the jet mass distribution, the background shape can be predicted from first principles.

Recently, W and Z tagging techniques have been used in measurements of diboson production at high p_T [605–607]. These measurements probe the EW production of two gauge bosons and the triple gauge boson coupling. These can also be used to probe for anomalous effects from BSM contributions to this coupling [608, 609]. The analyses are carried out in ℓ+jets final states, where the lepton is back-to-back with a tagged jet. The advantage of an isolated lepton is that QCD multijet production can be largely suppressed, leaving W+jet and $t\bar{t}$ production as the main backgrounds. This allows for the possibility to constrain these backgrounds sufficiently for a measurement of the small WW and WZ cross sections. The larger branching fraction in relation to fully leptonic final states allows for a higher reach in p_{T}. A WW+WZ cross section measurement by ATLAS at 8 TeV uses trimmed anti-k_{T} $R = 1.0$ jets to reconstruct the $\ell\nu J$ final state, where J stands for the single, large-R jet [605]. The jet mass distribution is used to fit the signal and background shapes to the data, as shown in Fig. 4.8 (right). Thanks to the trimming algorithm, the light quark and gluon jet background from W+jet production is observed to fall smoothly in m_{jet}. No selection on the substructure of the large-R jets is applied, such that the m_{jet} distribution is not sculpted. The W+jet normalisation is constrained to an uncertainty of 4% by the fit, but uncertainties in its modelling still result in the largest uncertainty in the measured cross section. The second largest uncertainty originates from the modelling of top quark production, which shows a resonant structure in m_{jet}. It is controlled by a dedicated sideband region. The cross section for WW+WZ production with $p_{\mathrm{T}} > 200$ GeV is measured in the fiducial region to be $30 \pm 11\,(\mathrm{stat}) \pm 22\,(\mathrm{syst})$ fb, in agreement with predictions at NLO precision. A similar measurement by CMS at 8 TeV [606] uses pruned CA $R = 0.8$ jets. In contrast to ATLAS, a selection on τ_{21} is applied, which suppresses the W+jet background by more than a factor of two. This comes at the cost of sculpting the distribution in m_{jet}, with large uncertainties in modelling the W+jet background. Overall, the sensitivity is very similar to the one by ATLAS, with an expected significance of about three standard deviations. A recent CMS measurement of WW+WZ production at 13 TeV in the $\ell\nu J$ final state uses soft-drop groomed anti-k_{T} $R = 0.8$ jets [607]. The jet substructure observables are corrected with the PUPPI algorithm for pileup mitigation. A similar selection on τ_{21} as in the 8 TeV analysis is performed. No attempt is made to model the W+jet background. Instead, it is derived from data in a control region using an extrapolation function (the α ratio method) [490, 492], with the benefit of reduced systematic uncertainties. While the WW+WZ cross section is not explicitly measured in this analysis, the most stringent limits to date on anomalous triple gauge couplings are derived.

4.2.2 Higgs Boson Production

Before the start of the LHC, the associated H boson production channels VH with the subsequent decay $H \to b\bar{b}$ have been considered to be poor search channels due to large backgrounds. The first study on high-p_{T} H production, introducing the mass drop tagger and filtering, has given hope to recover this channel [40]. However, the experimental challenges have turned out to be far more difficult than anticipated, and the expected significance of five standard deviations with data corresponding to 30 fb^{-1} has not been reached. Some of the experimental difficulties with respect to the assumptions made in Ref. [40] are higher background levels, uncertainties in the modelling of $t\bar{t}$, W+HF and Z+HF, worse jet mass resolutions resulting in overlapping distributions from VH and VV production, the small branching fractions of leptonic decays, which result in low event counts once a selection is made to suppress the main backgrounds, and systematic uncertainties related to b tagging and the jet reconstruction. Here, HF refers to heavy flavour and subsumes $b\bar{b}$, bc, b+light and $c\bar{c}$ production. While VH production with to $H \to b\bar{b}$ has been observed in resolved final states with a significance higher than five standard deviations by ATLAS [35] and CMS [36] using data from 2011–2017, corresponding to about 100 fb^{-1} per experiment, a first measurement using boosted H tagging has achieved a significance of 2.1 standard deviations using 137 fb^{-1} of data [610]. The measurement profits from a number of recent developments, most notably in subjet b tagging, where VR track-jets [187] are used to adjust the jet distance parameter dynamically to the decay kinematics [514, 515]. It is instructive to compare the result of the measurement to the phenomenological study from 2008 [40], as shown in Fig. 4.9. The jet mass distributions have been obtained in final states with one lepton and missing transverse momentum, consistent with the decay of a W boson in WH production. There are differences in the selected phase space, where e.g. the phenomenological study uses CA jets with $R = 1.2$ and $p_T > 200$ GeV and ATLAS uses anti-k_T jets with $R = 1.0$ and $250 < p_T < 400$ GeV, but the distributions can be compared qualitatively. In the study from 2008, the $t\bar{t}$ and VV backgrounds have been underestimated, while the signal efficiency is too optimistic. However, the predicted shape of the W+jets background is very similar to the one observed in the ATLAS measurement, which is remarkable. On the other hand, the shape of the $t\bar{t}$ background is very different, where its contribution below the WZ and H jet mass regions has been underestimated. Similar observations can be made when comparing the $Z \to \ell\ell$ and $Z \to \nu\nu$ channels. While it took a long time to obtain sufficient sensitivity in boosted VH production, this channel will play an important role in future studies, especially for differential cross section measurements.[3]

An important observation is that measurements of high-p_T H production through the gluon-gluon fusion (ggF) process provide sensitivity to the top quark Yukawa coupling. In ggF production, the H p_T is induced by the radiation of a hard gluon or quark, which resolves the loop induced contributions and allows the study of anomalous production mechanisms [611]. For a long time, the $H \to b\bar{b}$ channel has been considered impossible for ggF production due to the high background from QCD multijet production. However, with recent developments in jet substructure tagging, a first analysis in this channel has resulted in an observed significance of 1.5 standard deviations (with 0.7 expected) for $p_T > 450$ GeV [612], using 35.9 fb^{-1} of 13 TeV data. Similar to the VH analysis, the distribution in the jet mass, with soft drop grooming in this case, is used to search for a signal. The main challenge is the suppression of the QCD multijet background, while retaining the ability to predict its shape reliably. Anti-k_T $R = 0.8$ jets are selected based on the double-b tagger discriminant, which is by construction uncorrelated to m_{jet}. However, any further suppression of one-prong jets with jet substructure methods would result in a

[3] An area where H tagging has been indispensable for some time already are searches for heavy resonances decaying to VH or HH, which are described in Sect. 5.

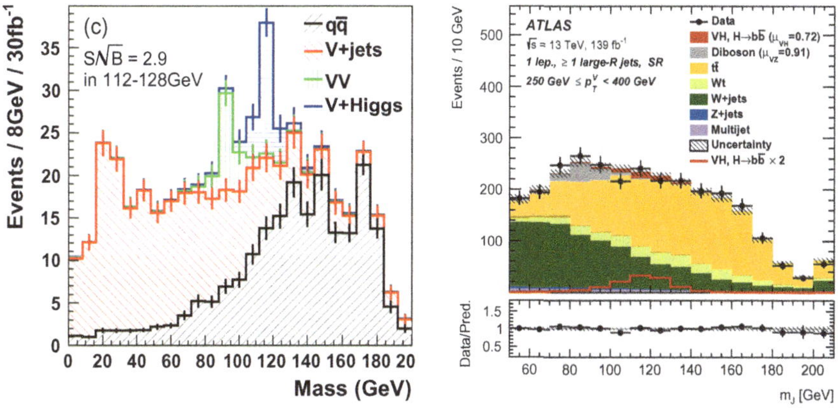

Fig. 4.9 Signal and background distributions in the jet mass for a SM H boson with a mass of 115 GeV, as obtained in a phenomenological study in 2008; taken from Ref. [40] (left). The same distributions as obtained by an ATLAS measurement in 2020, taken from Ref. [610] (right)

sculpting of the jet mass, which is difficult to model. To overcome this challenge, the DDT method [251] is used, where the $N_2^{(1)}$ ratio of generalised ECFs is transformed, such that the selection on $N_2^{(1),\mathrm{DDT}}$ yields a constant QCD background efficiency across the entire ρ and p_{T} range considered. A simultaneous fit to the distributions in m_{jet} for events passing and failing the double-b tagger requirement is then possible, allowing the extraction of the $H \to b\bar{b}$ and $Z \to b\bar{b}$ production cross sections and the determination of the normalisation and shape of the QCD multijet background. A complication arising in this method is that the distributions in m_{jet} are not exactly identical in the regions failing and passing the substructure selections. To take the residual differences into account, the pass-fail ratio is parametrised with a polynomial in ρ and p_{T}. Its free parameters are determined in the fit to the m_{jet} distributions. Besides the mentioned sensitivity to H+jet production in ggF,[4] Z+jets production has been observed with a significance of 5.1 standard deviations (5.8 expected). This is the first time the Z+jets process has been observed in a single-jet topology.

An update of this result using 137 fb^{-1} of 13 TeV data has been published recently by CMS [613]. Besides the larger dataset, the most important improvement with respect to the above analysis is the use of the deep double-b tagger. The algorithm is an improved version of the double-b tagger, based on a deep neural network. It improves the $H \to b\bar{b}$ tagging efficiency by a factor of about 1.5 for the same QCD misidentification probability [614]. This improvement comes at the cost of an anti-correlation at high tagger discriminator values and low m_{jet}, meaning that the jet mass distributions are different in the passing and failing regions. This difference needs to be accounted for, which in this case is done by deriving it in simulation and parametrizing residual differences between data and simulation using Bernstein

[4] Also other Higgs boson production mechanisms contribute to this analysis, but to a lesser degree of about 12% for weak vector boson fusion, 8% for VH and 5% for $t\bar{t}H$.

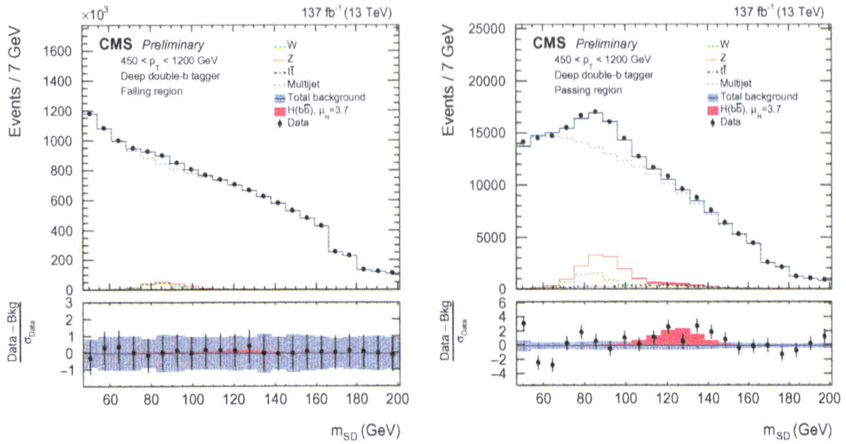

Fig. 4.10 Distributions of the soft drop jet mass for jets with $p_T > 450\,\text{GeV}$ in regions failing (left) and passing (right) the deep double-b tagger. The total background is obtained from a fit to the data with a signal+background model, where the signal strength for $H \to b\bar{b}$ production has been allowed to float freely. Taken from Ref. [613]

polynomials. While this is conceptually similar to the method used in the previous CMS analysis [612], the calculation of the pass-fail ratio from simulation represents an additional complication beset with systematic uncertainties. The soft drop m_{jet} distributions for events failing and passing the deep double-b tagger are shown in Fig. 4.10. A difference is apparent in the shapes of the multijet background at masses below about 80 GeV, originating from the anti-correlation of the deep double-b tagger with the soft drop jet mass. The expected backgrounds from W and Z+jets production are well reproduced in the data. The fitted signal strength for $H \to b\bar{b}$ is $3.68^{+1.58}_{-1.46}$, corresponding to a significance of 2.5 standard deviations (0.7 expected). The expected significance is equal to the one of the analysis using 35.9 fb^{-1} [612], due to an updated prediction of the H p_T spectrum [377]. The new H simulation results in about a factor of two less H boson events in the fiducial region. The expected significance increases to 1.7 standard deviations when using the previous H p_T simulation, highlighting the improvement in sensitivity by approximately a factor of two in relation to the previous analysis.

A similar measurement has also been performed by ATLAS, using 80.5 fb^{-1} of data [615]. ATLAS uses only b tagging information on the two subjets, which is shown to be uncorrelated from the trimmed jet mass for $70 < m_{\text{jet}} < 230\,\text{GeV}$. Below 70 GeV, also ATLAS observes a difference in the m_{jet} distribution for events passing and failing the double-b tagger requirement, such that this region is not considered in the analysis. Fitting the multijet background with an exponential function, where the exact functional form has been determined on events failing the double-b tagging requirement, the observed significance for $H \to b\bar{b}$ production is found to be 1.6 standard deviations.

While the H boson couplings to vector bosons and third generation quarks and leptons have been measured, its couplings to quarks and leptons of the second generation have not been experimentally accessible so far. For $H \to \mu\mu$, a significance of three standard deviations has been reached with 137 fb^{-1} of data [616]. The coupling to charm quarks is even more difficult to establish experimentally, even though the branching fraction $\mathcal{B}(H \to c\bar{c})$ is about a factor of 100 larger than $\mathcal{B}(H \to \mu\mu)$. However, with a value of 0.0288, $\mathcal{B}(H \to c\bar{c})$ is about 20 times smaller than $\mathcal{B}(H \to b\bar{b})$, and has much higher background levels at hadron colliders. A first attempt to study $H \to c\bar{c}$ at the LHC has been made by ATLAS in VH production [617]. The key to this analysis is a modification of the b tagging algorithm to target c quark jets, while rejecting b jets. Charm jet identification is particularly challenging as c hadrons have shorter lifetimes and decay to fewer charged particles than b hadrons. A c-jet efficiency of 41% is achieved for background efficiencies of 25% for b jets and 5% for light quark or gluon jets. In this analysis, the resolved topology has been targeted, without considering boosted H production. Using the distribution of the dijet mass of c-tagged jets, observed (expected) upper limits at 95% confidence level (CL) on the SM signal strength modifier of 110 (150^{+80}_{-40}) have been obtained. A recent analysis by CMS obtains a similar performance for small-R jet c tagging, where a working point corresponding to a c jet efficiency of 28% has been chosen with misidentification rates of 15% for b jets and 4% for light quark or gluon jets [618]. A multivariate discriminant using c tagging information and kinematic variables of the VH system in resolved final states as input is used for the $H \to c\bar{c}$ signal extraction. The observed (expected) upper limit on the signal strength modifier for SM VH production with $H \to c\bar{c}$ is 75 (38^{+16}_{-11}). In the same publication, CMS analyses boosted $H \to c\bar{c}$ production with $p_{\mathrm{T}} > 200$ GeV, where the signal is extracted via a binned maximum likelihood fit to the soft-drop mass distribution of large-R jets. Since the H production cross section for $p_{\mathrm{T}} > 200$ GeV amounts to approximately 5% of the total cross section only, it is paramount to choose R appropriately to keep the signal efficiency as high as possible. To meet this goal, anti-k_{T} jets with $R = 1.5$ are used (see Sect. 2.2.4). A deep neural network is trained for the identification of $H \to c\bar{c}$ decays, exploiting information related to jet substructure, flavour, and pileup simultaneously. The use of an adversarial training procedure results in a discriminator which is largely decorrelated from the jet mass, while preserving most of the method's discriminating power. The achieved $c\bar{c}$ tagging performance can be expressed in efficiencies for three working points of 23, 35, and 46% for $c\bar{c}$, with misidentification rates of 9, 17, and 27% for b jets and 1, 2.5, and 5% for light quark and gluon jets, respectively. The sensitivity is further improved with an event selection based on a multivariate discriminator derived from kinematic variables of the high-p_{T} VH system, uncorrelated to the signal jet mass and $c\bar{c}$ tagging discriminant. The statistical analysis uses distributions in the soft drop jet mass in nine categories (three lepton channels and three $c\bar{c}$ tagger working points). It results in an observed (expected) upper limit of 71 (49^{+24}_{-15}). It is a remarkable achievement that the boosted analysis has a comparable sensitivity as the resolved one, considering the smallness of the cross section in the boosted fiducial region. A significant fraction of events overlap between the resolved and boosted channels, which is fixed by a selection based on p_T of the associated V when combining the two results. The final observed upper limit at 95% CL on the SM signal strength modifier for VH, $H \to c\bar{c}$ is 70, with an expected limit of 37^{+16}_{-11}. The sensitivity exceeds previous expectations and can further be improved by analyses of the full available LHC data.

4.2.3 Top Quark Production

Measurements of top quark production provide important information on the dynamics of the heaviest particle in the SM. Through its mass, the top quark couples strongly to the H boson, resulting in its special role in EW symmetry breaking. Its large mass also implies that top quark production is a difficult multi-scale problem, where m_t can be of the order of top quark p_T, and hence can not be neglected. Precise predictions therefore require the inclusion of higher order terms and resummation corrections [619–623]. Differential measurements of top quark production allow for precise tests of QCD with heavy quarks, provide important input for the determination of PDFs [624, 625] and can be used to extract fundamental theory parameters, such as α_s and m_t [626, 627]. The reach in p_T of these measurements is limited by the onset of collimated decays, where selection efficiencies of standard reconstruction techniques start to decrease. In order to access the high-p_T regime, which provides also highest sensitivity to BSM effects [628–633], jet substructure techniques are indispensable. In addition, for certain BSM models interference effects with SM production are important and require precise differential measurements in order to resolve the dip-peak structure caused by the presence of new scalar particles [634–637].

A first differential measurement of the $t\bar{t}$ production cross section using top tagging has been performed on 8 TeV data by ATLAS [638]. The measurement is carried out in the ℓ+jets channel. The high p_T of the two top quarks implies a back-to-back topology of the collimated top quark decays. This results in standard lepton selection criteria to fail, because the lepton may get reconstructed inside the b jet from the $t \to bW \to b\ell\nu$ decay. Since the collimation, and thus the overlap, is expected to increase with increasing p_T, the efficiency for selecting leptons from the W decay decreases when requiring isolated leptons. In ATLAS and CMS, lepton isolation is defined by the relative p_T of particles in a fixed cone around the lepton four-vector [407, 639–643]. Usually, a selection based on this measure of isolation efficiently rejects leptons from weak decays within jets, at high selection efficiencies for leptons from W and Z decays. In semi-leptonic decays of boosted top quarks this isolation fails because of the b jet constituents falling inside the lepton isolation cone. In order to overcome this, a variable called mini-isolation [644] is used, which introduces a shrinking isolation cone size, proportional to p_T^{-1}. In high-p_T $t\bar{t}$ production, mini-isolation helps to retain high lepton reconstruction efficiency, while sufficiently suppressing backgrounds from QCD multijet production. It should also be noted that once a lepton has been reconstructed within a jet, an overlap removal

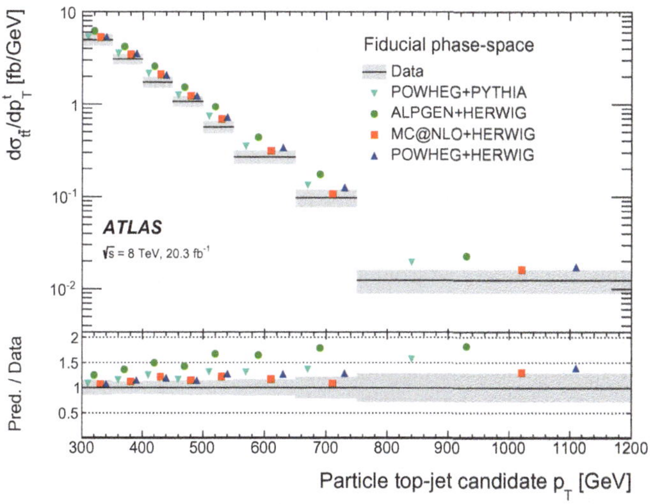

Fig. 4.11 Fiducial differential cross section of $t\bar{t}$ production as a function of $p_{\rm T}$ of the particle level top quark. Taken from Ref. [638]

needs to be applied in order to avoid double counting of the lepton momentum [527]. The fully hadronic top quark decay $t \to bW \to bq\bar{q}$ is reconstructed with a single, trimmed, anti-$k_{\rm T}$ $R = 1.0$ jet with $p_{\rm T} > 300\,\text{GeV}$. The Tagger III is used (see Sect. 3.5.4) to identify t jets, with a rather loose requirement on the trimmed jet mass of $m_{\rm jet} > 100\,\text{GeV}$. The largest background from W+jets production is estimated by exploiting the expected charge asymmetry in the production of W^+ and W^- bosons in pp collisions [645, 646]. The data are unfolded at the particle and parton level. While the particle level has the advantage of being theoretically well-defined in terms of stable, colour-neutral particles, the parton level cross sections can be readily compared to calculations in perturbative QCD, even though its definition is afflicted by theoretical ambiguities and depends on the event generator used for correcting the data. Differential cross sections are reported up to top quark $p_{\rm T}$ of 1200 GeV, as shown in Fig. 4.11. The measurement extended the reach of available measurements at that time by 400–700 GeV [647, 648]. A softer $p_{\rm T}$ spectrum is observed in data than predicted by the NLO+PS simulation POWHEG, where the trend persists when varying model parameters affecting the additional radiation in the event.

A similar measurement using 8 TeV data has been performed by CMS [649]. The lepton selection is based on a two-dimensional (2D) selection, requiring either $\Delta R(\ell, \text{closest small-}R\text{ jet}) > 0.5$ or $p_{\rm T}^{\rm rel} > 25\,\text{GeV}$, where $p_{\rm T}^{\rm rel}$ is the component of the lepton $p_{\rm T}$ perpendicular to the axis of the closest small-R jet. Note that in this measurement small-R jets are anti-$k_{\rm T}$ $R = 0.5$ jets, such that the first of the two requirements resembles an isolation selection. However, leptons in close proximity to small-R jets are kept if their relative $p_{\rm T}$ is large with respect to the nearest jet. This effectively removes leptons from weak hadronic decays, which have small

$p_{\mathrm{T}}^{\mathrm{rel}}$ because of the small hadron masses involved, while leptons from W decays are kept. This 2D lepton selection shows better performance in boosted $t\bar{t}$ decays against QCD multijet production compared to mini-isolation. With the 2D selection, a QCD rejection rate of more than 99% is achieved for a selection efficiency of 91% for muons, compared to a rejection rate of 60% for mini-isolation at the same efficiency [650]. In this measurement, boosted hadronic t decays are identified using the CMSTT for large-R jets with $p_{\mathrm{T}} > 400$ GeV. The dominant sources of background are W+jets, single top quark in the tW channel and multijet production. The contribution from the latter is derived from a control region, obtained by inverting the 2D lepton selection. The normalisations of the other backgrounds are derived through a maximum-likelihood fit, where their shapes are modelled using simulation. In this fit, also the $t\bar{t}$ signal yield and t tagging efficiency are obtained simultaneously. The cross sections, unfolded at the particle level, show a similar trend as in the ATLAS measurement, where the NLO+PS prediction from POWHEG and the multi-leg prediction from MADGRAPH have a harder p_{T} spectrum. A better description of the data up to the highest p_{T} of 1200 GeV is obtained with MC@NLO interfaced to HERWIG for the PS. In addition, the data at the parton level are compared to fixed-order predictions in perturbative QCD at NNLO [651] and approximate next-to-next-to-next-to-leading-order (aNNNLO) [620] accuracy. While the NNLO calculation shows very good agreement with the data, the aNNNLO calculation predicts a significantly too hard p_{T} spectrum.

A particular challenge is the measurement of $t\bar{t}$ production at high top quark p_{T} in the all-hadronic channel, which resembles a dijet topology. The overwhelming amount of background from QCD multijet production is difficult to model and technically unfeasible to simulate with accurate statistical precision. Hence, jet substructure methods enabling a background estimation from data are the only possibility to carry out measurements in this channel at the LHC. In a CMS analysis using 8 TeV data [652], the $t\bar{t}$ signal is identified by requiring two subjet b-tagged large-R jets with $p_{\mathrm{T}} > 400$ GeV, where the leading jet in p_{T} has to have a three prong structure with $\tau_{32} < 0.55$. The distribution in the pruned jet mass of the leading jet is used to perform a maximum likelihood fit to estimate the $t\bar{t}$ cross section in bins of p_{T}. The τ_{32} selection, in conjunction with $p_{\mathrm{T}} > 400$ GeV, leads to a sculpting of the m_{jet} distribution for light quark and gluon jets, resulting in a peak approximately at the same position as the signal peak. In order to reliably estimate the multijet background, sidebands in the jet mass with $100 < m_{\mathrm{jet}} < 140$ GeV and $250 < m_{\mathrm{jet}} < 400$ GeV are used, as well as events failing the τ_{32} requirement. With the help of these sideband regions, the multijet background can be interpolated into the signal region with $140 < m_{\mathrm{jet}} < 250$ GeV, taking correlations between m_{jet} and τ_{32} fully into account. In order to achieve higher accuracy, also events with zero and one subjet b tag are considered in this analysis. The final differential cross section unfolded at the parton level agrees well with the results obtained in the ℓ+jets channel, albeit with larger uncertainties by factors of about two to three.

The increase of $\sqrt{s}$ from 8 to 13 TeV in 2015 has offered a unique possibility for $t\bar{t}$ measurements in the highly boosted regime. While the total cross section for $t\bar{t}$ production increased by a factor of three, the cross cross section for top quark

$p_T > 400$ GeV increased by more than a factor of 10. The accumulated data in 2015, corresponding only to 3.2 fb^{-1}, lead already to an improved statistical precision for these measurements. ATLAS has performed a differential measurement in the ℓ+jets channel [653] using these data. The measurement uses a very loose top tagging selection based on τ_{32} and trimmed jet mass, with varying selection criteria as a function of p_T, resulting in an approximately flat top tagging efficiency of 80% for $p_T > 400$ GeV. The obtained cross sections at the particle level are overall well modelled by the SM predictions. The least-well described distribution is the p_T spectrum of the leading-p_T top quark, which is observed to be softer than the predictions, similar to the observations in 8 TeV data.

The large dataset collected in the years 2015–2018 offers unprecedented precision for studies at high top quark p_T. Recently, first analyses using about 36 fb^{-1} of data have been completed, corresponding to an increase in the number of events at high p_T of more than a factor of 15 in relation to the 8 TeV measurements. Both, ATLAS and CMS have performed measurements in the ℓ+jets and all-hadronic final states. In ATLAS, the all-hadronic analysis [654] uses the same top tagging algorithm as the ℓ+jets analysis based on 3.2 fb^{-1} [653], but with tighter selection criteria, because of the higher background levels. The resulting top tagging efficiency is 50% for $p_T > 500$ GeV, with a misidentification rate for light quark and gluon jets of 6% at $p_T = 500$ GeV, increasing to 10% at $p_T = 1000$ GeV. To further suppress non-top SM backgrounds, b tagging on small-R jets, matched to the large-R jets, is used. In the ℓ+jets channel [655], background suppression is not as crucial, such that a compromise between background rejection and modelling uncertainties is made in order to achieve the best total precision. The analysis uses reclustered large-R jets [431], improving the precision of the jet reconstruction by using calibrated inputs to the clustering. The trimmed mass of the reclustered jet is used for t tagging, where the selection of $120 < m_{\text{jet}} < 220$ GeV results in an efficiency of about 60%. This results in a total non-$t\bar{t}$ background of about 15%, which is estimated mostly from simulation. Only a small contribution from multijet production is estimated from a control region with a looser lepton selection. The unfolded results in the all-hadronic and ℓ+jets channels are consistent; the shapes of distributions are well described by NLO+PS calculations, but the measured total cross section is lower than the prediction by about 30% for $p_T > 500$ GeV. The large dataset also allows for the first time to perform double-differential cross section measurements in boosted final states. State-of-the-art fixed-order NNLO calculations [651, 656] describe the data very well, as shown in Fig. 4.12, where the ratio of data to predictions at NNLO accuracy are shown for measurements in resolved and boosted final states in the ℓ+jets channel [655], also highlighting the gain from the boosted measurement. The NNLO prediction provides a very good description of the data from top quarks produced at rest up to $p_T = 2$ TeV. This provides a very stringent test of the SM description of $t\bar{t}$ production. Measurements of differential cross sections for boosted $t\bar{t}$ production have also been reported by CMS in the ℓ+jets and all-hadronic channels using 36 fb^{-1} of data [657]. In the all-hadronic channel, a selection on the soft drop jet mass is used of $120 < m_{\text{jet}} < 220$ GeV, identical to the trimmed mass selection in the ATLAS analysis. In addition, a NN has been trained on $t\bar{t}$ and QCD jets with τ_1, τ_2 and τ_3 as

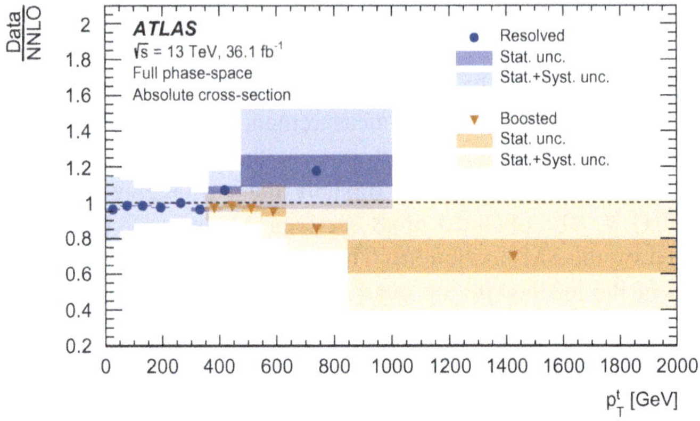

Fig. 4.12 Ratios of the measured $t\bar{t}$ differential cross-sections to NNLO predictions in resolved and boosted topologies as a function of top quark p_{T}, taken from Ref. [655]

inputs. The top tagging in the ℓ+jets channel is based solely on τ_{32}. The advantage of using only N-subjettiness variables for t tagging in this measurement is that these are uncorrelated to the subjet b tagging discriminator, which is used to define signal and control regions. The unfolded cross sections are in agreement with the results from ATLAS and show the same offset in normalisation of about 30% with respect to predictions at NLO accuracy matched to parton showers.

A novelty is the measurement of the forward-backward asymmetry A_{FB} in the process $q\overline{q} \to t\bar{t}$, made made possible by jet substructure methods. The parton-level A_{FB} is defined as the relative difference of the production cross sections of top quarks in the forward and backward hemispheres defined in the centre-of-mass frame, relative to the incident quark direction. At LO, A_{FB} is predicted to be zero, while NLO effects generate positive values for A_{FB} in $q\overline{q}$-initiated subprocesses with a value of 0.095 ± 0.007 [658]. BSM modifications of the top quark-antiquark-gluon vertex or the presence of heavy states coupled to top quarks can lead to modifications in A_{FB} [659, 660]. First measurements of A_{FB} at the Tevatron have received considerable attention since these have reported a value somewhat larger than the NLO prediction [661, 662]. However, more recent measurements at the Tevatron are consistent with the SM [663–665]. At the LHC, A_{FB} is difficult to access since $t\bar{t}$ production is dominated by gluon-gluon fusion processes and the direction of the quark initiating the process is unknown. Measurements of the top quark charge asymmetry [645, 666–670] do not separate $q\overline{q}$ from qg and gg initial states, and therefore do not achieve the same sensitivity to BSM effects as A_{FB}. An important observation is that the scaled longitudinal momentum of the $t\bar{t}$ system in the laboratory frame, x_F, reaches higher values for $q\overline{q}$ initiated events relative to qg and gg. The direction of the $t\bar{t}$ system is strongly correlated with the direction of the initial quark, because the gluon and anti-quark PDFs have a lower average longitudinal momentum fraction than quark PDFs. Events with high x_F thus provide a sample enriched with $q\overline{q}$ initiated events, where the initial parton direction is known to a good approximation.

Since high x_F implies high energies, the fraction of $q\overline{q}$ initiated events is large for highly boosted top quarks. Jet substructure methods are therefore the key for precise measurements of A_{FB} at the LHC. A first measurement of the linearised version of $A_{\mathrm{FB}}^{(1)}$, which approximates the full asymmetry within a few percent, has been carried out by CMS using 36 fb^{-1} of 13 TeV data [671]. Events are classified into t tagged, W tagged and fully resolved $t\overline{t}$ events, using the soft drop jet mass and τ_{32}. The measurement of $A_{\mathrm{FB}}^{(1)}$ is done by a template fit to a triple-differential distribution in the reconstructed value of x_F, the mass of the $t\overline{t}$ system and the top quark scattering angle in the centre-of-mass frame. The highest sensitivity to $q\overline{q}$ initiated $t\overline{t}$ production is observed in t tagged events, as expected. The result of the measurement is $A_{\mathrm{FB}}^{(1)} = 0.048^{+0.095}_{-0.087}\,(\mathrm{stat})^{+0.020}_{-0.029}\,(\mathrm{syst})$, which is less precise than the Tevatron combination [665], but is dominated by the statistical uncertainty and can be improved with future measurements. This measurement also places stringent constraints on top quark anomalous chromoelectric and chromomagnetic dipole moments, which would modify the production of the $t\overline{t}$ system.

Chapter 5
Direct Searches for New Physics

Abstract Heavy resonances, as predicted by theories extending the SM, result in highly energetic particles and very collimated decays. Already since the first collision data have been studied at the LHC, jet substructure methods have been an integral part in analyses searching for unknown effects. With higher centre-of-mass energies and the availability of larger datasets, their relevance has multiplied. Nowadays, jet substructure methods permeate numerous analyses aiming at very different final states. This chapter provides an overview of searches for new physical phenomena where substructure techniques have been indispensable and will play a major role in the future.

Jet substructure taggers have been a critical component in searches for BSM effects from the beginning of data analyses at the LHC, unlike SM measurements, where jet substructure has started to play a role much later. The reason lies in the fact that the expected cross sections of SM and BSM processes differ fundamentally. The partonic cross sections for the production of SM particles fall as $1/\hat{s}$ once the scale of the process is much larger than the mass of the SM particles involved, where $\hat{s}$ is the partonic centre-of-mass energy. Convoluted with the steeply falling PDFs for increasing fractional parton momentum x, this results in a very small cross section at high p_T, relative to the total production cross section. For example, the cross section for $t\bar{t}$ production with top quark $p_T > 350$ GeV is about 13 pb [655], compared to the total cross section of about 830 pb at $\sqrt{s} = 13$ TeV [627, 672, 673]. The statistical power of the data is largest for low-p_T SM processes. When the recorded dataset becomes larger, the statistical precision becomes sufficient to probe high-p_T SM production cross sections. Hence, measurements at high p_T are usually performed later in the lifetime of a particle collider. In contrast to this, the production of a narrow, heavy BSM resonance with a mass m_X beyond 1 TeV will result in a negligible cross section at low p_T, while the production cross section will be dominated by events with $p_T \approx m_X/2$, where p_T denotes here the transverse momentum of the resonance decay products. Already at $\sqrt{s} = 7$ and 8 TeV, cross sections larger than 1 pb for the production of resonances with $m_X > 1$ TeV have been predicted. Since the hadronic channels of W, Z, H and t decays are only accessible with substructure techniques

R. Kogler, *Advances in Jet Substructure at the LHC*, Springer Tracts in Modern Physics 284, https://doi.org/10.1007/978-3-030-72858-8_5

in this regime, searches for BSM effects have been the main application of these techniques very early in the experimental programme of the LHC.

A wealth of results has been produced in the last decade, probing a parameter space of BSM models inaccessible without jet substructure. Because of this large number of results, it is neither possible nor instructive to discuss all of them in detail here. The most recent results are described below, with a focus on results obtained with 13 TeV data. Older ones are mentioned where appropriate, but no attempt at completeness is made. Since no discovery has been reported, the results of these searches are typically given as exclusion limits in terms of upper cross section limits or lower limits on the mass of the heavy resonance. No attempt is made to interpret these results in terms of specific BSM models. Conclusions from these interpretations depend on the BSM model chosen and on the choice of free parameters. Any study of a complete model would require a combination of results from different final states, which is beyond the scope of this book. Instead, possible BSM signatures will be discussed in terms of the experimental signatures, with a focus on the experimental techniques needed. After all, the task of the experimental collaborations is to cover all conceivable and improbable final states, with the highest sensitivity achievable.

5.1 Diboson Resonances

The discovery of the Higgs boson marks the first discovery of a resonance decaying to two electroweak gauge bosons, WW and ZZ. Since m_H is of similar magnitude as m_W and m_Z, the two gauge bosons are produced with low p_{T} on average at the LHC. However, BSM models with extended scalar sectors predict the presence of additional Higgs bosons, which can be very heavy, with masses beyond 1 TeV. In general, most extensions of the SM predict the existence of additional scalar or vector bosons with potentially large couplings to gauge bosons. Examples are two-Higgs-doublet models [674–679], models with a additional gauge group structures resulting from a dynamical electroweak symmetry breaking [680–686], Little Higgs models [687–695] or models with large extra dimensions [696–703]. They all have in common that their experimental signatures include diboson final states, with two highly-energetic scalar or vector bosons, produced back-to-back. The existence of a light SM Higgs boson also resulted in new search channels for BSM resonances. Depending on the exact model and the choice of its free parameters, decays to VH or HH can have the highest branching fractions and thus lead to the best sensitivity. In the following, results of searches for heavy resonances decaying to VV, VH and HH are discussed in all-hadronic and ℓ+jets channels, where jet substructure techniques play a central role in the identification of the V and H bosons. A recent review on searches for BSM physics in diboson final states, including leptonic channels, $\gamma\gamma$ and gg can be found in [704].

5.1.1 WW, WZ and ZZ Resonances

In the all-hadronic decay mode, diboson resonances with masses beyond 1 TeV appear as an highly-energetic dijet system. The only possibility to suppress the large background from QCD jet production are jet substructure taggers. Since the jet mass resolution is not sufficient to resolve W and Z jets, searches in the all-hadronic final state can not distinguish between WW, WZ and ZZ resonances. An important aspect of all-hadronic channels are trigger constraints. Because of the high rate of multijet production, high p_{T} thresholds have to be implemented in order to keep the rate at which events are recorded and stored, at a manageable level. In ATLAS, the single-jet trigger required a jet to have $p_{\mathrm{T}} > 360\,\mathrm{GeV}$ in 2015, which has been continually raised to a final threshold of 440 GeV in 2018, due to the increasing instantaneous luminosity. Besides single-jet triggers, CMS also employed triggers based on the trimmed jet mass of large-R jets in order to lower the p_{T} thresholds, while maintaining a similar rate. For example, an additional requirement $m_{\mathrm{jet}} > 30\,\mathrm{GeV}$ allowed to lower the p_{T} threshold from 500 to 360 GeV for the single-jet trigger. Because of these trigger thresholds, analyses using data with $\sqrt{s} = 13\,\mathrm{TeV}$ usually start at dijet masses of $m_{\mathrm{jj}} > 1.1\,\mathrm{TeV}$ or higher, such that sensitivity to resonances with masses above 1.2 TeV is obtained.

In the two most recent all-hadronic VV searches at 13 TeV by ATLAS [494, 705], a W/Z tagger is used based on D_2, track multiplicity n_{trk}, and trimmed jet mass. In the earlier search, based on data corresponding to $36.7\,\mathrm{fb}^{-1}$ [494], the jet mass is reconstructed using the combined mass, which is a weighted average of calorimeter and tracking measurements where the weights are inversely proportional to the square of the jet mass resolution of the corresponding mass terms [427, 525]. In the more recent analysis, based on the full available Run 2 data with $139\,\mathrm{fb}^{-1}$ [705], TCC [410] are used for the reconstruction of jet substructure variables. The resolution in D_2 for TCC jets is largely improved compared to jets clustered from topoclusters, where at $p_{\mathrm{T}} \approx 2\,\mathrm{TeV}$ an improvement by a factor of about two is achieved. For V tagging, the improvement in D_2 resolution far outweighs the slight degradation in mass resolution below 1.5 TeV (see Sect. 3.2). The variable n_{trk} is used for additional discrimination against gluon jets, which have a higher charge multiplicity on average than quark jets (see Sect. 4.1.4). The V tagging selection based on m_{jet}, D_2 and n_{trk} is optimised either for a constant signal efficiency of 50% [494] or for the best overall sensitivity [705]. The latter approach results in an increasing efficiency from about 20% at $p_{\mathrm{T}} = 500\,\mathrm{GeV}$ to about 60% at $p_{\mathrm{T}} = 4\,\mathrm{TeV}$. This is achieved by p_{T}-dependent selection criteria on m_{jet}, D_2 and n_{trk}. The reason for the increase in efficiency with increasing p_{T} is the steeply falling background, which results in less background events at high p_{T}, such that a higher signal efficiency with smaller background suppression is beneficial. The search is performed as a typical bump-hunt [706], where a narrow resonance peak is expected to appear on a steeply falling V-tagged dijet mass spectrum. Since the background originates dominantly from dijet production, a background parametrisation identical to the ones used in dijet searches can be used [707]. Both ATLAS searches have found the parametrisation

$$f(x) = p_1(1-x)^{p_2 - \xi p_3} x^{-p_3} \tag{5.1}$$

to adequately model the background, where $x = m_{jj}/\sqrt{s}$, p_i are free parameters and ξ is a constant, introduced to minimise the correlation between p_2 and p_3.

The all-hadronic CMS analysis on 36.7 fb^{-1} of data [437] follows the same overall strategy. The V tagging algorithm used is less complicated, based on a selection using PUPPI soft-drop mass and τ_{21}. In order to retain high signal efficiency at high m_{jj} and sufficient background rejection at low m_{jj}, two V tag categories are defined. These are based on the τ_{21} selection, where low purity (LP, $0.35 < \tau_{21} < 0.75$) and high purity (HP, $\tau_{21} < 0.35$) tags have about the same signal efficiency. The HP selection has a higher background rejection than the LP tag. Taken together, the signal efficiency is close to 99%. The optimal use of the LP and HP tags is obtained by classifying events into LP+HP and HP+HP, depending on the V tag of the two leading large-R jets in the event. Two additional categories are introduced, where only one jet is V tagged with LP or HP, to achieve sensitivity to resonances decaying to qW and qZ. In total, 10 exclusive categories are formed and the same background parametrisation as used by ATLAS is found to describe the data. While the full functional form is needed for the qV categories, only two parameters are sufficient to describe the VV categories, obtained by setting $p_2 = \xi = 0$ in (5.1).

The results from the ATLAS and CMS analyses are shown in Fig. 5.1. The background distribution falls less steep in the ATLAS analysis than in CMS due to the p_T-dependent V tagging selection. Overall, ATLAS has a higher background suppression, resulting in less events, but the sensitivities are very similar when comparing analyses based on the same amount of data, i.e. [437, 494]. The data are very well described by the background parametrisations, without any hints of contributions from resonant signals.

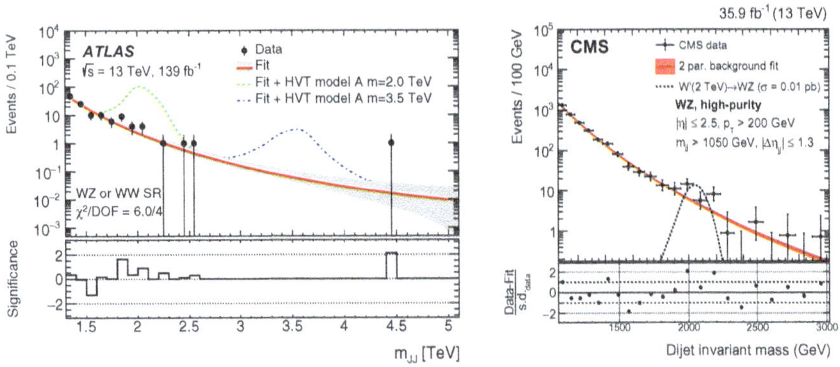

Fig. 5.1 Dijet mass distributions for two V-tagged jets in ATLAS (left) and CMS (right), with background functions obtained from fits to data. The expected signal distributions for a heavy resonance decaying to WW or WZ would result in a peak on top of the falling background. Taken from [705] (left) and [437] (right)

An improvement of the background estimation has been developed by CMS in an all-hadronic search based on 77.3 fb^{-1} of data [254]. Instead of searching for a peak in the m_{jj} distribution only, advantage is taken of the fact that the signal peaks in three observables simultaneously: in m_{jj} and in m_{jet} of the two leading jets. The full spectrum in m_{jet} is fitted for the two leading jets, such that no selection in m_{jet} is needed. The V tag is based solely on the decorrelated τ_{21}^{DDT}. This improves the signal efficiency by about 20% and helps to better constrain the background shape because of the higher statistical precision, especially at high m_{jj}. In this approach, templates for non-resonant and resonant backgrounds in m_{jet} have to be built, where non-resonant contributions entail smoothly falling distributions in m_{jet}, while resonant contributions from W/Z+jets production and $t\bar{t}$ production will result in a peak in one or both distributions of m_{jet}. Since all backgrounds result in falling distributions in m_{jj}, this approach allows to constrain the main backgrounds by a fit of these templates to data, while obtaining high sensitivity to resonant signals. This method yields an improvement in sensitivity of up to 30% relative to the one-dimensional fit in m_{jj}. Even though this search is based on a dataset only half as large as the one used by the most recent ATLAS search in this channel [705], the limits are better by up to a factor of three at high masses, highlighting the power of this method.

Searches for diboson resonances in ℓ+jets final states can be performed in three different channels, $WV \to \ell\nu J$, $ZV \to \ell\ell J$ and $ZV \to \nu\nu J$, where J stands for the V-tagged large-R jet. Each of these comes with its particular advantages and challenges. The $\ell\nu J$ final state has the second-highest branching fraction for WW decays with 17.0%, when including $\tau^- \to e^-\bar{\nu}_e\nu_\tau$ and $\tau^- \to \mu^-\bar{\nu}_\mu\nu_\tau$ decays and their charge conjugates. Only the all-hadronic channel has a higher branching fraction of 45.4%, but much higher background levels. The same is true for WZ resonances, where the single-lepton final state has a branching fraction of 17.7% compared to 48.9% for the all-hadronic final state. The presence of an isolated lepton allows for triggering these events with lepton triggers with thresholds as low as 25–30 GeV. This results in lowest reconstructed diboson masses of about 500 GeV, extending the lower reach of the all-hadronic analyses. For ZV resonances, the $\nu\nu J$ final state has the highest branching fractions after the all-hadronic final state, with $\mathcal{B}(ZZ \to \nu\nu J) =$ 14.0% and $\mathcal{B}(WZ \to \nu\nu J) = 13.5\%$. Recording these events is usually achieved with triggers requiring significant missing transverse momentum, $p_{\mathrm{T}}^{\mathrm{miss}}$, which have much higher thresholds than lepton triggers.[1] Dilepton signatures from $ZV \to \ell\ell J$ decays have the smallest branching fractions with 4.7% for ZW and 4.9% for ZZ. While dilepton triggers could be used for these events, usually single-lepton triggers are employed. For these, trigger paths without lepton isolation requirements are available, with the advantage of a stable efficiency if the two leptons from the highly boosted Z decay are within each other's isolation cone. The higher lepton p_{T} thresholds of these triggers compared to dilepton triggers with isolation requirements is inconsequential for heavy resonance searches.

[1] The $p_{\mathrm{T}}^{\mathrm{miss}}$ denotes the magnitude of the two-component vector $\mathbf{p}_{\mathrm{T}}^{\mathrm{miss}}$, which is calculated as the negative vectorial sum of the transverse momenta of electrons, muons, small-R jets, and unassociated tracks in ATLAS [708], or of all PF candidates in CMS [709].

In searches in the $\ell\nu J$ final state, the presence of an isolated lepton results in the QCD multijet background being largely suppressed, such that the most important backgrounds are $t\bar{t}$ and W+jets production. This leads to less stringent requirements on V tagging selections relative to all-hadronic analyses. It also offers a number of possibilities to estimate the SM backgrounds. In an early analysis of 13 TeV data in the $\ell\nu J$ final state, CMS uses the pruned jet mass to define a signal-depleted sideband region with $40 < m_{\text{jet}} < 65$ GeV [492]. The main background from W+jets production is obtained from this sideband region, by scaling the reconstructed diboson mass distribution m_{VV} in data with a transfer function $\alpha(m_{VV})$. The function $\alpha(m_{VV})$ is constructed from the ratio of probabilities to observe an event at a given value of m_{VV} in the signal region to the one in the control region. This ratio is obtained from the simulation of W+jets production and is largely insensitive to modelling uncertainties. However, when scaling the data in the sideband region, the expected background from $t\bar{t}$ production has to be subtracted, which introduces additional uncertainties. Also, the shapes of the probability density functions in the sideband and signal regions are affected by modelling uncertainties. In an analysis of 36.1 fb^{-1} of data, ATLAS considers weak vector-boson fusion (VBF) in addition to the usual gluon-gluon fusion (ggF) and $q\bar{q}$ production modes [495]. The VBF process $pp \to VVjj$ is characterised by the presence of two jets with large separation in rapidity, originating from the initial state quarks from which a vector boson is radiated. This leads to a total of twelve signal regions, categorised by VBF or ggF/$q\bar{q}$ production, HP or LP V tags, WW or WZ signals, and merged or resolved final states. The background estimation uses the simulated shapes of the twelve m_{VV} distributions, which are dominated by W+jets and $t\bar{t}$ production. Given an accurate modelling of these backgrounds, this has the advantage of constraining modelling uncertainties in a simultaneous fit of all backgrounds and a possible signal contribution to the data. In order to improve the precision of this procedure, W+jets and $t\bar{t}$ control regions are defined by large-R jets failing the V tagging requirements and the presence of additional b-tagged small-R jets. Similar to the all-hadronic final state, it is possible to make use of the full information contained in the m_{jet} and m_{VV} distributions to improve the precision of the background estimation. A CMS analysis in the $\ell\nu J$ final state, based on 35.9 fb^{-1} of data, has used conditional probabilities of m_{VV} as a function of m_{jet} to model the SM backgrounds [710]. The signal is modelled by a conditional probability of the signal mass as a function of m_{VV} and m_{jet}. The two-dimensional fit to data in the m_{VV}-m_{jet} plane results in an extended reach in signal mass and better sensitivity compared to the $\alpha(m_{VV})$ method.

A diboson search in the $\nu\nu J$ final state has been carried out by CMS, using 35.9 fb^{-1} of data [711]. The requirement of $p_{\text{T}}^{\text{miss}} > 200$ GeV results in a stable trigger efficiency of about 96% and a negligible amount of background events from QCD multijet production. Similar to the ℓ+jets analysis, HP and LP V-tagging categories are used to achieve an optimal signal-to-background ratio for low and high resonance masses. The main difference to analyses in leptonic final states is the unknown momentum in z direction of the $Z \to \nu\nu$ decay. Hence, the transverse mass m_{T}^{ZV} is used instead of the diboson mass. The main background in this search are events from Z+jet production with $Z \to \nu\nu$, and W+jet production with $W \to \ell\nu$, where the

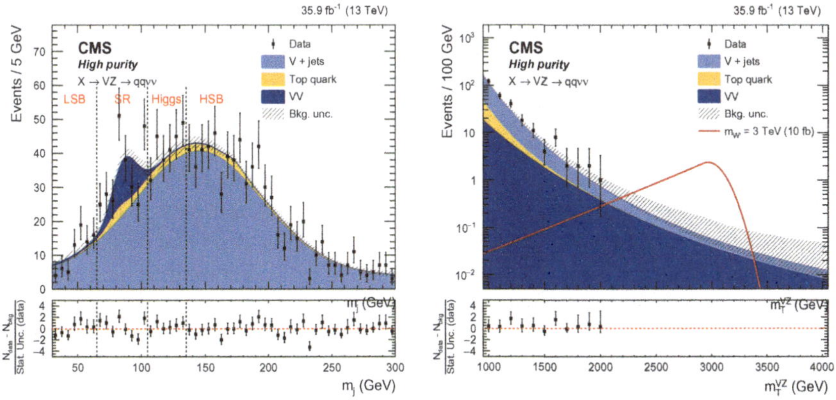

Fig. 5.2 Jet mass distribution in events with a single large-R jet, balanced by $p_{\mathrm{T}}^{\mathrm{miss}}$ (left). The distribution is obtained for HP V-tagged jets, defined by $\tau_{21} < 0.35$. The ZV signal region (SR), ZH signal region (Higgs) and two sideband regions are depicted by horizontal lines. Distribution of m_{T}^{ZV} in the SR for HP V-tagged jets (right). The V+jets background shapes are obtained with the α method from the sideband regions in m_{jet}. Taken from [711]

lepton is outside the detector acceptance or has been misreconstructed. These backgrounds are estimated with the α method, using sideband regions in m_{jet}, as shown in Fig. 5.2 (left). The distribution in m_{jet} for HP V-tagged jets shows a pronounced peak at about 150 GeV for light quark and gluon jets, originating from the HP requirement of $\tau_{21} < 0.35$. It also shows a peak from non-resonant SM VV production, indicating the potential of this final state for a measurement of the VV production cross section. The distribution in m_{T}^{ZV} in the signal region is shown in Fig. 5.2 (right). The smoothly falling background from V+jets production is obtained from events in the m_{jet} sideband regions multiplied by $\alpha(m_{\mathrm{T}}^{ZV})$, to account for small kinematical differences between the signal and sideband regions. A hypothetical signal with a mass of 3 TeV and a production cross section of 10 fb would show as a pronounced peak in the m_{T}^{ZV} distribution. A similar analysis $\nu\nu J$ final state is performed by ATLAS, also considering VBF production, and with a background estimation based on simulation [712]. The sensitivities and results obtained are very similar between the two analyses.

Diboson searches in the $\ell\ell J$ final state have the smallest branching fraction of the searches considered here, but restricting the dilepton system to the Z boson mass results in a very clean selection, with backgrounds mostly from Z+jets and VV production. Together with the low trigger thresholds, this allows to probe small resonance masses starting from 500 GeV, where the final state can be reconstructed using two small-R jets. At higher resonance masses than about 1 TeV, the final state becomes boosted and better sensitivity is obtained by reconstructing the V decay with a large-R jet. Care has also to be taken in the reconstruction of the dilepton system at large boosts, where the leptons are within each other's isolation cone. In analyses using 36 fb^{-1}, ATLAS uses mini-isolation for this purpose [712], whereas

CMS rejects the second lepton from the calculation of the isolation variable [713]. The analysis strategies in these searches are very similar to the ones in the $\ell\nu J$ and $\nu\nu J$ final states described above. Searches in this channel result in the best sensitivity at low ZV resonance masses, but beyond 2–3 TeV the sensitivity decreases and other channels with higher branching fractions are better.

A recent ATLAS result, using the full 13 TeV dataset with 139 fb^{-1}, includes all three dilepton+jets final states $\ell\nu J$, $\nu\nu J$ and $\ell\ell J$ [714]. A recurrent neural network is used to classify events into ggF/$q\overline{q}$ and VBF topologies. A similar methodology is employed as used in previous analyses based on 36 fb^{-1}. A combined interpretation of the results gives the best sensitivity for VV searches to date, with no signal above the expected backgrounds. Resonance mass exclusions are obtained at 95% CL of 4.3 and 3.9 TeV for a W' and Z' in the heavy-vector triplet (HVT) model B, respectively. Compared to previous searches at 7 and 8 TeV by ATLAS [715–717] and CMS [490, 518, 718, 719], the new results based on 13 TeV data improve the mass exclusion limits by more than 2 TeV.

5.1.2 *WH* and *ZH* Resonances

Diboson resonances decaying to VH offer a rich phenomenology because of the different H decay channels. In searches targeting masses above 1 TeV, the $H \to b\overline{b}$ channel offers the best sensitivity in most cases because of its large branching fraction and thus high signal efficiency in a region where the background falls steeply. In jet substructure analyses of VH resonances, subjet-b and double-b tagging improves the background rejection at a given signal efficiency compared to the V taggers employed in VV searches.

The largest branching fraction is obtained for the all-hadronic decay channels, with $\mathcal{B}(WH \to q\overline{q}b\overline{b}) = 39.2\%$ and $\mathcal{B}(ZH \to q\overline{q}b\overline{b}) = 40.6\%$. While the $H \to WW^{(*)} \to q\overline{q}q\overline{q}$ decay is technically also part of the all-hadronic channel, it has not been considered in analyses so far even though it has a relatively large H branching fraction of 9.8%. The reason is the use of b tagging for the identification of boosted H bosons, which has a small efficiency for the decay into four quarks. In analyses targeting the $H \to b\overline{b}$ decay, contributions from the four-quark decay are largely suppressed. New strategies would need to be developed to include this decay.

ATLAS [493] and CMS [720] have analysed data corresponding to 36 fb^{-1} in search for VH resonances in the all-hadronic final state. The use of double-b tagging is central in both analyses. ATLAS uses p_{T}-dependent trimmed jet mass selections for the identification of W, Z and H jets. The selections overlap, such that the jet with the higher mass is assigned to be the H jet. The H jet has to have either one or two ghost-associated b jets, where the latter results in better sensitivity for resonances with masses below about 2.5 TeV. Above 2.5 TeV, H jets with one b tag provide higher sensitivity because the fragmentation products of both b quarks merge into a single track-jet. The V jets have to satisfy a p_{T}-dependent D_2 selection, designed to achieve a constant signal efficiency in V-jet p_{T} of about 50%. CMS uses a categorisation into loose and tight V and H tags, defined by a selection on τ_{21} and the double-b tagging discriminator. In addition, the soft drop jet masses have to be within $65 < m_{\text{jet}} < 105$ GeV for a V tag and $105 < m_{\text{jet}} < 135$ GeV for an H tag. Jets tagged as V bosons are further categorised in W and Z jets, with $65 < m_{\text{jet}} < 85$ GeV and $85 < m_{\text{jet}} < 105$ GeV, respectively. Even though the W and Z jet mass peaks cannot be fully resolved, this classification allows a partial discrimination between WH and ZH resonances. The signal regions corresponding to these two final states can be readily combined in a statistical analysis of the results, as these correspond to independent data. In ATLAS, the WH and ZH signal regions have an overlap of approximately 60% due to the overlapping jet mass selections, such that these can not be combined easily. In both analyses, the multijet background is estimated by one-dimensional, monotonically decreasing functions, similar to the parametrisations used in VV resonances searches, given in (5.1). In addition, ATLAS constrains the normalisation of the multijet background from control regions with no b-tagged track-jets associated with the H jets. The two analyses achieve very similar sensitivities, despite the very different choices made in the design of the analyses. Upwards fluctuations in the excluded upper limits at 95% CL seen in one experiment correspond to downward fluctuations in the other, such that overall no significant excess or deficiency relative to the expected limits are observed. Recently, ATLAS has updated the VH resonance search in the all-hadronic final state using 139 fb^{-1} of 13 TeV data [721]. The analysis relies on the same strategy as the previous analysis, but the $H \to b\bar{b}$ identification has been improved by the use of VR track-jets, which result in higher efficiency at high p_{T} at constant background rejection. The larger data set and improved H tagging results in an improvement in the expected upper cross section limits of factors between 3 and 5, which translates into an extended mass reach by about 500 GeV.

Searches in dilepton+jets final states include the channels $\ell\nu J$, $\ell\ell J$ and $\nu\nu J$, similar to VV resonance searches, but with the fragmentation products of a $b\bar{b}$ pair merged into the large-R jet. In analogy to VV resonance searches, the dilepton+jets VH channels extend the mass range to lower values thanks to the leptonic V decay. ATLAS has considered resolved and merged H decays into $b\bar{b}$ in an analysis of 36.1 fb^{-1} of 13 TeV data [722], which succeeds an analysis of 8 TeV data [723] and supersedes an early result based on 13 TeV data [724]. For events which satisfy both the resolved and merged selections, priority is given to the resolved category with two identified small-R b jets over the merged category with a large-R H jet. This provides higher sensitivity to resonances with a mass near 1 TeV. This cascading selection results in sensitivity for resonance masses as low as 200 GeV up to 5 TeV. In a similar analysis by CMS [725] on the same amount of data, only merged final states are considered, probing resonance masses above 800 GeV. Unlike in the analysis by ATLAS, the background estimation is based on data from control regions with the α method, instead of using simulated SM backgrounds. The sensitivities of the two analyses are very similar and the versatility of the final states allows to constrain several BSM models. Besides placing limits on the HVT model, also the parameter space of two-Higgs doublet models and an extended two-Higgs doublet model with a

DM candidate [726] are probed. The sensitivity shows a large improvement compared to earlier analyses by CMS using 8 TeV [491] and 13 TeV [727] data.

In an analysis by CMS, also using about $36\,\mathrm{fb}^{-1}$, hadronic τ decays (τ_h) are probed in the context of *VH* resonances [728], a strategy which has been developed on 8 TeV data [729]. This analysis searches for resonances in $b\bar{b}\tau\tau$, where the *V* or *H* can decay either to $b\bar{b}$ or $\tau\tau$. Besides these direct decays, the analysis is also sensitive to $VH \to b\bar{b}VV$ with $VV \to \tau\tau\nu\nu$ decays. The analysis is further probing resonant HH production, as explained in the following section. The central element in this analysis is the reconstruction of a boosted *V* or *H* boson decaying to one or two τ_h candidates. Because of their close angular proximity, the two τ_h decays are reconstructed starting with a large-*R* CA jet with $R = 0.8$ and $p_\mathrm{T} > 100\,\mathrm{GeV}$. The CA jet is declustered, where in each step the mass drop condition of (2.39) is tested with $y_\mathrm{cut} = 2/3$. Once two subjets satisfying the mass drop condition are found, these are used as seeds in the standard τ identification algorithm [730, 731]. The τ_h candidate with highest p_T is selected with an efficiency of 50–60%. If two τ_h candidates are found, the isolation of the second τ_h candidate is relaxed to achieve a 70–80% efficiency, with a misidentification rate of less than 0.1%. The analysis considers $\ell\tau_\mathrm{h}$ and $\tau_\mathrm{h}\tau_\mathrm{h}$ categories, where the former includes leptonic τ decays of the sort $\tau^- \to \ell^-\bar{\nu}_\ell\nu_\tau$. In the $\ell\tau_\mathrm{h}$ category, the τ_h decay products are removed from the lepton isolation cones to ensure a stable reconstruction efficiency for highly collimated decays. After the selection of an *H* boson candidate decaying to $\ell\tau_\mathrm{h}$ or $\tau_\mathrm{h}\tau_\mathrm{h}$, events are classified based on the substructure of the other high-p_T large-*R* jet in the event. This results in twelve categories, depending on whether the other jet is classified as an HP or LP *W*, *Z* or *H* jet. The principal SM background after this selection is *V*+jet production, which is estimated with the α method. The subdominant contributions from $t\bar{t}$ and single top quark production are taken from simulation. Due to the smaller branching fraction $\mathcal{B}(H \to \tau\tau)$, the analysis is not as sensitive as the analysis in the $\ell\nu J$, $\ell\ell J$ and $\nu\nu J$ final states, but it provides complementary information and probes *VH* and HH resonances simultaneously.

5.1.3 HH Resonances

Pair production of *H* bosons is an important process in the SM. It probes the triple-*H* coupling, which is predicted to exist by the BEH mechanism with strength λ. While the size of λ can be inferred by measurements of m_H, a direct measurement represents a stringent test of the consistency of the theory. The parameter λ determines the form of the Higgs field potential, which generates the *H* self-coupling after EWSB. The exact form of this potential can have important cosmological implications, for example on the vacuum stability [732, 733]. The production cross section of $pp \to HH$ is predicted to be as small as $33.5^{+2.4}_{-2.8}$ fb at 13 TeV [138]. To access such a small cross section, all experimentally accessible *H* branching fractions need to be considered. A recent review on the relevant signatures, as well as the theoretical and experimental status of HH production can be found in [734]. The largest contribution to the total cross section comes from events where the H pair is produced at the kinematic threshold, such that boosted final states do not play a major role in the search of SM HH production. However, in several BSM scenarios the existence of massive resonances decaying to HH may enhance the cross section at high H p_T significantly and result in a peak in the di-Higgs mass distribution, m_{HH}. Early searches for heavy resonances decaying to HH have been performed by ATLAS [735–738] and CMS [739–741] at 8 TeV and on a small subset of 13 TeV data.

The highest sensitivity for high resonance masses is achieved by all-hadronic analyses targeting the $HH \to b\bar{b}b\bar{b}$ final state [742], with a branching fraction of 33.8%. In these analyses, QCD multijet production constitutes the largest background with 80–95%, depending on the number of b-tagged jets or subjets, with the remainder being $t\bar{t}$ events. The shape of the multijet background in the reconstructed m_{HH} signal region is obtained from signal-depleted sideband regions with less b-tagged jets. Care has to be taken to correct for differences in jet kinematics, which are expected because requiring b tags generally affects the jet kinematics. ATLAS has carried out an analysis using 36.1 fb^{-1} of 13 TeV data [743], considering resolved final states with four b-tagged small-R jets, and merged final states with two H-tagged large-R jets. The signal and sideband regions are defined by circles in the reconstructed H masses. In the resolved category, the four small-R jets with the highest b-tagging score are paired to construct two Higgs boson candidates. Requirements derived from the kinematics of the H decay are used to select the best pairing. This leads to a sculpting of the reconstructed H mass distributions, as shown in Fig. 5.3 (left). Even though the HH signals feature a more pronounced peak (not shown), the similarity between the signal and background distributions leads to a reduction in sensitivity. In contrast, in the merged category the background shapes are not sculpted, as shown in Fig. 5.3 (right), where trimming shifts the jet mass to lower values for light quark, gluon and b jets. The structure at 170 GeV in leading m_{jet} originates from merged t jets in $t\bar{t}$ production, which are absent in the resolved category. In the merged category, in addition to the requirements on the H masses, the signal regions are defined by the number of b-tagged track jets, ghost-associated to the large-R jets. The background shapes in m_{HH} are obtained from data, selected with one less b-tagged jet than in the corresponding signal region. Kinematic corrections due to the b tagging requirement are derived from distributions with loose m_{jet} requirements. The resulting m_{HH} distributions are fitted with a smoothly falling power-law function to model the multijet and $t\bar{t}$ backgrounds. The background normalisation is given by the b tagging pass-fail ratio, which is calculated in the sideband region (the outer ring in Fig. 5.3). The procedure is validated in the control region (the intermediate ring in Fig. 5.3), and after finding good agreement between the data and the background estimation, applied in the signal region.

In a similar analysis by CMS, only the fully merged final state with two H-tagged large-R jets is considered [744]; the resolved final state is covered by another analysis [745]. An H jet is defined through the soft drop jet mass, $105 < m_{\text{jet}} < 135$ GeV, $\tau_{21} < 0.55$, and a loose or tight selection on the double-b tagging discriminator. The reduced dijet mass is used as final discriminant, where the soft-drop masses of the H tagged jets are replaced by m_H. This removes fluctuations in the jet masses, leading

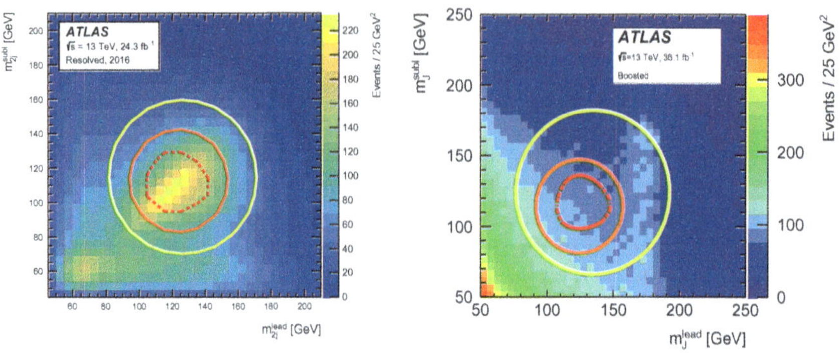

Fig. 5.3 Background distributions in planes of the leading versus sub-leading reconstructed H boson mass in the resolved (left) and merged (right) category. In the resolved category, the H masses are calculated from each of the best pairings of four b-tagged small-R jets. In the merged category, the masses are calculated from the trimmed jet masses of two large-R jets. The inner, dashed circle defines the signal region, the intermediate circle the control region and the outermost circle the sideband region. Taken from [743]

to an improvement of 8–10% in the dijet mass resolution. The background estimation procedure is similar to the one used in the ATLAS analysis, but with an important difference. Instead of correcting the jet kinematics of each jet for differences due to b tagging, the pass-fail ratio is calculated as a function of m_{jet}. This method has the advantage that any background shape can be extrapolated from a sideband to the signal region, simultaneously predicting its normalisation. The method proceeds as follows. The leading jet mass is used to define the signal and sideband regions. In the sideband regions, with $m_{\text{jet}} < 105$ GeV and $m_{\text{jet}} > 135$ GeV, the pass-fail ratio of the double-b tagging algorithm is calculated in bins of m_{jet}. The form of the pass-fail ratio is parametrised with a second-order polynomial, which has been found to model the shape sufficiently well. The resulting prediction of the pass-fail ratio in the m_{jet} signal region is then used to scale events in this region, failing the double-b tagging requirement. These scaled events constitute the background prediction in the signal region, and can be compared to data passing the double-b tagging requirements. After examining the signal region, the prediction of the pass-fail ratio can be compared to the values calculated directly in the signal region, and good agreement is found.

The searches described above target final states with either four b-tagged small-R jets, or with two H-tagged large-R jets, covering fully resolved and fully merged topologies, respectively. The obvious gap are final states where one H boson can be reconstructed by two small-R jets and the other one with a single H-tagged large-R jet. An analysis by CMS using 35.9 fb^{-1} of 13 TeV data targets these semi-resolved events [746]. A single H-tagged jet is required, with the same tagging criteria as in the fully merged analysis. The resolved $H \to b\bar{b}$ decay is reconstructed from the pair of b-tagged small-R jets with the highest sum of b-tagging discriminator values. The mass of this pair is required to be $90 < m_{b\bar{b}} < 140$ GeV. The analysis proceeds analogously to the fully merged analysis, also using the pass-fail ratio as a function of m_{jet} to estimate the multijet background. The results are statistically combined with the fully merged analysis, leading to a significant improvement in the search sensitivity for resonances with masses between 750 and 2000 GeV. In addition, signals from the non-resonant production of HH are also accessible, which is not the case for fully merged final states since such production typically results in an HH invariant mass that is lower than that of a postulated resonance signal.

Another channel considered is $HH \to b\bar{b}\tau\tau$ [728], which has been described above in the context of VH searches. After reconstructing one H candidate through collimated $\ell\tau_{\text{h}}$ or $\tau_{\text{h}}\tau_{\text{h}}$ decays, the other H candidate is obtained by a requirement on the soft drop jet mass, $105 < m_{\text{jet}} < 135$ GeV, and the presence of one or two b-tagged subjets. Due to the large branching fraction $\mathcal{B}(H \to \tau\tau) = 6.3\%$, in combination with a clean signature, resulting in background events mostly from V+jets and top quark production, this channel is comparable in sensitivity at high resonance masses to all-hadronic searches in the $HH \to b\bar{b}b\bar{b}$ channel. ATLAS has developed a new reconstruction method for boosted $\tau_{\text{h}}\tau_{\text{h}}$ pairs and employed it in a search for $HH \to b\bar{b}\tau\tau$ using 139 fb^{-1} of 13 TeV data [747]. The constituents of large-R jets with $p_{\text{T}} > 300$ GeV are clustered into anti-k_{T} subjets with $R = 0.2$. The two p_{T}-leading subjets are used to identify the di-τ system, where information from the calorimeter clusters, tracks and vertices associated with the subjets are used as input to a BDT. This reconstruction achieves signal efficiencies of about 60% for a background rejection of 10^4. The analysis is not as sensitive as the CMS analysis using 35.9 fb^{-1} [728], because leptonic τ decays are not considered. However, the new boosted $\tau_{\text{h}}\tau_{\text{h}}$ reconstruction promises to improve future searches in the $H \to \tau\tau$ channel.

A new channel in HH searches has been very recently explored by CMS, targeting the decay $HH \to b\bar{b}WW^{(*)}$, with the subsequent decay $WW^{(*)} \to q\bar{q}\ell\nu$ [748]. Both H bosons are reconstructed in a single large-R jet, but in the case of the $WW^{(*)}$ decay, a $q\bar{q}$ jet with a nearby lepton is required. The lepton is identified using mini-isolation, which performs better for boosted $q\bar{q}\ell\nu$ decays than the two-dimensional selection based on $p_{\text{T}}^{\text{rel}}$ and ΔR used in top quark analyses (see Sect. 4.2.3). The $q\bar{q}$ jet is required to have a small angular separation to the lepton with $\Delta R < 1.2$, and two subjets with $p_{\text{T}} > 20$ GeV, obtained from the soft drop algorithm. The $q\bar{q}$ jet has to satisfy $0.55 < \tau_{21} < 0.75$ for the LP category and $\tau_{21} < 0.55$ for the HP category. Note that no mass selection is imposed on this jet, as the W initiating this jet may be off-shell. The $b\bar{b}$ H jet is classified based on the subjet b tagging discriminator values. The analysis is conducted via a maximum likelihood fit to the m_{jet}–m_{HH} mass plane, where m_{jet} denotes the soft drop jet mass of the $b\bar{b}$ H jet (called $m_{b\bar{b}}$ in [748]). The background templates are modelled as conditional probabilities of m_{jet} as a function of m_{HH}.

Since a parametric function that models the full m_{HH} range for background events is difficult to obtain, the templates are obtained from simulated events using the kernel density estimation [749]. The background model is tested in two dedicated control regions, one for $t\bar{t}$ and one for W+jets and QCD multijet production. The background templates are fit to the two-dimensional mass planes in the control regions, with the

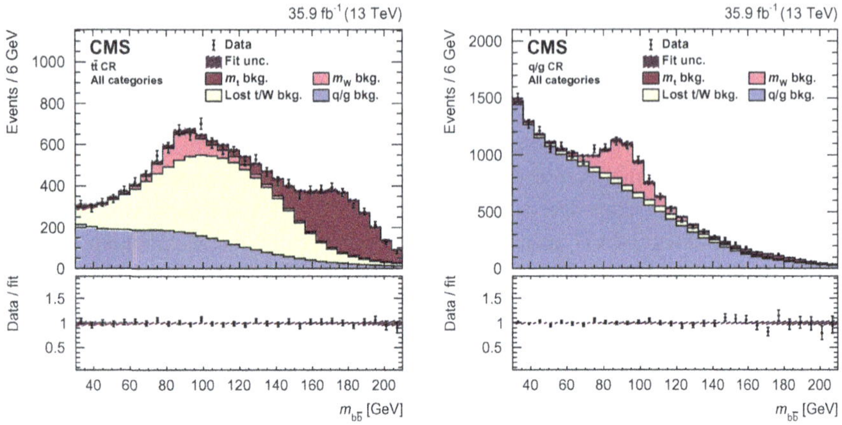

Fig. 5.4 Results of template fits to data in the $t\bar{t}$ (left) and W+jets/multijet (left) control regions, projected in the soft drop jet mass $m_{b\bar{b}}$. The background templates are shown by different colours and are compared to data. Taken from [748]

results shown Fig. 5.4. Excellent agreement between data and the sum of the four templates is found, where the templates are built to describe fully merged t (m_t bkg.) and W jets (m_W bkg.), q/g jets (q/g bkg.), and semi-merged t jets where one of the quarks from the W decay is not reconstructed within the large-R jet (lost t/W bkg.). The sensitivity of this analysis is about a factor of two weaker than obtained in the fully merged $H \rightarrow b\bar{b}b\bar{b}$ analysis, but with a number of possible future improvements comparable sensitivity is in reach.

5.1.4 Combinations

Realistic BSM models, postulating the existence of heavy resonances, predict these to decay into several different channels. For example, in the HVT model [750, 751] the nearly mass degenerate $W'^{\pm}$ and Z' have equal branching fractions for the decays $W' \rightarrow WZ$, $W' \rightarrow WH$, $Z' \rightarrow WW$ and $Z' \rightarrow ZH$. These bosonic signatures may be accompanied by non-negligible branching fractions into leptonic channels like $W' \rightarrow \ell\nu$ and $Z' \rightarrow \ell\ell$, or decays into quark-antiquark pairs, depending on the choice of free parameters. In order to obtain the best sensitivity for these models, and either discover them or place the most stringent experimental constraints on the viable parameter space, a combination of searches in all possible final states is necessary. A combination of VV searches has been performed by ATLAS using 20.3 fb^{-1} of 8 TeV data [752], and a later one uses 3.2 fb^{-1} of 13 TeV data [753]. A combination by CMS extends these analyses by including VV and VH final states based on 19.7 fb^{-1} at 8 TeV and 2.7 fb^{-1} at 13 TeV [754]. The most recent combinations by ATLAS [755] and CMS [756] are based on searches with approximately 36 fb^{-1}

of 13 TeV data. These enlarge the scope of previous combinations by including the leptonic channels $\ell\nu$ [757, 758] and $\ell\ell$ [759, 760], thus adding sensitivity to models with large fermionic couplings.

The difficulty in combining different analyses in very different channels and final states lies in ensuring full orthogonality between the analyses, i.e. statistical independence of all signal regions such that events in data are considered exactly once. The challenge here is that sideband and control regions of one analysis might be part of the signal region of another analysis. In addition, the same definitions of jet substructure observables and lepton identification variables have to be used in all analyses considered, such that no unwanted and unknown correlations are introduced. Precise results also require knowledge of correlations between sources of systematic uncertainties. The feasibility of a combination including a large number of individual results requires careful planning and coordination prior to the execution of the individual searches. Note that when designing the analyses, the optimal sensitivity should be achieved for the combined result and not for individual analyses. In many cases, optimality in a single channel does not result in optimality of the combination.

The result of the latest CMS combination [756] of diboson final states is shown in Fig. 5.5 in terms of upper cross section limits at 95% CL on the production of a heavy W' (left) and Z' (right). At the highest masses up to 4.5 TeV, analyses in the $\ell\nu q\bar{q}$ and all-hadronic final states have the highest sensitivity. At intermediate masses, final states with $p_{\mathrm{T}}^{\mathrm{miss}}$ contribute, where analyses in $\nu\nu q\bar{q}$ and $\nu\nu b\bar{b}$ final states show comparable or better sensitivity to the all-hadronic and single-lepton channels. At resonance masses of 1 TeV and below, dilepton final states result in the best sensitivity. The small background in multi-lepton final states has been exploited by ATLAS, where the inclusion of the fully leptonic searches $WW \to e\nu\mu\nu$ [761], $WZ \to \ell\nu\ell\ell$ [762], and $ZZ \to 4\ell$ [763] improves the combined limits at low masses [755]. As a result of these combinations, W' and Z' resonances below masses of 4.3 and 3.6 TeV are excluded in the HVT model. Constraints are also derived in the plane of lepton versus diboson coupling strengths.

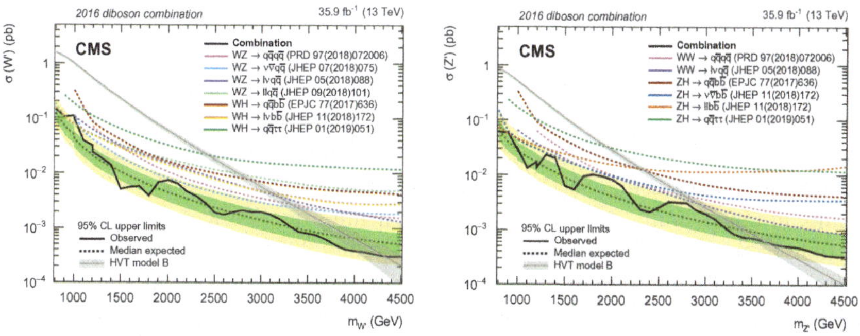

Fig. 5.5 Observed and expected upper limits at 95% CL on the W' (left) and Z' (right) cross sections as a function of the resonance mass. The statistical combination of the VV and VH channels is shown, together with the expected limits of the individual channels. Taken from [756]

Similar to VV and VH resonances, also HH resonance searches are combined in dedicated analyses [764, 765]. These combinations include mostly searches for resonances below masses of 1 TeV, with sensitivity to the SM HH production cross section. The best sensitivity at low masses is achieved by the $b\bar{b}\gamma\gamma$ analyses [766, 767]. At masses above 1 TeV, the $b\bar{b}b\bar{b}$ [743, 744] and $b\bar{b}\tau\tau$ [728] final states dominate the combined results. For spin-0 resonances decaying to HH with masses above 2 TeV, production cross sections of 6 fb or higher can be excluded at 95% CL. It should be noted that the $b\bar{b}WW^{(*)}$ analysis, with $WW^{(*)} \rightarrow q\bar{q}\ell\nu$ [748], has not been considered in these combinations so far. Further improvements are expected from its inclusion.

5.1.5 $V\gamma$ and $H\gamma$ Resonances

A special class of diboson resonances are resonances decaying to photons, where the decay is usually loop-mediated. A prominent example of such a diphoton decay is $H \rightarrow \gamma\gamma$, with a branching fraction of only 0.23%. Nevertheless, this channel was one of the discovery channels. Even smaller is the branching fraction to $Z\gamma$ with 0.15%, which has not been observed to date [768]. In BSM models, high-mass $Z\gamma$ resonances appear as loop-mediated decays as well. Examples are compositeness models, like the little Higgs model [769], extra-dimension models [770, 771] or extensions of the SM with an additional scalar field [772]. Some models of technicolour predict $W\gamma$ resonances in addition to $Z\gamma$, originating from a triplet of technirhos [773]. These BSM theories can be generalised by an effective model introducing colour-neutral spin-0 or spin-2 states, denoted by X, decaying to $Z\gamma$. Spin-1 W' and Z' can also be introduced, with the loop-induced W' and Z' decays to $W\gamma$ and $H\gamma$ [774], respectively. The branching fraction for the $H\gamma$ decay is typically very small with $\mathcal{B}(Z' \rightarrow H\gamma) = O(10^{-5})$ and smaller, but if the photon is not a single photon but a cluster of photons from the decay of one or more highly boosted light particles, much larger branching fractions are obtained [774]. Since these decay cascades would lead to the same detector signature as a single photon, searches for $H\gamma$ resonances might lead to discoverable BSM effects.

Early searches for $W\gamma$ and $Z\gamma$ resonances have focussed on resonance masses below 1 TeV and leptonic V decays. Examples are searches by ATLAS and CMS using 7 and 8 TeV data [775, 776] and small samples of 13 TeV data [777]. A search by ATLAS using 36.1 fb^{-1} of 13 TeV data [778] in the $Z \rightarrow \ell\ell$ channel results in upper limits on the product of branching fraction and production cross section, $\sigma(pp \rightarrow X)\mathcal{B}(X \rightarrow Z\gamma)$, between 88 and 2.8 fb for masses in the range 250–2.4 TeV. The reach in resonance mass can be improved by considering $Z \rightarrow \nu\nu$ decays, resulting in a mono-γ signature, which is a common signature in searches for dark matter (see Sect. 5.5). An example is the interpretation of an ATLAS search for dark matter [779] as a search for a $Z\gamma$ resonance. This search extends the reach in the $Z\gamma$ resonance mass up to 5 TeV. The observed upper limits on $\sigma(pp \rightarrow X)\mathcal{B}(X \rightarrow Z\gamma)$ are between 26 and 43 fb for a mass range of 2–5 TeV.

While the mass reach by this search is higher, the limits at about 2 TeV are about an order of magnitude weaker than the ones from dilepton final states, which can be attributed to larger backgrounds and the inability to reconstruct the resonance mass in the $\gamma + p_{\mathrm{T}}^{\mathrm{miss}}$ final state.

These disadvantages can be alleviated by considering hadronic Z and W decays. These result in larger signal efficiencies than leptonic channels due to the larger branching fractions, and therefore in a larger reach in resonance mass. Jet substructure methods can help to suppress the reducible backgrounds from γ+jet production, and the reconstruction of the resonance mass helps to improve the sensitivity compared to the $\gamma + p_{\mathrm{T}}^{\mathrm{miss}}$ final state. The first search to consider V tagging in $Z\gamma$ resonances has been performed by ATLAS on 3.2 fb^{-1} of 13 TeV data [780]. The search includes the $Z \to \ell\ell$ and $Z \to q\overline{q}$ channels, taking advantage of the better signal-to-background ratio in the $Z \to \ell\ell$ channel for low resonance masses and the higher sensitivity of the $Z \to q\overline{q}$ channel at high masses. The V tagger uses the trimmed jet mass, n_{trk} and a p_{T}-dependent selection on D_2 to identify $Z \to q\overline{q}$ decays, merged into large-R jets with $p_{\mathrm{T}} > 200$ GeV. The reconstructed $Z\gamma$ mass distribution is fit with the power-law function of (5.1) to model the smoothly falling background. The hadronic channel extends the reach of the analysis to 2.75 TeV, where the observed upper limit on $\sigma(pp \to X)\mathcal{B}(X \to Z\gamma)$ is about 10 fb. An analysis by CMS using 35.9 fb^{-1} also considers leptonic and hadronic Z decays [781]. Dilepton final states are selected using mini-isolation for the lepton identification to retain signal efficiency at high $Z\gamma$ masses. Three different categories of large-R jets are used in this search: untagged, τ_{21} tagged and b tagged. Jets in the untagged category have $p_{\mathrm{T}} > 200$ GeV and a pruned jet mass $75 < m_{\mathrm{jet}} < 105$ GeV. Jets in the τ_{21}-tagged category pass a requirement of $\tau_{21} < 0.45$ in addition to the requirements of the untagged category, and jets in the b-tagged category are required to have a b-tagged subjet in addition to the τ_{21}-tagged category. These three categories are made mutually exclusive by first testing for a b-tagged jet, then for a τ_{21}-tagged jet and lastly for an untagged jet. The untagged category improves the sensitivity at intermediate and high masses compared to a previous CMS analysis considering only b- and τ_{21}-tagged jets [782]. The b-tagged category improves the sensitivity by exploiting $Z \to b\overline{b}$ decays with smaller SM backgrounds. The product of acceptance times signal efficiency increases from 7 to 9% in the untagged category and from 3 to 9% in the τ_{21}-tagged category for signal masses between 0.65 and 4 TeV. In all three categories, the background shape in the reconstructed $Z\gamma$ distribution can be described by a power-law function with two free parameters controlling the shape. The leptonic channels have better sensitivity for $Z\gamma$ resonance masses below 2 TeV, while above this value the hadronic channels are more sensitive. A combination results in limits on $\sigma(pp \to X)\mathcal{B}(X \to Z\gamma)$ ranging from 50 to 0.3 fb for masses between 0.35 and 4 TeV, which considerably improves the limits by the ATLAS search on the same amount of data considering only leptonic final states [778].

The hadronic V decay channels are the target of a search by ATLAS using 36.1 fb^{-1} of 13 TeV data [783]. Similar to the CMS search, this search makes use of subjet b-tagging and categorises events according to the requirements on subjet b tagging, D_2 and trimmed jet mass. In addition to these three categories, a fourth category is introduced for large-R jets failing all requirements. This last category improves the sensitivity for masses larger than 4 TeV and extends the mass reach up to 6.8 TeV. The analysis has comparable sensitivity for $Z\gamma$ resonances to the CMS search with the same amount of data [781]. Limits on spin-1 $W\gamma$ resonances are derived as well and are equivalent to the results for $Z\gamma$. The b-tagged category is used to search for an $H\gamma$ resonance for the first time. In this case, the jet mass range is changed to $93 < m_{\text{jet}} < 134$ GeV and no selection on D_2 or n_{trk} is applied to achieve high signal efficiency. The signal efficiency for b-tagged jets in the $H\gamma$ search is 25% for masses of 1 TeV, decreasing to 7% at 3 TeV. The decrease in efficiency can be attributed to a b-tagging inefficiency for highly boosted H jets with ghost-associated $R = 0.2$ track-jets (see Sect. 3.5.3). The obtained upper limits on $\sigma(pp \to Z')\mathcal{B}(Z' \to H\gamma)$ vary between 10 and 4 fb for resonance masses between 1 and 3 TeV. In a dedicated search for $H\gamma$ resonances by CMS [784], the double-b tagging algorithm is used, which is designed for boosted H tagging. Together with a selection on the PUPPI-corrected soft drop jet mass of $110 < m_{\text{jet}} < 140$ GeV, a signal efficiency is achieved of 36% at a mass of 1 TeV and 20% at 3 TeV. The sensitivity of the search for masses above 2 TeV is further improved by the addition of an untagged category. The gain in signal efficiency, together with the improved suppression of SM backgrounds by the double-b tagger for low masses, compared to the ATLAS analysis [783] results in a significant gain in sensitivity by factors between 2 and 10. The analysis reports upper limits of 25 fb for a mass of 720 GeV, 5.3 fb for 1 TeV and 0.4 fb for 3.25 TeV.

The latest result in search for an $H\gamma$ resonance is reported by ATLAS using 139 fb^{-1} [785]. The analysis has been optimised for this channel, making use of the centre-of-mass subjet reconstruction for subjet b tagging for the first time. Two categories are defined for this search, based on single and double b-tagged large-R jets. The combined signal efficiency is about 19% for a mass of 1 TeV, increasing to 30% at 3 TeV. For masses below 2.7 TeV double-b tagged jets have higher efficiency, for masses larger than that single-b tagged jets have higher efficiency. At 4 TeV the efficiency of double-b tagged jets is still 10%. The new subjet b tagging approach results in an excellent signal-to-background ratio even at the highest masses. This constitutes a significant improvement relative to the first ATLAS analysis in this channel. The reconstructed $H\gamma$ distributions are shown in Fig. 5.6 for the single and double b-tagged categories. The background in each category is modelled by a function with three free shape parameters, validated in control regions without subjet b and jet mass requirements. The data are well described by the background-only fit, and the two categories are combined to set upper limits on $\sigma(pp \to Z')\mathcal{B}(Z' \to H\gamma)$. The analysis achieves the best limits to date, of 10 fb for a mass of 720 GeV, 3 fb for 1 TeV and 0.24 fb for 3.25 TeV. It is an excellent example on how improvements in jet substructure tagging can improve the sensitivity of searches at the LHC.

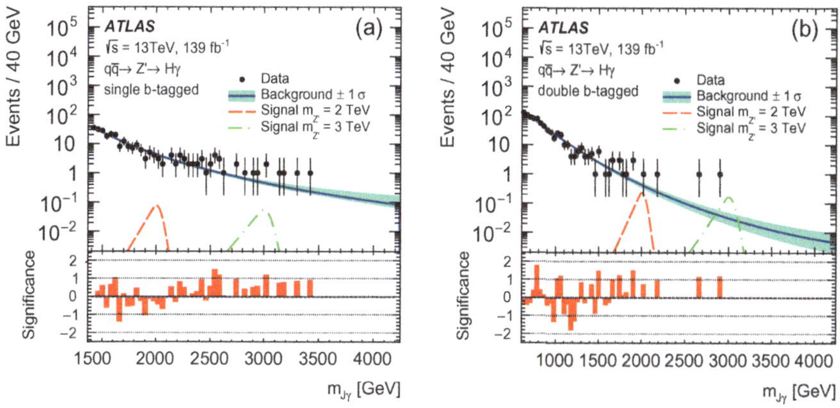

Fig. 5.6 Distributions in the reconstructed $H\gamma$ mass obtained by ATLAS from events with a high-p_T photon and an H-tagged jet. Events are categorised into H jets with a single b tag (left) and a double b tag (right). The background is modelled by a parametric function with three free parameters. Taken from [785]

5.2 Resonances Coupling to Third Generation Quarks

In certain BSM models, the couplings of heavy resonances to third generation quarks is enhanced. This includes Kaluza-Klein excitations of gluons [39, 786] or gravitons [787, 788], massive colour-singlet W' and Z' bosons [789–793], colourons [794–796], axigluons [797, 798], and pseudoscalar Higgs bosons [628, 679, 799]. In some models of warped extra dimensions [699], resonance decays into bosons or third generation quarks can be important, depending on the nature of the resonance and the choice of free parameters [800]. For heavy neutral resonances, decays to $b\bar{b}$ would be observable in dijet searches with b-tagged jets (see for example [801]), but decays to $t\bar{t}$ result in distinct final states. Dedicated analyses are needed for optimal sensitivity, reconstructing the $t\bar{t}$ system from its visible decay products. The decay of a narrow resonance would be observable as a peak above the falling spectrum in the mass of the $t\bar{t}$ system, $m_{t\bar{t}}$. Searches of this kind are probing the highest energy scales in $t\bar{t}$ production, thus also testing our models used to simulate these final states. Charged resonances would preferably decay to $t\bar{b}$ or its charge conjugate, if the couplings to first and second generation quarks are suppressed. While also leptonic decays of the kind $W' \to \ell\nu$ are possible, these are forbidden for purely right-handed couplings of the W' boson if the right-handed neutrino is heavier than a few GeV [802]. Similar to $t\bar{t}$ resonance searches, a structure in the mass of the $t\bar{b}$ system is searched for.

5.2.1 $t\bar{t}$ Resonances

Searches for resonances with masses larger than 1 TeV decaying to $t\bar{t}$ have been the first applications of top tagging algorithms at the LHC. Interest in models predicting resonant $t\bar{t}$ production has been fuelled by measurements of the forward-backward asymmetry A_{FB} in $p\bar{p} \rightarrow t\bar{t}$ at the Tevatron (see Sect. 4.2.3). As a consequence, the first searches at the LHC have already been performed with 7 TeV data recorded in 2011 [803, 804]. Even though no excess has been found in these early analyses, searches for resonant $t\bar{t}$ production have been carried out on each dataset recorded by ATLAS and CMS. These searches can give testimony of the improvements in terms of sensitivity due to developments in the field of jet substructure.

The $t\bar{t}$ system can be classified into all-hadronic, ℓ+jets and di-leptonic final states. In the searches described here, decays of the type $W \rightarrow \tau\nu$ are included through leptonic τ decays, and are part of the ℓ+jets final state. Hadronic τ decays are not considered. The three distinct final states have very different reducible backgrounds, while the irreducible background from $t\bar{t}$ production is common to all. In addition to $t\bar{t}$ production, also single top quark production in the tW channel in association with an additional b quark contributes to the irreducible background. In the all-hadronic channel, with a branching fraction of 45.4%, multijet production constitutes the largest reducible background. This background originates mostly from light quark and gluon jets produced via the strong force, but also includes hadronic decays of W and Z bosons produced in association with jets. Compared to the multijet background, other backgrounds from processes not involving top quarks are negligible. In the ℓ+jets final state with a branching fraction of 34.1%, the largest reducible background originates from W+jets production with the decay $W \rightarrow \ell\nu$. The dilepton channel has the smallest branching fraction of 6.4%. The dominant reducible background in this channel is Z+jets production with $Z \rightarrow \ell\ell$. Due to the small branching fraction, this channel contributes mostly at masses below 1 TeV. Above this value, searches in the all-hadronic and ℓ+jets channels achieve higher sensitivity.

Analyses in the all-hadronic channel rely heavily on top tagging, as this is the only possibility to reduce the large multijet background. At 7 TeV, CMS has used the CMSTT to identify boosted t jets [803]. In addition to topologies with two t jets, also events with one t jet and one W jet have been considered, where the W jet is paired with a small-R jet in angular proximity. The multijet background is estimated using the mistag probability, which is obtained in a sample with two high-p_{T} jets without t-tagging requirements. The fraction of events with one t-tagged jet defines the mistag probability, which depends on p_{T}. A sideband region is defined by requiring a dijet topology with one t-tagged jet. Events in this sideband region are weighted by the mistag probability as a function of p_{T} of the non-tagged jet. Since these jets have on average a smaller jet mass than t-tagged jets in the signal region, their jet mass is set to a random number drawn from the simulated jet mass distribution in the signal region. The obtained distribution in $m_{t\bar{t}}$ constitutes the background estimation for the signal region with two t-tagged jets. A similar procedure is applied for semi-merged events. In the signal region, the multijet background is about an order of magnitude larger than the irreducible background from $t\bar{t}$ production. The same is observed for semi-merged events, which result in a similar sensitivity as fully merged events for resonance masses of about 1 TeV. For masses of 1.5 TeV and higher, the signal efficiency for the semi-merged selection is lower by a factor of four and more. Since later analyses in the all-hadronic channel focus on high resonance masses, semi-merged events are not considered any longer.

The sensitivity of the analysis could be largely improved by a follow-up analysis, using four times more data recorded at 8 TeV [805]. The same techniques as in the 7 TeV result are employed, particularly the CMSTT is used to select two t-tagged jets [805]. Relative to the 7 TeV result, the sensitivity improves because of the increased $\sqrt{s}$ and the larger dataset, while the ratio of reducible multijet to the irreducible $t\bar{t}$ background is the same. Advanced top tagging algorithms allowed for an improvement on these results using the same data [806]. The performance of the CMSTT is enhanced by using subjet b tagging and a selection of $\tau_{32} < 0.7$. In addition, the HEPTopTagger, applied on CA jets with $R = 1.5$, is used to gain sensitivity to low mass resonances in the all-hadronic channel. Events failing the t tagging selection based on the CMSTT, are tested for two HEPTopTagger t-tagged jets. The use of subjet b tagging and τ_{32} makes the background estimation for multijet production more complicated. While the overall strategy is the same, correlations in the variables $m_{t\bar{t}}$, p_{T}, τ_{32} and the subjet b tagging discriminant are observed. The mistag probability in a given bin of $m_{t\bar{t}}$ is therefore parametrised by the three variables p_{T}, τ_{32} and the value of the subjet b tagging discriminator. For events selected with the HEPTopTagger, the background estimation is obtained from a sideband region where the jet mass and pairwise subjet masses are inverted. The use of better top tagging techniques results in a large suppression of the multijet background, as shown in Fig. 5.7. This background constitutes only about 30% of the total background for events with two b-tagged subjets, and increases to about 70% for one b-tagged subjet. The better background suppression improves the 95% CL upper cross section limits on the production of a narrow resonance by nearly 50% for masses between 1 and 2 TeV.

In analyses of 13 TeV data, CMS uses the soft drop mass, τ_{32} and subjet b tagging applied on the soft drop subjets to identify t jets. First results in the all-hadronic channel have been obtained on a dataset corresponding to 2.6 fb^{-1} [807]. The same algorithm is used in the latest CMS analysis in this channel using 35.9 fb^{-1} [808]. The algorithm shows a very similar performance as the CMSTT with τ_{32} and subjet b tagging. The multijet background is obtained from sideband regions, similar to the approach developed for the 7 and 8 TeV analyses. The only difference is that the mistag rate is parametrised as function of p instead of p_{T}. This has been shown to give more stable results because the Lorentz boost is roughly constant in a given interval of p, whereas it depends on the rapidity when considering an interval in p_{T}. Events are classified into six categories based on the number of subjet b tags and the rapidity difference between the jets. This allows to retain signal efficiency at high masses, where the SM background decreases rapidly. The six categories have different background compositions, which allows to constrain the $t\bar{t}$ background in a simultaneous fit to data. This is exploited in a combination with analyses in the

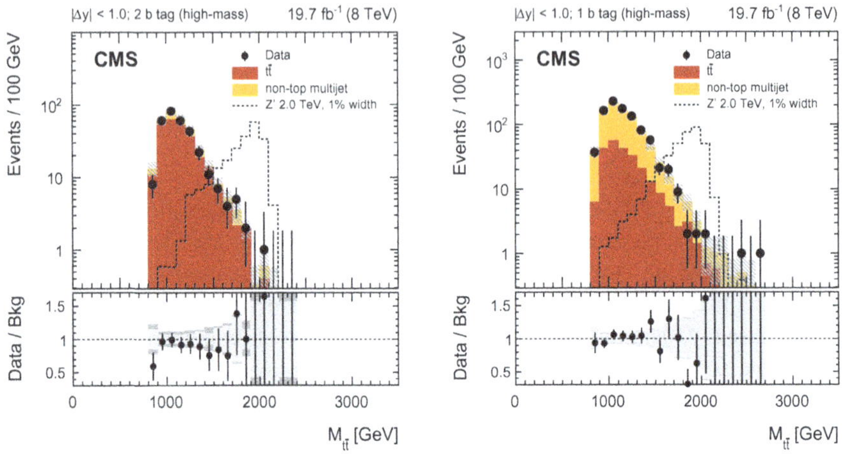

Fig. 5.7 Measured distributions in $m_{t\bar{t}}$ obtained from events in the all-hadronic channel with two t-tagged jets using the CMSTT with $\tau_{32} < 0.7$, and two (left) or one (left) b-tagged subjet. The jets are required to be central, with small rapidity difference $|\Delta y| < 1.0$. Taken from [756]

ℓ+jets and dilepton channels, described below. The combination allows for an in situ determination of top tagging data-to-simulation corrections, which would otherwise result in large uncertainties in the signal efficiency. The largest systematic uncertainty at high $m_{t\bar{t}}$ originates from the modelling of $t\bar{t}$ production, where uncertainties related to renormalisation and factorisation scale variations have a large impact. An analysis by ATLAS based on 36.1 fb^{-1} of 13 TeV data in the all-hadronic channel considers resolved and fully merged final states [809]. The analysis in the resolved final state is carried out using the "buckets of tops" algorithm [810]. In this algorithm, small-R jets are assigned to groups, referred to as buckets. This algorithm ensures a smooth transition from resolved to semi-merged final states, as no requirement is imposed on the number of jets to reconstruct a top quark. In the boosted analysis, the trimmed jet mass, subjet b tagging and τ_{32} are used to identify t-tagged jets. The values of τ_{32} from the leading jets are combined into a single likelihood ratio, which is used to categorise events into loose, medium or tight t-tagged events. The multijet background is obtained from sideband regions, defined by the loose and medium categories with one or no b-tagged subjets. The shapes of the $m_{t\bar{t}}$ distributions are extrapolated to the signal regions with the help of transfer factors which are also derived from data. Similar proportions of reducible and irreducible backgrounds as in the 13 TeV analyses by CMS are obtained. Because the resolved and boosted selections are not mutually exclusive, the results of the two analyses are not combined in the statistical interpretation. Instead, upper limits are calculated using the analysis with higher sensitivity. The transition, where the expected sensitivity is better for the boosted analysis compared to the resolved analysis is at resonance masses of about 1 TeV.

A very recent analysis by ATLAS in the all-hadronic channel uses the full 13 TeV dataset with 139 fb^{-1} [811]. The identification of t jets is performed with a deep neural network [525], which uses high-level discriminants of anti-k_T $R = 1.0$ jets as input, such as jet p_T and mass, τ_i, splitting scales and energy correlation functions. A working point of the algorithm is used, corresponding to an efficiency of of 80% over the full range of top quark p_T considered. The misidentification probability for light quark and gluon jets is approximately 3% at $p_T = 500$ GeV, increasing to 8% at $p_T = 3$ TeV. The analysis is performed in two categories, defined by the number of b-tagged ghost-associated track-jets. The high efficiency of the t tagging algorithm, together with the high statistical power of the data, allows for the first time in a $t\bar{t}$ resonance search to estimate the background from a fit to data with a parametrised function of the form

$$f(x) = p_1(1-x)^{p_2} x^{p_3 + p_4 \log(x)}. \tag{5.2}$$

This power-law function has one more free parameter than the functional form used in VV searches, (5.1). The additional flexibility is needed because of the different shapes of the multijet and $t\bar{t}$ background distributions. The form of (5.2) has been established in control regions, where data and simulated events have been mixed in order to achieve sufficient statistical power. The fit to the data in the signal region with two b-tagged subjets is shown in Fig. 5.8. The background fit describes the data well over the full range, $1.4 < m_{t\bar{t}} < 7$ TeV, without a trace of a resonant signal. The advantage of estimating the background with an analytic function lies in its independence from modelling uncertainties of the $t\bar{t}$ background. This analysis therefore provides an independent test of previous results obtained, using simulation for the prediction of the $t\bar{t}$ background. The disadvantage of the method lies in the steeply

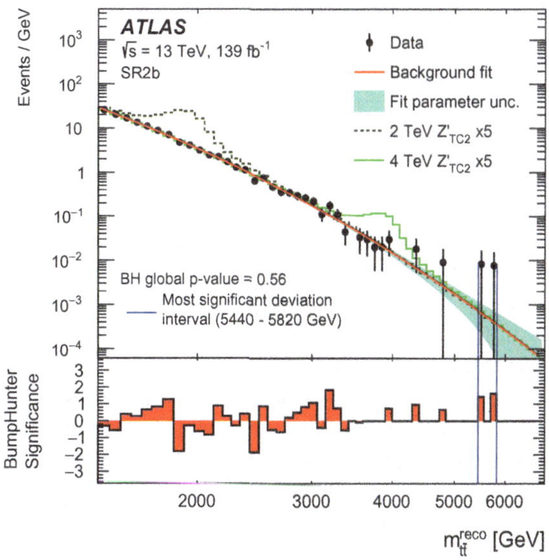

Fig. 5.8 Observed $m_{t\bar{t}}$ distribution in ATLAS data for events with two t-tagged jets, each associated with a b-tagged subjet. The result of a background fit and two possible signal distributions are shown as well. Taken from [811]

falling background, causing the function to be less constrained in the high mass region. This leads to a spurious-signal uncertainty, which describes a bias observed when testing the signal+background model against pseudo data generated under the background-only hypothesis. Intrinsic limitations, associated with the fit range and number of free parameters, create a systematic difference between the estimated and real backgrounds, producing a spurious signal. The corresponding uncertainty is approximately 30–40% up to 4 TeV, but increases rapidly to 200% at 5 TeV, where the lack of data can not constrain the background model sufficiently well. Nevertheless, the advanced top tagging algorithm in conjunction with the parametric background model provide an improvement of 65% in the expected cross-section upper limit at 4 TeV compared to the previous ATLAS analysis [809], when using the same data set.

Searches for resonant $t\bar{t}$ production in the ℓ+jets channel have been performed using 7 and 8 TeV data by ATLAS [527, 812, 813] and CMS [805, 806, 814]. While ATLAS uses t tagging already in the first analysis in this channel [813], CMS has adopted a cascading selection, where small-R jets are assigned to the leptonic or hadronic top quark decay [805, 814]. In a given event, all permutations of jets are tested. The final assignment is chosen by the permutation minimising a χ^2-metric, consisting of squares of differences between the reconstructed and expected top quark masses. No assumption on the number of small-R jets assigned to the reconstructed top quark hypotheses is made, such that a smooth transition between resolved and boosted final states is obtained. In order to probe resonance masses of 500 GeV and below, both ATLAS and CMS have dedicated analyses in resolved final states [527, 805, 812, 814]. In addition, dedicated analyses have been performed in search for scalar resonances by ATLAS [815] and CMS [816], taking into account interference with SM $t\bar{t}$ production. ATLAS [527] and CMS [806] have explored the use of a cascading selection for boosted and resolved final states. The boosted topology is defined by a t-tagged jet opposite of the selected lepton. At high resonances masses, the lepton is in close proximity to the b jet from the t decay, and is identified with mini-isolation in ATLAS and a two-dimensional isolation requirement in CMS (see Sect. 4.2.3). If an event fails the boosted selection, it is reconstructed using a resolved selection based on the χ^2-metric from above. This approach has been shown to improve the sensitivity of the searches by about 30% with respect to the χ^2 assignment of small-R jets without t tagging.

Analyses at 13 TeV in the ℓ+jets channel have been performed by CMS using data corresponding to 2.6 fb^{-1} [807] and 35.9 fb^{-1} [808], and by ATLAS with 36.1 fb^{-1} [817]. In these analyses, the same t tagging algorithms have been employed as in the corresponding all-hadronic channels, to allow for a combination of the results. An important difference is the omission of subjet b tagging. In the ℓ+jets channel, the largest irreducible background is W+jets production, which can be sufficiently suppressed by substructure and kinematic requirements, as well as b tagging on the small-R jet from the leptonic top quark decay chain. This leaves $t\bar{t}$ as the dominant background, which can not be suppressed by a requirement on a b-tagged subjet. Since such a selection would reduce the signal and background efficiency by the same amount, it would lead to a lower overall sensitivity. Because of the non-isolated lepton selection, the background from multijet production is larger than in analyses with an isolated lepton. This background is larger in the electron than in the muon channel, because of larger misidentification probabilities for electrons. In ATLAS, the multijet background is estimated from a sideband region, obtained with a looser set of lepton identification criteria and small $p_{\mathrm{T}}^{\mathrm{miss}}$ [817]. In CMS, the multijet background is reduced to a negligible level by a selection of $p_{\mathrm{T}}^{\mathrm{miss}} > 50$ and 120 GeV for the muon and electron channels, respectively [808]. The shape of the W+jets background is obtained from simulation in ATLAS and CMS. While ATLAS determines its normalisation from the W charge asymmetry, CMS reduces this background with a dedicated BDT, trained on the kinematic variables of jets and the lepton, as well as global event shapes. The output distribution of the BDT is used to separate events into the signal region and a W+jets control region. The measured $m_{t\bar{t}}$ distribution in the μ+jets channel for events with one t-tagged jet are compared between ATLAS and CMS in Fig. 5.9. Note that for ATLAS events with one b-tagged jet are shown, whereas in CMS events have been selected by the W+jets BDT, which includes b tagging information. The observed distributions are very similar between ATLAS and CMS, in normalisation as well as in shape and background composition. Both measurements are described very well by the simulation of W+jets and $t\bar{t}$ production, which is also observed in the electron channel and other signal categories and control regions. When interpreting these results in terms of 95% CL upper limits on the production cross section of a heavy resonance, the achieved sensitivities are comparable between ATLAS and CMS.

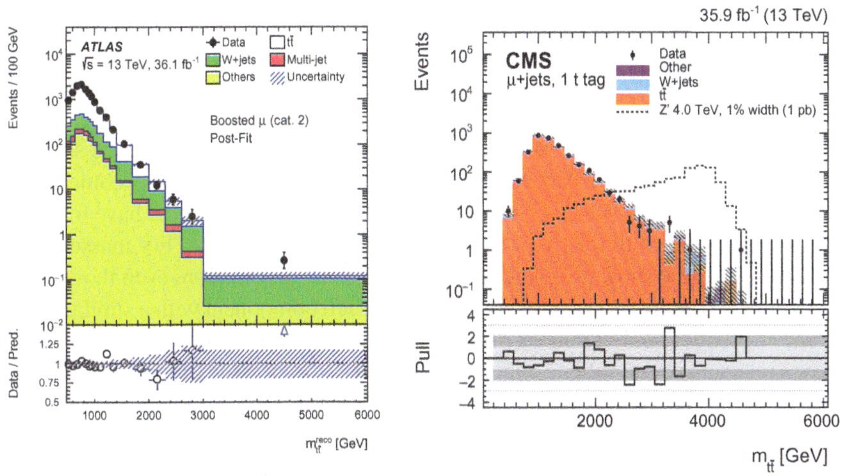

Fig. 5.9 Reconstructed $m_{t\bar{t}}$ distribution in ATLAS (left) and CMS (right) for events in the μ+jets channel with one t-tagged jet. The background distributions are obtained from simulation, except for the multijet background in ATLAS, and are shown after a likelihood-fit to data under the background-only hypothesis. Taken from [817] (left) and [808] (right)

Searches for $t\bar{t}$ resonances in the boosted dilepton channel have been performed by CMS using 8 [806] and 13 TeV [808] data. In these analyses, the presence of two neutrinos complicates the reconstruction of the $t\bar{t}$ system. Studies have shown that S_T, defined as the scalar sum of p_{T} of jets, leptons and $p_{\mathrm{T}}^{\mathrm{miss}}$, provides higher sensitivity in this channel than $m_{t\bar{t}}$. The large mass of the resonance leads to a collimated system from each top quark decay, consisting of a lepton and a b jet. To account for the overlap between the lepton and the b jet, two-dimensional isolation criteria are used, based on $p_{\mathrm{T}}^{\mathrm{rel}}$ and $\Delta R(\ell, \mathrm{j})$, where the latter is the angular distance between the lepton and the nearest small-R jet. The distribution in $\Delta R_{\mathrm{sum}} = \Delta R(\ell_1, \mathrm{j}) + \Delta R(\ell_2, \mathrm{j})$ is found to have high discrimination power between signal and background, where ℓ_1 and ℓ_2 denote the p_{T}-leading and p_{T}-subleading leptons in the event. Lepton-jet pairs from resonance decays populate the low-ΔR_{sum} region. Events are separated into signal and control regions, defined by $\Delta R_{\mathrm{sum}} < 2$ and $\Delta R_{\mathrm{sum}} > 2$, respectively. The S_T distribution in the signal region is used to search for a signal in ee, $e\mu$ and $\mu\mu$ final states. The expected signal distributions are broader compared to distributions in $m_{t\bar{t}}$ in the ℓ+jets channel, resulting in a lower sensitivity for high mass resonances. This is exacerbated by the smaller branching fraction in this channel. However, for resonances with masses of 1 TeV or smaller, this channel achieves similar sensitivity as the ℓ+jets channel.

The best sensitivity is achieved by a combination of analyses in different decay channels of the $t\bar{t}$ system. This allows for the most precise determination of the SM backgrounds, as well as for better constraints on experimental and theoretical uncertainties. Such combinations have been performed by CMS considering the all-hadronic and ℓ+jets final states [805, 807], as well as all three $t\bar{t}$ decay channels considered in searches [806, 808]. In these combinations, the dilepton and ℓ+jets channels contribute in approximately the same amount to limits for masses below 1 TeV. Above 1.5 TeV, the sensitivity from the all-hadronic and ℓ+jets channels is about the same.

The off-shell component in the production of high mass resonances is strongly enhanced by the available parton luminosity at lower $m_{t\bar{t}}$. This effect becomes more apparent for wider resonances. So far, LHC searches for $t\bar{t}$ resonances have focussed on narrow signals with relative widths Γ/M between 1 and 30%. For masses up to about 3 TeV, these show pronounced peaks in the $m_{t\bar{t}}$ distribution, even though signals with $\Gamma/M > 10\%$ have already a large off-shell component with a similar shape as SM $t\bar{t}$ production. For masses of 5 TeV and higher, only signals with $\Gamma/M \approx 1\%$ show a resonant structure. The searches presented here have only limited sensitivity to broad, high mass resonances. For example, the ATLAS analysis presented in Fig. 5.8 has remarkable sensitivity to narrow resonances with $\Gamma/M \approx 1\%$, but it is blind to non-resonant enhancements of the $t\bar{t}$ production cross section at high $m_{t\bar{t}}$. In general, recent searches at 13 TeV are already limited by systematic uncertainties for wide, high-mass resonances. However, recently these resonances or general enhancements in the high energy tails of $t\bar{t}$ production have received considerable theoretical attention. Examples are five-dimensional BSM models with Kaluza-Klein excitations that result in an increase of the continuum production of $t\bar{t}$ [818], or low-mass axion-like mediators in s-channel processes mediating $t\bar{t}$ production [819]. Future searches for these non-resonant BSM models will need dedicated techniques to constrain the SM $t\bar{t}$ background in the tails of the distributions in order to achieve optimal sensitivity.

5.2.2 $t\bar{b}$ Resonances

Heavy charged gauge bosons W', decaying to $t\bar{b}$ (and charge conjugates), can be searched for in ℓ+jets and all-hadronic final states. Single top quark production in the s-channel constitutes the same signature and is part of the non-resonant background of this search. In fact, for a W' boson with left-handed couplings there is interference between SM s-channel $t\bar{b}$ production and $t\bar{b}$ production through an intermediate W'. This interference needs to be taken into account, by simulating resonant and non-resonant $t\bar{b}$ production simultaneously. In the leptonic channel, with a branching fraction of 25.3%, a lepton is produced in close angular proximity to the b jet from the t decay. The resulting lepton-jet-$\mathbf{p}_{\mathrm{T}}^{\mathrm{miss}}$ system is balanced by a high-p_{T} b jet. The $\ell\nu b\bar{b}$ final state can also be produced by $t\bar{t}$ and W+$b\bar{b}$ production, which constitute irreducible backgrounds in this search. In the all-hadronic channel, with a branching fraction of 67.4%, the t and b jet pair resembles a high-mass dijet system. Hence, multijet production is the largest background in this channel, with an irreducible component from $t\bar{t}$ production.

The first searches for $W' \rightarrow t\bar{b}$ have been performed with 7 TeV data in the ℓ+jets channel by ATLAS [820] and CMS [821]. These early searches considered isolated leptons and one or two b-tagged jets. Masses below about 1.85 TeV could be excluded, considering right handed couplings and a branching fraction into $t\bar{b}$ of one. Similar analyses on 8 TeV data by ATLAS [822] and CMS [823] improved this limit to 2.05 TeV. It should be noted that the ATLAS analysis requires two b-tagged jets, which results in a much smaller signal efficiency for high resonance masses, such that the ATLAS upper cross section limits are about a factor of 7 weaker than the ones by CMS. Analyses in the all-hadronic final state have been made possible with the advent of top tagging algorithms, and have been performed by ATLAS [824] and CMS [825] on 8 TeV data. ATLAS uses a top tagging algorithm based on the splitting scale $\sqrt{d_{12}}$, and the N-subjettiness ratios τ_{32} and τ_{21}. No selection on the jet mass is applied. This choice results in a t-tagging efficiency of about 50% with a suboptimal misidentification rate of about 10% for light quark, b and gluon jets. The background is estimated by a fit of a parametric function to data, where an exponential function has been found to describe the shape of the background distribution. The CMS analysis uses the CMSTT with $\tau_{32} < 0.55$ and subjet b tagging. The chosen working point results in a t-tagging efficiency of about 25% with a misidentification rate of 0.3%, which is more than an order of magnitude better than in the ATLAS analysis. The background is estimated from data in sideband regions, obtained by inverting individual steps of the substructure selection. The final m_{tb} distribution is described well by the predicted backgrounds from multijet, $t\bar{t}$ and to a lesser degree, single t production. Due to the better t tagging algorithm, the background is about an order of magnitude smaller compared to the ATLAS analysis. Consequently, the obtained upper cross section limits are better by a factor of two. A combination with the ℓ+jets channel results in an exclusion of a W' boson with right-handed couplings below 2.15 TeV.

For analyses in the ℓ+jets channels at 13 TeV, ATLAS and CMS have changed the lepton selections to non-isolated leptons. Both experiments have performed analyses of 36 fb^{-1} of data, where ATLAS uses mini-isolation for the lepton selection [826] and CMS uses the two-dimensional ΔR, $p_{\mathrm{T}}^{\mathrm{rel}}$ selection [827]. The CMS analysis is optimised for high mass signals, with higher p_{T} thresholds on leptons and jets compared to the ATLAS analysis. In turn, ATLAS achieves sensitivity for W' masses starting at 0.5 TeV, whereas the CMS analysis starts at 1 TeV. Besides this, the two analyses strategies are similar, and both analyses estimate the SM backgrounds from simulation. The CMS analysis achieves higher sensitivity at high masses, because of a categorisation of the signal regions based on the p_{T} of the four-vector sum of the two p_{T}-leading small-R jets. The achieved sensitivities are comparable, where ATLAS achieves better sensitivity for masses below 2 TeV, and CMS above this value. The mass exclusion limits improve by nearly 1.5 TeV compared to the 8 TeV analyses, to 3.6 TeV for a right-handed W'.

A search in the all-hadronic channel using 36.1 fb^{-1} of 13 TeV data has been performed by ATLAS [521]. The $t\bar{b}$ final state is reconstructed using one small-R b-tagged jet with $p_{\mathrm{T}} > 420$ GeV, which is required to have a large angular separation to the t-tagged jet, also with $p_{\mathrm{T}} > 420$ GeV. The shower deconstruction (SD) algorithm [257, 258] is used to identify t jets. The input to the SD algorithm are subjets, which are obtained by reclustering the selected large-R jet with the k_{T} algorithm. The clustering is stopped once the splitting scale is larger than 15 GeV, at which point the remaining protojets are used as subjets. The six p_{T}-leading subjets are used as input for the SD tagger. Large-R jets are only considered further if they have at least three subjets and two or more subjets have a combined invariant mass between 60.3 and 100.3 GeV, and by adding one or more subjets a total jet mass between 132 and 212 GeV is obtained. These jets are passed to the SD algorithm, and its output is used to define two working points. The loose working point has a t jet efficiency of 80% with a misidentification rate of 4–10%; the tight working point has an efficiency of 50% with a misidentification rate of 1.3–3.3% for p_{T} between 0.45 and 1.3 TeV, respectively. The lose and tight working points, together with the requirement of the small-R jet being b-tagged or not, define six regions. Depending on the presence of a b-tagged small-R jet, overlapping with the t-tagged jet, six more regions can be defined. Out of these twelve regions, the two regions with the tight t-tagging requirement and one b-tagged jet opposite to the t jet, define the signal regions. A third signal region is obtained from loose t-tagged jets, but two b-tagged jets in the event. The multijet background is estimated from the other eight regions, where one region is used to validate the procedure. The multijet background makes up more than 90% of the background in two signal regions, and 75% in the third, with the remainder being $t\bar{t}$. The measured m_{tb} distributions are described by the estimated background within the uncertainties over the full accessible range between 1 and 6 TeV. The sensitivity achieved by this analysis is nearly identical to the sensitivity of the ℓ+jets analysis for masses above 1.5 TeV. This can be considered a great achievement, considering the difficulties in the identification of highly boosted t and b jets.

In several BSM models, there exists a relation between W' and Z' resonances. An important example are non-universal left-right models, where new right-handed gauge bosons couple predominantly to the third generation [828, 829]. These models have been introduced to explain the deviation in the $Z \to b\bar{b}$ decay asymmetry observed at LEP [107], and have recently been studied in the context of flavour anomalies in the b sector [830, 831]. Searches for $t\bar{t}$ and $t\bar{b}$ resonances at the LHC help to constrain the parameter space of these models [832].

5.3 Vector-Like Quarks

Hypothetical quarks, for which the left- and right-handed chiral components transform in the same way under the SM electroweak symmetry group, are often referred to as vector-like quarks (VLQs). While a fourth chiral generation of quarks has been excluded by the measurements of H mediated cross sections [833, 834] and by considerations of the vacuum stability [835], VLQs are a viable possibility for a fourth generation of quarks [836, 837]. Gauge-invariant mass terms of VLQs can be included in the SM without generating them through interactions with a scalar field, such that VLQs evade constrains from H measurements. Electroweak precision measurements and the absence of flavour changing neutral currents constrain the possible mixings of VLQs with SM quarks, such that a dominant mixing with third generation quarks emerges [838, 839]. Assuming the SM symmetry groups and a scalar sector with only $SU(2)_L$ doublets, VLQs can only appear in two singlets, three doublets or two triplets [840, 841]. The four quarks in these multiplets are the two third-generation partners T and B with electric charges of $+2/3$ and $-1/3$, and two exotic quarks X and Y with electric charges of $+5/3$ and $-4/3$, respectively. Possible decays of the heavy quark mass eigenstates are into t/b and W, Z or H. The branching fractions depend on the mixing parameters with SM quarks and the multiplet representation. The three benchmark points for the ratios of branching fractions of up-type quarks $\mathcal{B}(T \to Wb) : \mathcal{B}(T \to Zt) : \mathcal{B}(T \to Ht)$ are $1 : 0 : 0$, $2 : 1 : 1$ and $0 : 1 : 1$. These cover scenarios with minimal and maximal mixings for all multiplet representations [837]. Conversely, for down-type quarks the benchmark points for the ratios $\mathcal{B}(B \to Wt) : \mathcal{B}(B \to Zb) : \mathcal{B}(B \to Hb)$ are $1 : 0 : 0, 2 : 1 : 1$ and $0 : 1 : 1$. In experimental analyses, often not only these benchmark points are considered but constraints are derived for all possible combinations of branching fractions under the assumptions of $\mathcal{B}(T \to Wb) + \mathcal{B}(T \to Zt) + \mathcal{B}(T \to Ht) = 1$ and $\mathcal{B}(B \to Wt) + \mathcal{B}(B \to Zb) + \mathcal{B}(B \to Hb) = 1$. For quarks with electric charges of $+5/3$ and $-4/3$, the only possible decays are $X \to W^+t$ and $Y \to W^-b$. Since all VLQs have quark-like triplet colour charges, they can be produced in pairs in pp collisions through diagrams involving the strong interaction. For a VLQ of mass $m_{\mathrm{VLQ}} = 1.2$ TeV, the pair-production cross section at the LHC with $\sqrt{s} = 13$ TeV is of the order of 10 fb. Single production of VLQs is mediated by the weak interaction through t-channel exchange of a W or Z boson. Since these processes depend on the size of the TbW and TtZ vertices (with similar vertices for B), the cross sections show a pronounced model dependence, but can be larger than the pair production cross sections for high VLQ masses, $m_{\mathrm{VLQ}} \gtrsim 1$ TeV. Maximum values allowed by constraints from electroweak precision observables can be as large as 100 fb at 1.2 TeV [837].

The phenomenology of VLQs has been studied extensively [842–849], because VLQs appear in many extensions of the SM. VLQs are required to induce electroweak symmetry breaking if the H is a pseudo-Goldstone boson, as in Little Higgs [850] or Composite Higgs [851–854] models. In partial-compositeness theories of flavour, VLQs emerge as fermionic resonances [855], in extra-dimension models they appear as quarks propagating in the bulk of these dimensions [856, 857], and they are also present in grand unified theories [858]. A model independent framework for studying the phenomenology of VLQs has been introduced in [859], and many searches are based on the effective Lagrangian defined there.

5.3.1 *Pair Production*

Pair production of VLQs in pp collisions occurs through SM-like interactions with either a gluon or a photon in the s-channel exchange. These diagrams only depend on the strong and electromagnetic couplings, such that a model-independent production cross section is obtained. However, there are diagrams contributing to pair production of VLQs, that depend on the VLQ-quark-vector boson (QqV) vertex, namely s- or t-channel diagrams with a weak vector boson as mediator. These lead to a dependence of the cross section on the value of the QqV coupling. If VLQs couple to first-generation SM quarks, the t-channel diagram can be dominant for high m_{VLQ} above 1.7 TeV, because it is the only diagram with two valence quarks in the initial state [861]. If, however, VLQs couple to second- and third-generation SM quarks preferentially, the relevance of these diagrams decreases and the pair production cross section is approximately independent of the QqV coupling [862]. Pair production of VLQs results in a plethora of possible final states. Models with non-negligible mixings with SM quarks result in decay chains involving two third-generation quarks and two SM bosons. For example, the pair production $pp \to T\overline{T}$ can lead to the intermediate states $bW\bar{b}W, bW\bar{t}Z, bW\bar{t}H, tZ\bar{t}Z, tZ\bar{t}H, tH\bar{t}H$ and their charge conjugates, as illustrated in Fig. 5.10. Considering the many possible decays of the W, Z and H, final states with only jets or one to four leptons are possible. A similar situation is obtained for $pp \to B\overline{B}$. Since it is not feasible to cover all possibilities in dedicated experimental analyses, the ATLAS and CMS Collaborations have focussed either on searches assuming 100% branching fraction into one of the three possible decays, i.e. $bW\bar{b}W$, $tZ\bar{t}Z$ and $tH\bar{t}H$ in the case of $pp \to T\overline{T}$, or final states characterised by the number of leptons. In fact, this approach covers all conceivable possibilities once at least one analysis per decay channel has been

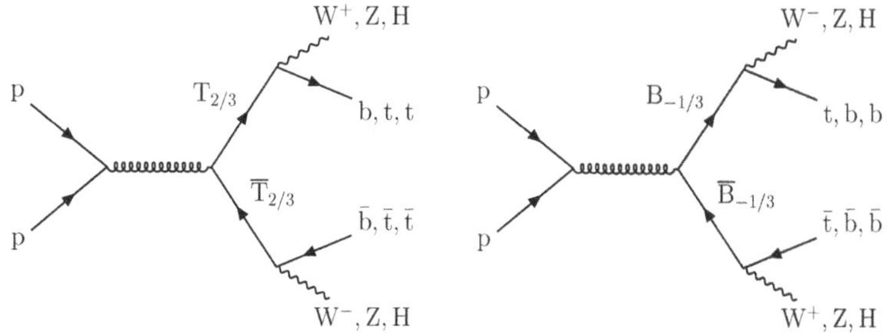

Fig. 5.10 Example leading-order Feynman diagrams for $T\overline{T}$ (left) and $B\overline{B}$ (right) production with subsequent decays. Taken from [860]

carried out. Often more than one analysis is sensitive to a VLQ decay, for example analyses in the all-hadronic and ℓ+jets final states complement each other. The best overall sensitivity to the complete model is then achieved by a combination of the experimental results. A prevalent feature of all VLQ searches are decays of highly boosted t, Z, W and H, such that in nearly all LHC analyses at 13 TeV jet substructure techniques are being used.

The situation was different when only 8 TeV data were available. ATLAS performed analyses in final states with leptons and small-R jets, searching for $T\overline{T}$ and $B\overline{B}$ production [863–866]. The sensitivity of these searches reaches lower limits on T masses of 715–950 GeV at 95% CL, where the W, Z, H and t are not too strongly boosted, such that their decays can be reconstructed with separate small-R jets. The first search using jet substructure methods is an inclusive search for $T\overline{T}$ production in final states with one, two or three leptons, carried out by CMS using 8 TeV data [867]. The search uses CA $R = 0.8$ jets for W and t tagging in the single-lepton channel to categorise events and as input to a BDT. No attempt is made to reconstruct specific decays, but the general character of the search, where a potential signal is searched for based on the BDT discriminator distribution, results in sensitivity for all three T decays. Besides improving the sensitivity at high masses, jet substructure methods allowed for VLQ searches in all-hadronic final states. The first search in the all-hadronic channel targeted the $T \rightarrow Ht$ decay, where the HEPTopTagger and H tagger with subjet b tagging are used for the suppression of the large multijet background [868]. An analysis targeting the $T \rightarrow Wb$ decay in the ℓ+jets and all-hadronic channels used pruned CA $R = 0.8$ jets, complemented with mass drop, to identify W jets [869]. A combination of T searches achieves mass limits between 770–920 GeV [869]. These mass limits are comparable to the ones from ATLAS, but at higher m_{VLQ} the upper cross section limits are considerably better because of the use of jet substructure in the CMS searches. Similar results have been obtained for $B\overline{B}$ and $X\overline{X}$ production at 8 TeV, with mass limits between 740 and 900 GeV for B [870] and 740–840 GeV for X [864, 866, 871].[2]

[2]For completeness, note that the limits on $T \rightarrow Wb$ can be interpreted as limits on Y, since the decays $T\overline{T} \rightarrow (W^+b)(W^-\overline{b})$ and $Y\overline{Y} \rightarrow (W^-b)(W^+\overline{b})$ are indistinguishable in searches.

The lower mass limits of about 900 GeV set the target for searches at 13 TeV, where the focus is on VLQs with masses of 1 TeV and above. An early CMS analysis of 13 TeV data, corresponding to 2.3 fb^{-1}, obtains mass limits for X between 990 and 1020 GeV, depending on its chirality [872]. The limits are based on a combination from same-sign dilepton and ℓ+jets final states. An update of this analysis by CMS, using 35.9 fb^{-1} of data, improved these bounds to 1.3 TeV [873]. In this analysis, the use of jet substructure methods results in a significant improvement compared to the analysis with 2.3 fb^{-1}. In the ℓ+jets channel, t tagging based on the soft drop jet mass and τ_{32} is employed. Large-R jets failing the top tagging requirement are tested for a W tag using the pruned jet mass and τ_{21}. While in the analysis with 2.3 fb^{-1} of data the dilepton channel has better sensitivity than the ℓ+jets channel, additional categories in the ℓ+jets channel with one W and one t-tagged jet result in better sensitivity of the ℓ+jets channel in the newer analysis. A similar result is obtained by ATLAS, where the multi-lepton analysis [874] achieves better sensitivity for X masses below 1 TeV, and an analysis in the ℓ+jets channel [875] achieves higher sensitivity above 1 TeV thanks to the usage of W tagging based on the trimmed jet mass and D_2. ATLAS obtains a mass limit of 1.35 TeV for $\mathcal{B}(X \to Wt) = 1$, compatible with the limit by CMS.

The presence of multiple weak vector bosons in the decay cascades of pair-produced T or B VLQs leads to a large branching fraction into ℓ+jets final states. This is exploited in a number of analyses by ATLAS and CMS, where the presence of an identified lepton with high p_{T} suppresses SM multijet production. Jet substructure techniques are used to suppress backgrounds from t, $t\bar{t}$ and V+jet production. Due to the hadronic activity from the large number of jets produced, usually mini-isolation or a two-dimensional lepton isolation criterion is used for the lepton reconstruction, similar as in searches for $t\bar{b}$ resonances. The first search at 13 TeV has been performed by CMS on 2.3 fb^{-1} of data [876]. This analysis is optimised for $T \to Ht$ and $T \to Wb$ decays, while being inclusive about the decay of the other VLQ in the event. Two distinct channels in the ℓ+jets final state are considered, based on the presence of a H- or W-tagged large-R jet, where the H jet identification considers either one or two b-tagged subjets. The analysis has been updated using 35.9 fb^{-1}, also adding final states with three leptons [860]. The highest sensitivity is obtained by a combination with the search in the same-sign dilepton final state [873]. The results are inclusive in the $T\overline{T}$ and $B\overline{B}$ decay channels and sensitivity for all branching fraction combinations is achieved. Masses of T and B quarks below 1140 and 910 GeV are excluded at 95% CL, respectively, for any combination of $\mathcal{B}$s.

ATLAS has taken a different approach and has performed four dedicated analyses in the ℓ+jets final state, each targeting a specific VLQ decay: $Wb + X$ [877], $Wt + X$ [875], $Z(\nu\nu)t + X$ [878] and $Ht + X$ [879]. In the Wb and Wt analyses, W jets are identified with the trimmed jet mass and D_2. The W jet, lepton, $p_{\mathrm{T}}^{\mathrm{miss}}$ and additional jets are used to reconstruct m_{VLQ}, where jets are assigned to the leptonic and hadronic W candidates. The permutation which minimises the mass difference of the two reconstructed VLQ candidates is used. This reconstruction works well for the targeted decay, where it produces distinct peaks for signal events. For decays where this algorithm fails to reconstruct the correct VLQ mass, it nonetheless provides sufficient separation power between signal and background events. In the Wt analysis, events failing the selection criteria, e.g. not enough jets are found or no jet has been W tagged, can not be reconstructed with this algorithm. Instead of discarding these events, they are passed to a BDT in order to retain signal efficiency. The analyses result in mass limits of 1.35 TeV for $\mathcal{B}(T/Y \to Wb) = 1$ [877] and $\mathcal{B}(B/X \to Wt) = 1$ [875]. The $Z(\nu\nu)t$ analysis targets final states with one lepton and $p_{\mathrm{T}}^{\mathrm{miss}} > 350\,\mathrm{GeV}$ from the $Z \to \nu\nu$ decay. In addition, two large-R jets with trimmed jet masses above 60 and 80 GeV are selected, representing the other VLQ decay in the event. The backgrounds are already sufficiently suppressed in the signal region, such that no specific assumptions on the second VLQ decay need to be made and no explicit jet tagging is applied. Mass limits of 1.16 TeV for $\mathcal{B}(T \to Zt) = 1$ are obtained [878]. The most involved of the four analyses is the analysis targeting $Ht + X$ final states [879], which also includes a $p_{\mathrm{T}}^{\mathrm{miss}}$+jets final state to improve the sensitivity for $Z(\nu\nu)t + X$ decays. Orthogonality to the $Z(\nu\nu)t$ analysis in the ℓ+jets final state is achieved by a lepton-veto. This analysis uses re-clustering of large-R jets [431] and defines t jets by a trimmed jet mass larger than 140 GeV and at least two subjets; H jets are required to have $105 < m_{\mathrm{jet}} < 140\,\mathrm{GeV}$ and either exactly two subjets if $p_{\mathrm{T}} < 500\,\mathrm{GeV}$ or one or two subjets if $p_{\mathrm{T}} > 500\,\mathrm{GeV}$. Selected events are categorised into different regions depending on the small-R jet multiplicity, the b-tagged jet multiplicity, and the H and t-tagged jet multiplicity. Overall, this results in 34 signal regions and 26 validation regions, which are used to constrain the SM background predictions. In each of these signal regions, the S_T distribution (called effective mass in the publication, m_{eff}) is used as sensitive variable to search for a signal. The total event yields in the signal regions are summarised in Fig. 5.11, where also the possible contribution from $T\overline{T}$ production is shown, where the T is assumed to be in an electroweak doublet with $\mathcal{B}(T \to Zt) = \mathcal{B}(T \to Ht) \approx 0.5$ and $\mathcal{B}(T \to Wb) = 0$. The best sensitivity is observed in signal regions with one or more H- or t-tagged large-R jets and (partially overlapping) b-tagged small-R jets. The analysis reports the strongest mass limits up to date from a single analysis of 1.43 TeV for $\mathcal{B}(T \to Ht) = 1$.

There is one exception to the inclusive approach followed by CMS in the ℓ+jets final state. The similarity of $T\overline{T} \to WbW\overline{b}$ with $t\overline{t} \to WbW\overline{b}$, as well as the dominant decays $T \to Wb$ and $Y \to Wb$, have prompted CMS to perform a dedicated analysis in the ℓ+jets channel [880]. Because the $\ell\nu bq\overline{q}\overline{b}$ final state can be fully reconstructed, it is possible to constrain the reconstructed neutrino and jet momenta[3] by a kinematic fit [881, 882], to achieve optimal resolution in the reconstructed VLQ mass. In order to take decays of highly boosted W bosons into account, the soft-drop mass of large-R jets is required to be between 60 and 100 GeV. Once a W jet has been found, its two soft-drop subjets are used as input to the kinematic fit instead of the corresponding one or two small-R jets. The kinematic fit results in an excellent relative mass resolution of about 7%, which compensates for the

[3] Also the lepton momentum is adjusted, but to a lesser degree, because the experimental resolution is better than for jets.

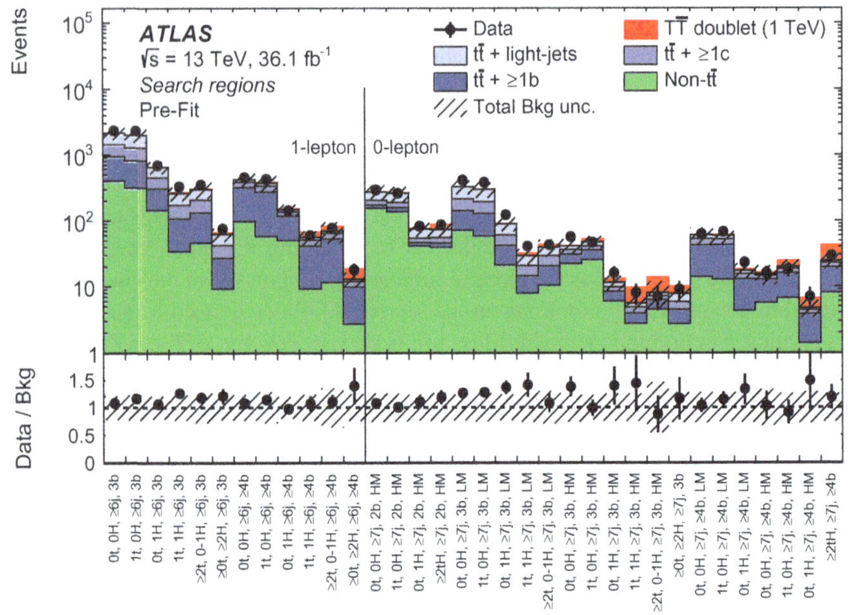

Fig. 5.11 Comparison between data and the background prediction for the total number of observed events in 34 signal regions in a search for $T\overline{T}$ production in ATLAS. The signal regions are defined by multiplicities of small-R jets and b-tagged jets, and large-R H- and t-tagged jets. Taken from [879]

necessary requirement of reconstructing the full $\ell\nu bq\overline{q}\overline{b}$ system. The achieved sensitivity and corresponding mass limits of 1.3 TeV are very similar to the ATLAS search for $T\overline{T} \to Wb + X$ [877].

VLQ decays involving Z bosons can be analysed in final states with two oppositely charged electrons or muons, consistent with coming from the decay of a Z boson. While the $Z \to \ell\ell$ branching fraction is small, the advantage of reduced backgrounds compared to single-lepton final states compensates the lower signal efficiency, and results in comparable sensitivity in this final state. In fact, exclusive decays with 100% branching fractions in $B \to Zb$ can not be investigated in analyses in ℓ+jets final states. ATLAS [883] and CMS [884] have analysed 36 fb^{-1} of data in the $Z \to \ell\ell$ channel, optimised for the decays $T\overline{T} \to Zt + X$ and $B\overline{B} \to Zb + X$. ATLAS uses a very inclusive selection based on large-R jets with trimmed jet mass greater than 50 GeV. Two signal regions are defined, both requiring a dilepton system with mass close to the Z boson and two b-tagged small-R jets. The first signal region selects two large-R jets with $p_T > 200$ GeV, with a high efficiency of reconstructing boosted W, Z, H and t. No other requirements on the jet substructure are needed, because the dilepton selection with two heavy jets reduces backgrounds from SM processes sufficiently. In order to retain signal efficiency, a second category with either no or one large-R jet is constructed. A third signal region selects events with three leptons, out of which two are consistent with coming from a Z boson decay. The highest sensitivity to $T\overline{T}$ production is obtained from the signal regions with two large-R jets and three leptons, whereas for $B\overline{B}$ production with $\mathcal{B}(B \to Zb) = 1$, the signal region with one large-R jet is better than the trilepton region and comparable in sensitivity to the selection with two large-R jets. The analysis by CMS follows a different strategy, being less inclusive in its jet substructure selection. A total of 16 signal regions are defined, based on the number of small-R b-tagged jets and large-R V-, H- or t-tagged jets. In order to maintain signal efficiency at low m_{VLQ}, also resolved events are considered. Doublets and triplets of small-R jets are built, which do not overlap with tagged large-R jets. Based on the mass of these pairings, V, H or t candidates are built. For $m_{\mathrm{VLQ}} < 1.2\,\mathrm{TeV}$ the resolved and substructure taggers are equally efficient in identifying signal events. For higher masses, the jet substructure taggers result in a more efficient signal selection. Note that this strategy for tagging resolved decays only works in the $Z \to \ell\ell$ channel, where multijet and $t\bar{t}$ backgrounds are small. The ATLAS and CMS analyses achieve comparable sensitivity to $B\overline{B}$ and $T\overline{T}$ production, and exclude VLQs below masses between 1 and 1.34 TeV.

The presence of a number of highly boosted vector bosons or top quarks facilitates searches for VLQ pair production in all-hadronic final states with low $p_{\mathrm{T}}^{\mathrm{miss}}$. Optimised jet substructure taggers are the key for achieving the best possible sensitivity, which depends on an efficient suppression of the large multijet background and at the same time high signal efficiency. Because of the large number of possible decay cascades from VLQ pair production, and the need for a dedicated and thorough background estimation for each measured distribution, it is advantageous to design an inclusive all-hadronic analysis. ATLAS and CMS have both performed a search for $T\overline{T}$ and $B\overline{B}$ production in the all-hadronic final state, using machine learning jet substructure taggers to suppress the multijet background. In the ATLAS analysis [885], VR jets are used (see Sect. 2.3.2), reclustered from calibrated small-R jets with $R = 0.4$. The VR jets are clustered using $\rho = 315\,\mathrm{GeV}$, $R_{\mathrm{min}} = 0.4$ and $R_{\mathrm{max}} = 1.2$, a parameter choice which is a compromise between the values for V decays ($\rho \approx 200\,\mathrm{GeV}$) and for t decays ($\rho \approx 600\,\mathrm{GeV}$), offering the possibility to reconstruct all relevant objects in the final state. Once VR jets have been clustered, trimming is applied, removing subjets of the VR jet if their p_{T} is less than 5% of the VR jet p_{T}. The VR jets are required to have $p_{\mathrm{T}} > 150\,\mathrm{GeV}$, $m_{\mathrm{jet}} > 40\,\mathrm{GeV}$ and $|\eta| < 2.5$. The subjets of the VR jet are used to train a DNN-based multi-class tagger, able to discriminate between V, H, t and background jets. The advantage of the small-R jet reclustering into VR jets is that the small-R jets have already been calibrated, leading to a precise momentum and mass reconstruction of VR jets, with uncertainties derived from the small-R jet uncertainties. These uncertainties can be propagated through the DNN to obtain uncertainties on the tagging efficiency for signal jets. The disadvantage is that the minimum distance between decay products that can be resolved is governed by the resolution parameter of the small-R jets, which is 0.4 in this case. At high p_{T}, this is too coarse to resolve the substructure of highly boosted hadronic decays, leading to a decreasing efficiency of the derived multi-class tagger. Twelve signal regions are defined based on the multiplicity of b-tagged small-R jets, V-, H-, and t-tagged VR jets, designed to cover all possible VLQ decays.

A matrix-element method [886] is used to separate signal from background events, and the analysis uses the logarithm of the signal likelihood distribution as final discriminating variable. The multijet background is estimated from sideband regions, obtained from a loosely tagged boson selection and large p_T^{miss}. The expected mass limits obtained by this analysis are 970 GeV for $\mathcal{B}(B \to Hb) = 1$ and 1010 GeV for $\mathcal{B}(T \to Ht) = 1$, which constitute the strongest channels of this analysis. The sensitivity of the analysis degrades for high m_{VLQ}, because the matrix-element method uses VLQ signals with a mass set to 900 GeV, leading to a decreasing separation power of the signal likelihood distribution with increasing m_{VLQ}. In the CMS analysis [887] a similar strategy is employed, where events with exactly four large-R jets are categorised based on the labels from the BEST algorithm [229]. The BEST algorithm runs on large-R jets, and boosts the jet constituents into four different rest frames under the assumption that the jet originates from a W, Z, H or t. The boost vector is formed by using the jet four-vector with the mass altered to be that of the particle under consideration, while keeping the jet momentum constant. In each rest frame, the jet constituents are used to calculate kinematic quantities including Fox-Wolfram moments [202, 203], aplanarity, sphericity, and isotropy, based on the eigenvalues of the sphericity tensor [43], and the jet thrust [44–46]. In addition, the soft drop jet mass, jet charge, τ_{32}, τ_{21}, and the CSVv2 [504] discriminator values calculated for the soft drop subjets in the original jet reference frame are used. In total, 59 quantities are used as inputs to a DNN, which labels jets as W, Z, H, t, b or light. All possible combinations of these labels, given to four jets, result in 126 independent signal regions with varying signal and background contributions. Regions with high multiplicities of W, Z, H and t have the best sensitivity to VLQ pair production. An advantage of the BEST algorithm is that it naturally includes subjet b tagging, making a matching between small-R and large-R jets unnecessary. In all signal regions with sufficient simulated events to model the dominant and sub-dominant background processes, the H_T distribution is used as discriminating variable between signal and backgrounds. The H_T is calculated from the scalar p_T-sum of the large-R jets. The multijet background in the signal regions is estimated by measuring the jet classification probability for each BEST label in a sample with exactly three large-R jets, which is dominated by multijet production with negligible signal contamination. These probabilities are used to calculate event weights in an untagged selection with exactly four large-R jets, considering every possible permutation of jets to arrive at a given event category. In the final result, uncertainties in the signal efficiency originating from the BEST classification are included through 11 independent nuisance parameters, one each for the classification and misclassification efficiencies for the five BEST labels, and a final one for the QCD categorisation efficiency. These nuisance parameters are allowed to float during the signal extraction, such that the signal efficiencies are determined in-situ in this measurement, with their prior values set to the expectations from simulation. The expected mass limits obtained by this analysis are 1170 GeV for $\mathcal{B}(T \to Ht) = 1$, 1100 GeV for $\mathcal{B}(T \to Zt) = 1$ and 950 GeV for $\mathcal{B}(B \to Wt) = 1$, which are the strongest channels of this analysis. The weakest sensitivity is observed for final states with two b jets, as obtained from $\mathcal{B}(T \to Wb) = 1$ and $\mathcal{B}(B \to Zb) = 1$, where the backgrounds from multijet production are large. In order to improve this situation, a dedicated analysis targeting the $WbWb$ and $ZbZb$ final states is performed in the same publication, achieving expected mass exclusion limits of 1070 GeV for $\mathcal{B}(T \to Wb) = 1$ and 1130 GeV for $\mathcal{B}(B \to Zb) = 1$, complementing the inclusive analysis with BEST.

Naturally, the best overall sensitivity to VLQs is obtained by combining dedicated analyses targeting different VLQ decays. Similar as in combinations of searches for diboson resonances, the feasibility of a combination relies on exclusive signal, control and validation regions. While it would be possible to include correlations between regions, the exact implementation and calculation of these correlations is technically very involved. This needs coordination and planning, already in the design stage of all analyses entering the combination. ATLAS has performed such a combination, including seven searches for the pair production of VLQs, all based on 13 TeV data corresponding to 36.1 fb^{-1} [888]. The analyses included are the four analyses in the ℓ+jets final state $Wb + X$ [877], $Wt + X$ [875], $Z(\nu\nu)t + X$ [878] and $Ht + X$ [879], the dilepton analysis $Z(\ell\ell)t/b + X$ [883], the multi-lepton analysis [874], and the all-hadronic analysis [885]. The analyses are either complementary by the final state, or complentarity is ensured by the use of consistent definitions of physics objects and jet substructure taggers. Only small adjustments had to be made to individual analyses, the largest one in the multi-lepton analysis, where events are removed with more than three leptons or events with a lepton pair having an invariant mass compatible with a Z boson decay. The loss in sensitivity by the individual analysis is compensated by the inclusion of the results from the $Z(\ell\ell)t/b + X$ search. The results of the combination are shown in Fig. 5.12, which also displays the expected 95% upper cross section limits of the individual analyses. The results are shown for a T and B in their electroweak singlet representations with branching fractions of approximately $\mathcal{B}(T \to Wb) = 0.5$, $\mathcal{B}(T \to Ht) = 0.25$ and $\mathcal{B}(T \to Zt) = 0.25$, and $\mathcal{B}(B \to Wt) = 0.5$, $\mathcal{B}(B \to Hb) = 0.25$ and $\mathcal{B}(B \to Zb) = 0.25$. The best sensitivity on $T\overline{T}$ production is achieved by the $Ht + X$ search, but the $Wb + X$ and $Z(\ell\ell)t/b + X$ significantly contribute to the overall result, too. The combination improves the individual limits by up to a factor of 1.7, resulting in the best constraints on $T\overline{T}$ pair production to date. For $B\overline{B}$ production, the best cross section limits for $m_{\text{VLQ}} > 900$ GeV are obtained from the $Wt + X$ search, while for $m_{\text{VLQ}} < 900$ GeV the $Z(\ell\ell)t/b + X$ and the multilepton analyses are more sensitive. The ultimate strength of a combination of VLQ searches is that sensitivity to all combinations of branching fractions is achieved, such that mass limits independent of the exact VLQ representation can be obtained. Under the assumption that there are no invisible decays, i.e. $\mathcal{B}(T \to Wb) + \mathcal{B}(T \to Ht) + \mathcal{B}(T \to Zt) = 1$ and $\mathcal{B}(B \to Wt) + \mathcal{B}(B \to Hb) + \mathcal{B}(B \to Zb) = 1$, mass limits above 1 TeV are obtained for any combination of branching fractions, as shown in Fig. 5.13. In particular, the observed lower mass limits for T are 1.31 TeV with 1.22 TeV expected, and 1.03 TeV for B with 0.98 TeV expected. At the time of writing, these limits constitute the strongest model-independent constraints on VLQs and it is conceivable that it will take a long time until these can be improved considerably. The weakest mass limits for T are observed for the singlet configuration with slightly lowered $\mathcal{B}(T \to Ht)$, where all decay channels contribute, but the dominant decays are $T \to Wb$ and

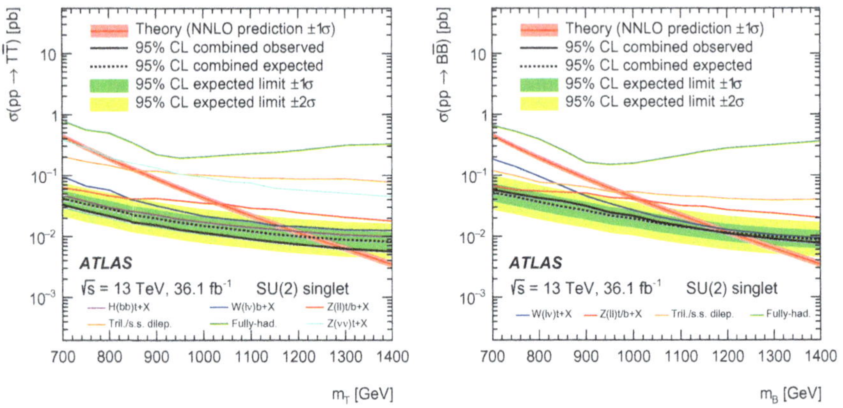

Fig. 5.12 Observed and expected upper limits at 95% CL on the production cross sections of $T\overline{T}$ (left) and $B\overline{B}$ (right) for the scenario of VLQs arranged in electroweak singlets. Expected limits from the individual analyses are shown by coloured lines. The combined result is shown in black with coloured bands corresponding to ±1 and ±2 standard deviations around the expected limit. Taken from [888]

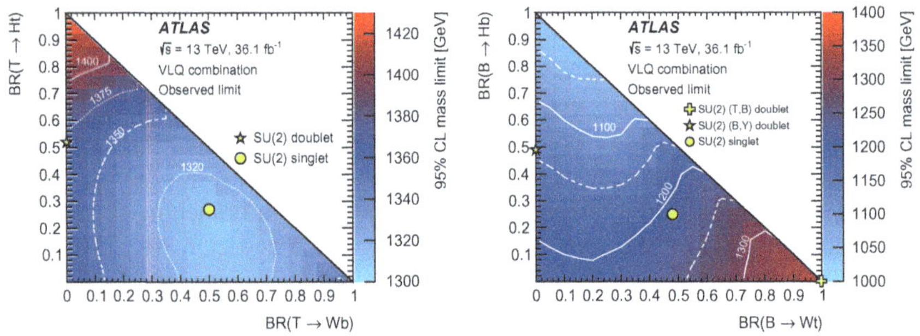

Fig. 5.13 Observed lower limits at 95%CL on the mass of a T (left) and B (right) as a function of branching ratios. The third possible branching ratio $\mathcal{B}(T \to Zt)$ and $\mathcal{B}(B \to Zb)$ is not shown, but can be inferred from the relations $\mathcal{B}(T \to Wb) + \mathcal{B}(T \to Ht) + \mathcal{B}(T \to Zt) = 1$ and $\mathcal{B}(B \to Wt) + \mathcal{B}(B \to Hb) + \mathcal{B}(B \to Zb) = 1$. Taken from [888]

$T \to Zt$. The situation is different for B, where the weakest mass limit is observed for $\mathcal{B}(B \to Hb) = 1$. This is not targeted by a dedicated analysis, but the strongest constraints come from the all-hadronic search.

Since the lowest sensitivity to VLQ pair production is in the $B\overline{B} \to bH\overline{b}H$ channel, the first VLQ search at the LHC using the full data recorded during 2016–2018 with 137 fb^{-1}, has been optimised for this decay [889]. Because of the large branching fraction for $H \to b\overline{b}$, the analysis is carried out in the all-hadronic final state. It targets B pair production in the $bHbH$, $bHbZ$ and $bZbZ$ channels, which result in six high-p_{T} quarks in the final state, where either six, four or two of them are b quarks. It improves on the previous results obtained in this final state [885, 887] by fully reconstructing the event kinematics, thereby allowing for a reconstruction of the B mass. The analysis is carried out in three different categories, defined by the small-R jet multiplicity with four, five or six jets, where two, three or four have to be b-tagged, respectively. In the categories with four or five jets, large-R jets are used to reconstruct the boosted H or Z, where one is required to be identified with the double b tagger. Events are reconstructed and sorted into three categories corresponding to the decay modes $bHbH$, $bHbZ$ and $bZbZ$ with a χ^2 estimator. The estimator is built for all possible jet combinations and compares the reconstructed Z and H masses with the expectations from simulation, as well as the mass difference between the two VLQs in the event. The multijet background is obtained by a fit of an exponential function to events in the three categories, before b- and double b-tagging selections are applied. This sample has a high rate of multijet background events, with a signal-to-background ratio more than two orders of magnitude smaller than in the signal regions. The exponential function is observed to describe the shape of the reconstructed VLQ mass distribution well for masses greater than 1 TeV. The parametrised shape is then multiplied by the misidentification rate of the b tagging requirements, obtained in a control region with small reconstructed VLQ mass. The misidentification rates are checked in validation regions with large values of the χ^2 estimator. The final result of this analysis consists of nine measured VLQ mass distributions for the three jet multiplicity regions and three decay modes. The highest sensitivity is obtained from the 4-jet, $bHbH$ category. The limits on the B mass are improved to 1570 GeV in the $\mathcal{B}(B \to Hb) = 1$ case, and to 1390 GeV in the $\mathcal{B}(B \to Zb) = 1$ case. The large improvement of about 400 GeV with respect to the all-hadronic analyses stems mostly from the specialisation to the particular decay mode, which allows for a reconstruction of m_{VLQ}. The larger data sample, where this analysis uses about four times the data, plays a role as well, but a smaller one. A naive extrapolation of the improvement in expected sensitivity of the two all-hadronic searches results in a higher mass limit by about 100 GeV from the increase in luminosity only. While the previous all-hadronic analyses use advanced machine learning methods, this analysis accomplishes high sensitivity without multi-dimensional classification algorithms. This shows that in some cases it is advantageous to search for a particular signature, instead of performing an inclusive analysis aiming at a large number of channels.

The LHC results with 13 TeV data move the allowed masses of all VLQs considered, B, T, X and Y, to values beyond 1.3 TeV. On the one hand, this is a remarkable achievement which relies very much on the development and commissioning of jet substructure methods by the experimental collaborations. On the other hand, this means that some BSM models where VLQs are an integral component become more unlikely to be realised in nature. However, non-minimal models, for example models with an additional scalar [890–892], a new quantum number which is conserved for VLQs [893], a new gauge symmetry group [894], or models with VLQs embedded in a two-Higgs-doublet model [895] can lead to significantly lower bounds on VLQ masses because of the presence of additional decay modes. There is still room left for these ideas to be explored theoretically and experimentally, which will also lead to new search strategies for the pair production of VLQs at the LHC.

5.3.2 Single Production

Single production of VLQs occurs through the weak interaction, with a t-channel exchange of a SM weak vector boson. The incoming quark is preferentially scattered under small angles, or large absolute values of rapidity, where the resulting jet is usually referred to as forward jet. In the case of VLQs coupling to third-generation quarks, the VLQ is produced in association with a b or t quark, which is on average produced with much smaller p_T than the decay products of the VLQ. While there are diagrams involving the exchange of an H boson, their contribution to the total cross section is small because of the smallness of the Yukawa coupling to first and second generation quarks. Therefore models with $\mathcal{B}(B \to Hb) = 1$ and $\mathcal{B}(T \to Ht) = 1$ predict vanishing production cross sections, because of the absence of charged and neutral currents with VLQs. In contrast to pair production, the production cross section depends on the size of the QqV coupling and therefore on the weak isospin and the representation of the VLQ. Because of the larger phase space for the production of a single heavy particle compared to two, single VLQ production can have larger production cross sections than pair production at high m_{VLQ}. The exact value of m_{VLQ} where the single production cross section is larger and becomes the better search channel depends on the unknown QqV coupling, but can be between 1 and 1.5 TeV. The largest allowed cross sections, given the constraints on VLQs from precision data, are in the range of 20–80 fb for $T\bar{b}j$, 7–55 fb for $B\bar{b}j$ and about 150 fb for $Y\bar{b}j$ production at $\sqrt{s} = 13$ TeV and $m_{\mathrm{VLQ}} = 1$ TeV [837]. The ranges indicate the cross sections obtained for VLQs in the singlet or doublet representations. VLQ production in association with a top quark involves initial gluon splitting $g \to t\bar{t}$ and has a much smaller cross section than b-associated production [896]. The cross sections for $T\bar{t}j$ and $X\bar{t}j$ are of comparable size, about 2 fb for $m_{\mathrm{VLQ}} = 1$ TeV [837, 897]. Note that $T\bar{t}j$ is the only production process for a T quark with very small coupling to a W boson. Only for $m_{\mathrm{VLQ}} > 2$ TeV, the cross section for VLQ single production in association with a top quark can be larger than the pair-production cross section. In some cases interference effects with SM processes need to be taken into account, for example in $pp \to Tj \to Wbj$, which results in the same final state as single top quark production. However, interference effects contribute only in cases where the width of the VLQ is large, with only a few percent change in the cross section when including these effects for a relative width of 10% [849]. Due to a certain degree of model dependence and a dependence on the QqV coupling, searches by the ATLAS and CMS Collaborations usually are performed without considering the associated b or t quark. The phenomenology of single VLQ production has been studied in several analyses, see for example [896, 898–904].

The first searches for singly produced VLQs have been carried out by ATLAS using 20.3 fb^{-1} of 8 TeV data. A search in the dilepton final state targets the channels $T \to Zt$ and $B \to Zb$, where the t and b quarks are reconstructed using b-tagged small-R jets [863]. The first 8 TeV search using jet substructure techniques targets the $B \to Wt$ channel, where either the W or t are reconstructed in a single large-R jet [905]. The large-R jet selection is not very constraining, where only the trimmed jet mass has to be larger than 50 GeV. This ensures that W and t jets can be captured, and the angular distance between the lepton, small-R jets and the selected large-R jet is used to determine if the large-R jet originates from a W or t decay. A further reduction of backgrounds is not necessary, as the analysis requires the presence of a forward jet with $2.5 < |\eta| < 4.5$, which suppresses SM backgrounds sufficiently. A different strategy is followed by an ATLAS analysis searching for the decay $T \to Wb$ in the ℓ+jets channel [906]. The analysis reconstructs the $W \to \ell\nu$ decay with an identified lepton and $\mathbf{p}_{\mathrm{T}}^{\mathrm{miss}}$, such that only a small-R jet from the b quark decay is expected in the final state. In order to suppress high-p_{T} $t\bar{t}$ background, a veto on large-R jets with a trimmed jet mass larger than 70 GeV is implemented. This results in an efficient rejection of $t\bar{t}$, reducing the $t\bar{t}$ background to about 20% of the total background in the signal region. The observed cross section limits on single VLQ production are translated into limits on the QqV couplings, but the achieved sensitivity from 8 TeV analyses is weaker than indirect constraints from electroweak precision data [837].

At a centre-of-mass energy of 13 TeV, single production of VLQs gains importance because the production cross sections increase by factors of 7 for $m_{\mathrm{VLQ}} =$ 1 TeV, up to 17 for $m_{\mathrm{VLQ}} = 2$ TeV compared to 8 TeV [837]. In the following, 13 TeV searches for the single production of up-type VLQs in the channels $T \to Ht$ and $T \to Zt$ will be discussed, where results for $T \to Wb$ will be covered as well. Down-type VLQs are searched for in the $B \to Hb$ and $B \to Wt$ channels.

A first search for single VLQ production at 13 TeV has been carried out by CMS and uses 2.3 fb^{-1} of data. It targets the $T \to Ht$ channel in the ℓ+jets final state. The analysis targets heavy VLQs with $m_{\mathrm{VLQ}} > 1$ TeV, and makes use of non-isolated lepton identification and H tagging of large-R jets. The H tagger uses $\tau_{21} < 0.4$ and the soft drop jet mass has to be between 90 and 160 GeV. The number of b-tagged subjets from the soft-drop algorithm defines the signal and control regions, together with the presence or absence of a forward jet. The T quark mass is reconstructed by combining the H jet with the lepton, $\mathbf{p}_{\mathrm{T}}^{\mathrm{miss}}$ and a small-R jet that matches the kinematics of a t decay best. The distribution of SM backgrounds in the reconstructed VLQ mass is obtained from a control region with one b-tagged subjet and without a forward jet. The assumption of similar background distributions in events with and without a forward jet is verified in a validation region with a forward jet, but with no b-tagged subjet. The analysis sets upper limits at 95% CL on the product of T production cross section and $\mathcal{B}(T \to Ht)$ between 0.2 and 1 pb for T masses of 1.8 TeV and 1 TeV, respectively. Note that this analysis is only sensitive to T production if also the couplings to bW and tZ are sizeable, because the production through a Higgs boson is strongly suppressed. The $T \to Ht$ channel can also be probed in all-hadronic final states, enabled by the large hadronic t and H branching fractions. A recent analysis by CMS uses 35.9 fb^{-1} of 13 TeV data and combines a resolved and a boosted selection [907]. Both selections target the $pp \to (bq\bar{q}')(b\bar{b})q$ final state, where the first three quarks originate from a t quark decay, the $b\bar{b}$ pair may come either from the decay of a H or Z boson, and the last quark is the scattered quark from the electroweak production. The similarity between the $H \to b\bar{b}$ and $Z \to b\bar{b}$ decays allows for an optimisation in the $T \to Ht$ and $T \to Zt$ channels within the same analysis. The resolved search is optimised for $m_{\text{VLQ}} < 1$ TeV and requires six small-R jets within $|\eta| < 4.5$, three of which have to be b-tagged. In this kinematic regime, the difficulty lies in the suppression of the abundant QCD multijet background and a method for its estimation, able to model the trigger turn-on. These challenges are met by a cascade of selection steps and a background extrapolation from regions with loosened b tagging requirements. The boosted analysis selects a dijet topology with two high-p_{T} large-R jets, where one is t-tagged and the other either Z- or H-tagged. The Z and H taggers use $\tau_{21} < 0.6$ and two b-tagged subjets. Jets have to have a pruned mass in the range 65–105 GeV to be Z tagged and 105–135 GeV to be H tagged. Similar requirements are imposed on t-tagged jets, with $\tau_{32} < 0.57$, one b-tagged subjet and a soft drop mass in the range 105–220 GeV. The overlap in the Z/H and t taggers results in an ambiguity when selecting jets, but increases the signal efficiency in this analysis where a confusion between the Z/H and t tagged jets does not matter. The main backgrounds are $t\bar{t}$ and multijet production, where the latter is estimated from control regions, obtained by reversing the tagging requirements. Upper cross section limits are reported for $\mathcal{B}(T \to Ht) = 1$, $\mathcal{B}(T \to Zt) = 1$ and $\mathcal{B}(T \to Ht) = \mathcal{B}(T \to Zt) = 0.5$, the two different production modes with either an associated b or t, and different fractional widths. The limits are based on whichever of the two selections, resolved or boosted, achieves the best estimated expected sensitivity. In general, for $m_{\text{VLQ}} \lesssim 1$ TeV the resolved analysis is better and for masses above 1 TeV the boosted analysis is better. The limits on the product of T production cross section and $\mathcal{B}(T \to Ht)$ are between 0.028 and 0.12 pb for T masses of 1.8 TeV and 1 TeV, respectively, thus improving the previous limits obtained in the ℓ+jets final state by nearly an order of magnitude. Comparable sensitivity is obtained for $T \to Zt$ and a mixture of $T \to Ht$ and $T \to Zt$.

The diversity of Z boson decays allows for searches for $T \to Zt$ also in leptonic final states, including charged leptons and neutrinos. Neutrinos from the $Z \to \nu\nu$ decay result in large $p_{\text{T}}^{\text{miss}}$, balanced by a boosted top quark. This particular topology, which does not have a counterpart in the SM, has been searched for by ATLAS using 36.1 fb^{-1} of 13 TeV data [908]. Large-R jets are identified as t jets based on the trimmed jet mass and τ_{32}. A loose working point is chosen with a t-tagging efficiency of 80% above p_{T} of 400 GeV, for high signal efficiency. Contrary to the all-hadronic analysis, tighter requirements on the t tagging are not needed because the background from SM multijet production can be reduced efficiently by a selection based on $p_{\text{T}}^{\text{miss}}$. This is achieved by $p_{\text{T}}^{\text{miss}} > 200$ GeV, and requiring the azimuthal angle between $\mathbf{p}_{\text{T}}^{\text{miss}}$ and the t-tagged jet to be large. In order to ensure well-reconstructed $p_{\text{T}}^{\text{miss}}$, the p_{T} of the t-tagged jet and $p_{\text{T}}^{\text{miss}}$ have to be balanced, and no small-R jets should be present close in azimuthal angle to $\mathbf{p}_{\text{T}}^{\text{miss}}$. This selection reduces the multijet background to a negligible level of <1% of the total background in the signal region, quite remarkable for an all-hadronic analysis. In addition to the presence of one t-tagged large-R jet with $p_{\text{T}} > 250$ GeV, one b-tagged small-R track-jet is required, which reduces the background from V+jets. The obtained limits on the product of production cross section and $\mathcal{B}(T \to Zt)$ are between 0.045 and 0.15 pb for $m_{\text{VLQ}} = 1.8$ and 1 TeV, respectively, comparable to the sensitivity of the all-hadronic search by CMS [907]. Searches in the dilepton final state, targeting $T \to Z(\ell\ell)t$,

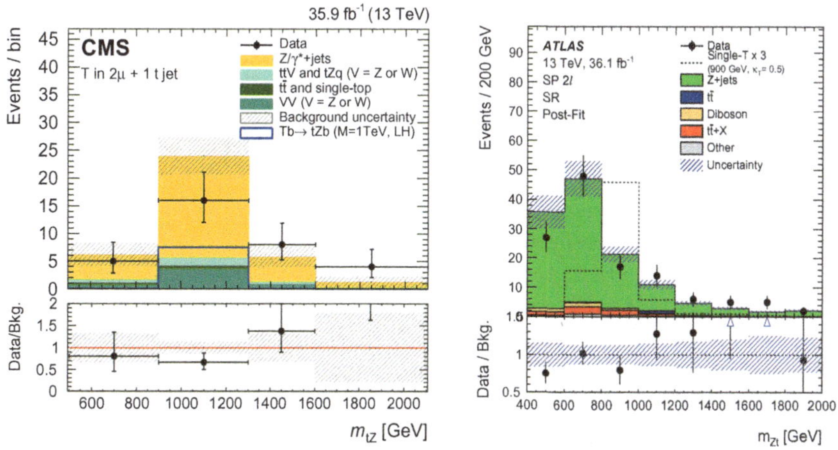

Fig. 5.14 Reconstructed T mass in CMS (left) and ATLAS (right), for events with two opposite-sign, same-flavour leptons with small angular separation, balanced in p_T by a t-tagged large-R jet. Taken from [910] (left) and [883] (right)

have been performed by CMS using 3.2 fb^{-1} [909] and 35.9 fb^{-1} [910], and by ATLAS using 36.1 fb^{-1} of data [883]. The CMS and ATLAS analyses follow a very similar strategy. In CMS, the t quark is reconstructed either by a t-tagged large-R jet, by a W-tagged large-R jet combined with a b-tagged small-R jet, or by three small-R jets, one of which has to be b-tagged. In ATLAS, only events with a t-tagged jet are considered for the signal region. The W tagging algorithm used in CMS relies on the pruned jet mass and τ_{21}. The t jets are identified using τ_{32} and the soft drop jet mass in CMS or τ_{32} and the trimmed jet mass in ATLAS. The T mass is reconstructed by summing the momenta of the Z boson candidate, given by two identified leptons, and the t candidate. The presence of a forward jet improves the sensitivity of the searches. The dominant background from Z+jets production is estimated from control regions with a veto on b-tagged small-R jets. The distributions in the reconstructed T mass obtained by CMS and ATLAS are shown in Fig. 5.14 for events with two opposite-sign, same-flavour leptons and one t-tagged jet. Note that CMS shows only events with two muons, whereas ATLAS shows events with two muons or two electrons. Also, the signal is multiplied by a factor of three in the ATLAS case. Summing the electron and muon channels in CMS, the signal and background yields are very similar in these two analyses, even though some details of the selections differ. The observed (expected) limits are also comparable and are about 0.04–0.05 pb (0.02–0.03 pb) and 0.08–0.1 pb (0.07 pb) for $m_{\rm VLQ} = 1.8$ and 1 TeV, respectively. These limits are the most stringent limits on single T production in pp collisions to date.

The single production of T and Y with decays to Wb are covered by searches with an isolated lepton, $p_T^{\rm miss}$ and a high-p_T b-tagged small-R jet in the final state. In this case, the QbW coupling is present in both, production and decay of the VLQ, such that the production cross section times branching fraction depends on the fourth power of this coupling. Interference effects with SM single top-quark production resulting in the same final state, $pp \rightarrow Wbq$, need to be taken into account. ATLAS [911] and CMS [912] have performed searches in this final state, where SM backgrounds are suppressed by requiring low jet multiplicity and the presence of a forward jet. The best limits are obtained by the ATLAS analysis, which is based on a data sample four times larger than the one used in the CMS analysis. The upper limits on the cross section times branching fraction are between 0.08 and 0.11 pb for $m_{\mathrm{VLQ}} = 1.8$ and 1 TeV, respectively.

Two searches have been performed by CMS targeting the single production of B, both based on 13 TeV data with 35.9 fb^{-1}. A search performed in the all-hadronic final state is optimised for the $B \rightarrow Hb$ channel, where the $H \rightarrow b\bar{b}$ decay is reconstructed by a H-tagged large-R jet [913]. The H jets are defined by the pruned jet mass in the range 105–135 GeV and by two b-tagged subjets, obtained with the soft drop algorithm. Events in the signal region require an H jet balanced by a high-p_{T} b-tagged small-R jet. Trigger requirements lead to a selection of $H_T > 950$ GeV, calculated from all small-R jets with $p_{\mathrm{T}} > 30$ GeV. Events are sorted into four categories, based on the presence of a forward jet and the value of H_T. The low-mass category with $H_T < 1250$ GeV shows higher sensitivity for signals with $m_{\mathrm{VLQ}} < 1500$ GeV, whereas the multijet background is reduced in the high-mass category with $H_T > 1250$ GeV, resulting in a better sensitivity for signals with $m_{\mathrm{VLQ}} > 1500$ GeV. The main background in this search is multijet production, with only 5–7% from $t\bar{t}$ production. Other SM processes give negligible contributions. The multijet background is estimated from three sideband regions, obtained by requiring only one b-tagged subjet and/or changing the soft drop jet mass to $75 < m_{\mathrm{jet}} < 105$ GeV or $m_{\mathrm{jet}} > 135$ GeV. For the method to work, the subjet b-tagging has to be uncorrelated from the soft drop jet mass, which has been verified using simulation. The analysis excludes cross sections times branching fractions above 0.07 and 0.4 pb for $m_{\mathrm{VLQ}} = 1.8$ and 1 TeV, respectively. The limits worsen by factors between 1.3 for $m_{\mathrm{VLQ}} = 1$ TeV and 2.1 for $m_{\mathrm{VLQ}} = 1.8$ TeV, when increasing the relative width of the B from 1% (narrow width approximation) to 30%.

A dedicated search for $B/X \rightarrow Wt$ is carried out in the ℓ+jets channel [914], where the lepton can originate either from the $t \rightarrow Wb \rightarrow \ell\nu b$ decay, or from the W boson from the B/X decay. The analysis selects leptons with $p_{\mathrm{T}} > 55$ GeV, which are identified with a two-dimensional isolation requirement in order to achieve high selection efficiency for decays of boosted t quarks (see Sect. 4.2.3). The analysis uses W and t tagging, based on the soft drop jet mass, τ_{21}, τ_{32} and subjet b tagging. Selected events are attributed to five categories, defined by the presence of a t tag, a W tag, two, one or no b-tagged small-R jets. In the t tag category, the VLQ mass is reconstructed from the four-vectors of the t-tagged jet, the lepton and $\mathbf{p}_{\mathrm{T}}^{\mathrm{miss}}$. In all other categories, it is reconstructed using combinations of small-R jets, where the best combination is chosen based on a χ^2 estimator. The data sample is divided into a signal region with a forward jet and a control region without one. The background distribution in the reconstructed VLQ mass in the signal region is estimated from the corresponding distribution in the control region. This allows for a data-driven background estimation of all SM backgrounds in this search. Residual differences

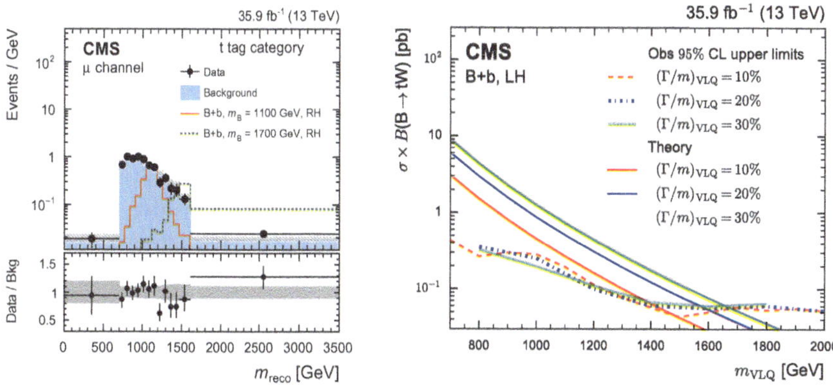

Fig. 5.15 Distribution in the reconstructed B mass in events with one t-tagged jet and a forward jet, where the SM background is obtained from a control region without a forward jet (left). Product of observed upper cross section limits and $\mathcal{B}(B \to Wt)$ as a function of m_{VLQ} for different B widths (right), for the single production of $B + b$. Taken from [914]

in the shapes of these distributions in the signal and control regions can arise from different background compositions due to the presence of a forward jet. The observed differences are small, with average values of 10%, and are corrected for by using factors derived from simulation. The distribution in the reconstructed VLQ mass is shown in Fig. 5.15 (left) for the t tag category in the μ+jets channel. The signal distributions for a B with right-handed couplings, produced in association with a b are shown as well, for two different values of m_{VLQ} with an assumed production cross section of 1 pb. Upper cross section limits times $\mathcal{B}(B \to Wt)$ on single production of $B + b$ and $X + t$ are derived by combining the five categories measured in the muon and electron channels. The observed (expected) limits for $B + b$ production with left-handed couplings and narrow width are between 0.04 pb (0.04 pb) and 0.3 pb (0.2 pb) for $m_{\text{VLQ}} = 1.8$ and 1 TeV, respectively. These limits can differ by 10–20% for VLQs with right-handed chirality or different widths, as shown in Fig. 5.15 (right). Similar exclusion limits are obtained for $X + t$ production. These limits are the most stringent limits on the single production of B and X to date.

No dedicated search for the single production mode $pp \to Bb \to Zbb$ exists, although limits are derived by the CMS search in the dilepton final state targeting $T \to Zt$, using 3.2 fb^{-1} of 13 TeV data [909]. The upper cross section limits are between 0.2 and 1 pb, and thus nearly one order of magnitude weaker than limits in other B decay modes. However, a recent study suggests that a dedicated analysis in $B \to Z(\ell\ell)b$ can result in competitive sensitivity compared with other decay channels [915].

Despite the theoretically well motivated dominant mixing with third-generation SM quarks, also couplings to light SM quarks are possible. This has been explored in a search by CMS, including the single production $pp \to Dq$, as well as pair production $pp \to Q\overline{Q}$, where D denotes down-type VLQs and Q up- and down-type VLQs [916]. The search uses 19.7 fb^{-1} of 8 TeV data and considers the single production decays channels $Dq \to Wqq$ and $Dq \to Zqq$, and the pair production channels $WqWq$, $WqZq$, $WqHq$, $ZqZq$ and $ZqHq$. The analysis is performed in the ℓ+jets and multi-lepton final states, where a kinematic fit is performed in the single-lepton final state to reconstruct the VLQ mass. In this fit, the subjets of W-, Z- or H-tagged large-R jets are used if the decay can not be resolved by small-R jets, similar to the strategy adopted in [880]. Observables sensitive to m_{VLQ} are used in the multi-lepton final states to discriminate between background and signal events. This analysis sets the most stringent mass limits on VLQs coupling to light quarks, ranging from 400 to 1800 TeV, depending on the electroweak coupling and the branching fractions into W, Z and H. The weakest constrains are obtained for large branching fractions into H, where this analysis has not been optimised for. Even when considering only pair production, the obtained limits are better by more than 150 GeV than results from searches by ATLAS at 7 [917] and 8 TeV [918], which did not include jet substructure methods.

The results commonly given in experimental searches are upper cross section limits. These are directly derived from the data and can be obtained in a fairly model-independent fashion. The only assumptions are the acceptance and experimental efficiency to observe a given final state, which are derived using explicit signal models. As long as the acceptance and efficiency are approximately unchanged, the cross section limits can be used to place constrains on any model predicting the same final state. However, an intermediate step is needed in order to derive the allowed parameter space of a given model. Cross sections have to be computed for a set of parameters and compared to the experimental results. For a phenomenological analysis, it is more convenient to have bounds on model parameters explicitly appearing in the Lagrange density, like couplings and masses. This eliminates the need for a computation of cross sections for various model parameters. A more general representation of the experimental results is more important for single production of VLQs than for pair production. In pair production, the QCD-induced cross section is independent of the electroweak representation and the corresponding coupling parameters. This allows for a reinterpretation of the results by treating the branching fractions as free parameters instead of the coupling parameters. In single production, the situation is more involved, because the couplings enter the production and decay simultaneously. A possible strategy is to use a simplified model with an effective Lagrangian, which can be written as [919]

$$\mathcal{L} = \sum_{\zeta,q,Q} \left(\frac{g_W}{2} \sum_V c^{Qq}_{\zeta,V} \overline{Q}_\zeta \gamma_\mu V^\mu q_\zeta + c^{Qq}_{\zeta,H} H \overline{Q}_{\zeta'} q_\zeta \right) + \text{h.c.} \tag{5.3}$$

where g_W is the weak coupling, Q denotes the VLQ fields, q the SM up- or down-type quark fields, ζ and ζ' are alternate chiralities and γ_μ are the usual gamma matrices. The free parameters of the Lagrangian are the VLQ masses m_{VLQ} and their couplings $c^{Qq}_{L/R,V/H}$ to the SM quarks by the exchange of gauge bosons V or the Higgs boson H. Once experimental results are expressed as bounds on the free parameters

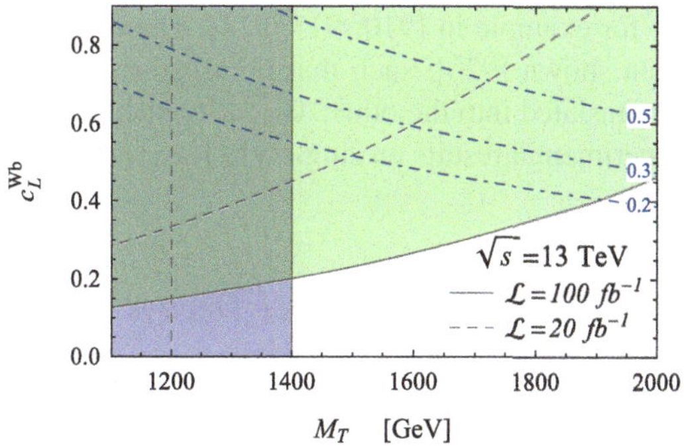

Fig. 5.16 Projected bounds, obtained in 2014, on m_{VLQ} of a T with $\mathcal{B}(T \to Wb) = 0.5$, presented for different values of the left-handed coupling $c_{L,W}^{Tb}$ to the bottom quark. The blue area on the left is excluded from pair production, the green area corresponds to the exclusion from b-associated single production. The dash-dotted blue lines show contours with relative VLQ widths of 20, 30 and 50%. Taken from [919]

m_{VLQ} and $c_{L/R,V/H}^{Qq}$, these can be mapped to explicit models through a one-to-one correspondence of the tree-level couplings. An example is shown in Fig. 5.16, where prospects of exclusion limits are shown for T searches in the m_{VLQ}–$c_{L,W}^{Tb}$ plane. Results from a search for pair production are compared to results from single production, showing the complementarity of the two searches. While pair production can exclude the full coupling space below a certain value of m_{VLQ}, single production can exclude higher values of m_{VLQ} for large couplings.[4] This also shows why searches for VLQ single production can not quote absolute limits on m_{VLQ}, but can only exclude regions in the mass-coupling parameter space. Care has to be taken in the interpretation of experimental results once the VLQ width becomes too large, where it has to be verified that the assumptions made in the design of the analysis are still valid, and the acceptance and efficiencies usually determined for relative width up to 30% are adequate. In these cases, where non-resonant production is dominant, SM measurements offer complementary sensitivity to direct searches [862]. Some experimental analyses by ATLAS have adopted the approach from (5.3), and have derived limits in the mass-coupling plane [883, 906, 908, 911]. CMS has followed a different approach, where the fact is exploited that for analyses mostly sensitive to a certain VLQ decay channel with small efficiencies for other decay channels, the excluded cross section becomes a function of the total VLQ width and not of the individual choices of the couplings [849]. The reason is that a different choice of couplings, resulting in the same decay width but in a different branching fraction, results only in a change in normalisation of the signal, and hence the cross section times branching fraction is insensitive to this change. All the information from the experimental analysis can therefore be presented in the plane of m_{VLQ} versus rela-

[4]Note that the bounds shown in Fig. 5.16 have been obtained by a projection of 8 TeV analyses in 2014, before data at 13 TeV have been available. Remarkably, the bounds reflect the current best-limits rather well.

tive width, as done for example in [910, 913, 914]. The equivalence of these two approaches can been shown [920], such that limits presented using one of these approaches can be translated into the other. This also enables a future combination of the variety of experimental results on single VLQ searches.

5.3.3 Production Through Resonance Decays

In ultraviolet-complete physical models, i.e. renormalisable theories with gauge bosons resulting from symmetry requirements, VLQs are not the only new heavy resonances in the particle content of the theory. The underlying symmetry groups require the presence of spin-1 resonances, as for example in minimal composite Higgs models [921, 922], where spin-1/2 VLQs appear naturally together with electrically neutral and charged spin-1 resonances ρ^0 and $\rho^{\pm}$ [844, 845, 923]. In models with extra dimensions [699], the lightest Kaluza-Klein excitation of the gluon, G^*, can couple to the lightest fermionic resonances, which are VLQs [632, 924, 925]. In general, a Higgs boson with a mass of 125 GeV requires fermions with masses $O(1\,\text{TeV})$ [926, 927], which are usually lighter than the spin-1 resonances. In such models, decays of the heavy resonances, generically called Z' and W', into VLQs are allowed. If the mass of these resonances is larger than $2m_{\text{VLQ}}$, decays into pairs of VLQs are possible with branching fractions of 60% and higher [925]. This usually results also in a large width of the spin-1 resonance, such that limits from the non-resonant pair production of VLQs can be reinterpreted by rescaling the production cross section [928]. If the mass of these resonances is smaller than $2m_{\text{VLQ}}$, but larger than $m_{\text{VLQ}} + m_q$, then mixed decays into a SM quark with mass m_q and a VLQ can be dominant [929]. These heavy-light decays include $W' \to T\overline{b}$ and $W' \to t\overline{B}$, complementing the light-light decays $W' \to t\overline{b}$ and heavy-heavy decays $W' \to T\overline{B}$. While the light-light decays are covered by searches for $t\overline{b}$ resonances and VLQ pair production, heavy-light decays need dedicated analyses at the LHC in order to achieve the best sensitivity. For the neutral Z' the situation is analogous, where the heavy-light decays $Z' \to B\overline{b}$ and $Z' \to T\overline{t}$ are not covered by searches for $t\overline{t}$ resonances and VLQ pair production, such that dedicated analyses for the production of single VLQs in resonance decays have to be carried out.

The first search for the production of a single VLQ in the decay of a heavy resonance has been performed by ATLAS using 19.5 fb^{-1} of 8 TeV data [930]. It targets the process $pp \to Z' \to B\overline{b} \to Hb\overline{b} \to b\overline{b}b\overline{b}$, as suggested in [931]. For large m_{VLQ} the $H \to b\overline{b}$ decay is reconstructed with a large-R jet with $p_{\text{T}} > 300$ GeV and trimmed jet mass in the range 90–140 GeV. The large-R jet is required to be matched to at least one b-tagged small-R jet. Two additional small-R jets have to be present in the event. If no H jet is found, the event is reconstructed using four small-R jets with p_{T} >50 GeV, three of which have to be b-tagged. In this case, the H boson candidate is reconstructed from the two jets with invariant mass nearest to 126 GeV and p_{T} of the dijet system larger than 200 GeV. The multijet background is estimated from control regions, obtained from events with exactly two b-tagged jets and events where the H boson mass is reconstructed outside the region 90–140 GeV. Five signal regions for the resolved and merged categories are defined, depending on the reconstructed m_{VLQ} and $m_{Z'}$. The observed numbers of events in the ten signal regions are used in the statistical evaluation, performed as a counting experiment. The analysis places upper limits on the product of cross section $\sigma(pp \rightarrow B\bar{b})$ and branching fractions $\mathcal{B}(B \rightarrow Hb)\mathcal{B}(H \rightarrow b\bar{b})$ for $m_{\mathrm{VLQ}} < m_{Z'}/2$ and $m_{\mathrm{VLQ}} = m_{Z'} - 250\,\mathrm{GeV}$ in a plane of m_{VLQ} versus $m_{Z'}$.

The case of a large coupling between the Z' and up-type quarks can be investigated in searches for $pp \rightarrow Z' \rightarrow T\bar{T}$. A promising decay channel is $WbWb$ [904], which differs from the kinematics of the $Z' \rightarrow t\bar{t} \rightarrow WbWb$ resonance search in two important aspects. The large mass of the T results in very different boosts of the two Wb systems. The top quark from the Z' decay can receive a large boost if the mass difference $m_{Z'} - m_{\mathrm{VLQ}}$ is large, while the boost of the T will be moderate at most for $m_{Z'}$ in the range 1.5–4 TeV and m_{VLQ} between 0.7 and 3 TeV. When considering the constraint $m_{Z'} < 2m_{\mathrm{VLQ}}$, Lorentz factors not larger than $\gamma = 1.5$ are realised for the T, such that the Wb system from its decay can not be reconstructed in a single jet. However, the W and b will be approximately back-to-back, with large p_{T} in the laboratory rest frame. The second aspect is that the Wb system from the T decay will have a mass close to $m_{\mathrm{VLQ}} \gg m_t$, such that the usual selection employed in $t\bar{t}$ resonance searches will result in a rejection of these events. Both aspects, the different boosts and different masses of the two Wb systems, result in an insensitivity of existing searches to this signal, despite the same final state. A dedicated search has been carried out by CMS in the all-hadronic final state using 13 TeV data with 2.6 fb^{-1} [932]. The analysis selects events with a three-jet topology, with one large-R t-tagged, one large-R W-tagged and one small-R b-tagged jet. The t and W tagging relies on the soft drop mass, and τ_{32} and τ_{21}, respectively. It is checked that the b-tagged small-R jet does not overlap with the two large-R jets. Two signal regions are defined, depending on the presence of a b-tagged subjet in the identified t jet. Both signal regions have approximately the same signal efficiency, with different background efficiencies and compositions. The subjet b tag reduces the multijet background by a factor of about four, such that the corresponding signal region has better sensitivity than the one without a subjet b tag. However, the latter still contributes to the overall sensitivity of this search and validates the multijet background estimation, which is obtained from sideband regions with vetoes on b-tagged jets and subjets. The uniquely identified decay particles of the signal decay chain allow for a reconstruction of $m_{Z'}$ and m_{VLQ}, where both masses could be determined in case of a potential signal in the data. The distributions in $m_{Z'}$ and m_{VLQ} are shown in Fig. 5.17 for the signal region with a subjet b-tag. The relative mass resolution for signal events is about 15%, such that pronounced peaks on the falling background would be visible. Since both distributions are obtained from the same events in data, only the distribution in $m_{Z'}$ is used to extract upper cross section limits on a potential signal.

While the all-hadronic search achieves high sensitivity for $T \rightarrow Wb$ decays, the channels $T \rightarrow Ht$ and $T \rightarrow Zt$ have been targeted by a dedicated search optimised for $pp \rightarrow Z' \rightarrow T\bar{T} \rightarrow Ht\bar{t}$ and $Zt\bar{t}$. The search has been carried out by CMS in the

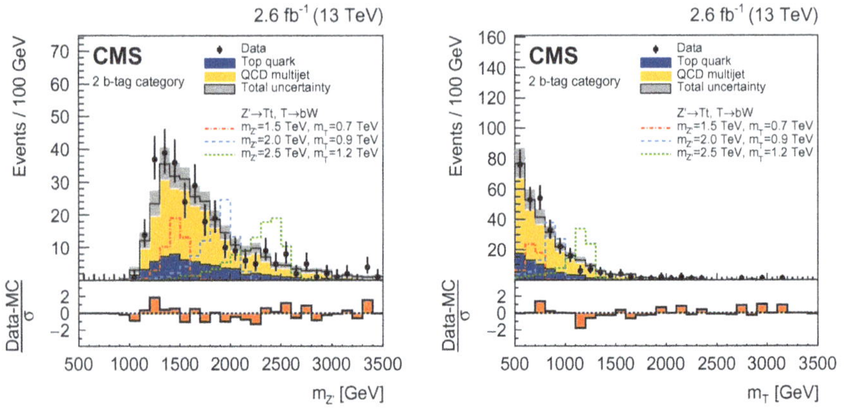

Fig. 5.17 Reconstructed $m_{Z'}$ (left) and m_{VLQ} (right) obtained in a search for $pp \to Z' \to T\bar{t}$ in the all-hadronic final state. The Z' is reconstructed using a t-, a W- and a b-tagged jet, while the T is reconstructed using the latter two jets. Taken from [932]

ℓ+jets final state using 35.9 fb^{-1} of 13 TeV data [933]. The presence of t, H, Z and W in the final state makes this search special in terms of single VLQ searches, where usually only two of these particles are produced in a given channel. The analysis considers events with one high-p_{T} lepton and a V- or H-tagged jet. In addition, events in the signal region are categorised depending on the presence of a t-tagged jet. The substructure taggers rely on the soft drop mass of large-R jets, where the mass regions for V, H and t tagging are 60–115, 100–150 and 150–220 GeV, respectively. The V-tagged jets have to fulfil $\tau_{21} < 0.5$, t-tagged jets $\tau_{32} < 0.57$, and H-tagged jets have to have either one (H_{1b}) or two (H_{2b}) subjet b tags. The overlap between the V and H taggers is resolved by giving priority to the H tagger for jets which fulfil both criteria, which results in an overall better sensitivity of this search. One t decay is reconstructed using the lepton, $\mathbf{p}_{\mathrm{T}}^{\mathrm{miss}}$ and an additional jet. The possibility to reconstruct the other t decay with a t jet depends on the boost of the two t quarks in the event, and thus on m_{VLQ} and the difference $m_{Z'} - m_{\mathrm{VLQ}}$. Events without a t-tagged jet are reconstructed using a combination of small-R jets, not overlapping with the V- and H-tagged large-R jet. All possible assignments of jets to the leptonic and hadronic t decay cascades are considered, and the hypothesis with the smallest difference of reconstructed and expected m_t is chosen. In the $Ht\bar{t}$ and $Zt\bar{t}$ channels there is an ambiguity which top quark is emitted by the Z' decay, such that m_{VLQ} can not be reconstructed. The reconstruction of $m_{Z'}$ is achieved by summing the four-momenta of the chosen $t\bar{t}$ system and the tagged V or H boson. Six signal regions are defined for each lepton flavour, categorised by a V-, H_{1b}- or H_{2b}-tagged jet and the presence or absence of a t-tagged jet, resulting in a total of 12 signal regions. The background is estimated from simulation, necessitating the measurement of efficiencies and misidentification rates of the three substructure taggers used. These measurements are performed in samples enriched in $t\bar{t}$ and multijet events.

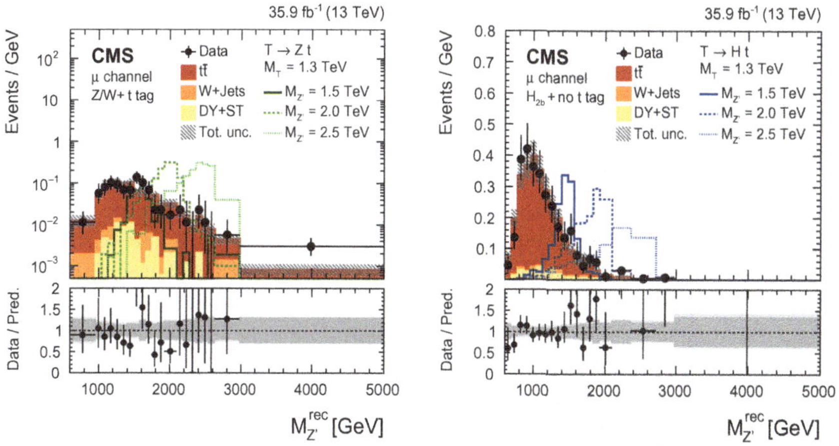

Fig. 5.18 Reconstructed $m_{Z'}$ obtained in a search for $pp \to Z' \to T\bar{t}$ in the ℓ+jets final state, in events with a V- and a t- tagged jet (left) and in events with an H-tagged jet (right). Taken from [933]

Differences in efficiencies between data and simulation are used to derive data-to-simulation correction factors, which are generally found to be compatible with unity within the uncertainties. In addition to these measurements, control regions enriched with the two main backgrounds, $t\bar{t}$ and W+jets, are used to validate the simulation and constrain systematic uncertainties in the modelling of these backgrounds. Two reconstructed $m_{Z'}$ distributions in the μ+jets channel are presented in Fig. 5.18, where the signals have been obtained for $m_{\rm VLQ} = 1.3$ TeV. The $Zt\bar{t}$ channel with a V tag and a t tag is shown, as well as the $Ht\bar{t}$ channel with an H_{2b} tag without a t tag. In the $Zt\bar{t}$ channel, the signal efficiency for $m_{Z'} = 1.5$ TeV is smaller than for signals with higher Z' masses because of the small mass difference $m_{Z'} - m_{\rm VLQ}$. This results in a t emitted from the Z' decay nearly at rest, such that only the t and Z from the $T \to Zt$ decay receive a large boost. Compared to signals with $m_{Z'} = 2$ and 2.5 TeV, there is only one boosted t instead of two, thus the selection efficiency is reduced by a factor of two in this category. In categories without a t-tagged jet, the efficiency is comparable for $m_{Z'}$ between 1.5 and 2 TeV. The efficiency for $m_{Z'} = 2.5$ TeV is smaller, because events with a t-tagged jet are more frequent and are reconstructed in the corresponding category. This search achieves the best sensitivity to production of $T \to Ht$ in a resonance decay and similar sensitivity to $T \to Zt$ as a non-resonant single VLQ search by CMS in the dilepton channel [910], which can be interpreted in this model as well. Upper limits on the product of cross section $pp \to Z' \to T\bar{t}$ and branching fraction $\mathcal{B}(T \to Ht, Zt, Wb)$ are derived. The simultaneous sensitivity to $T \to Ht$ and $T \to Zt$ results in the best constraints to date on models with a heavy gluon and on composite Higgs models, predicting $Z' \to T\bar{t}$ decays.

A search for the heavy-light decay of a W' has been performed by CMS in the all-hadronic final state, using 35.9 fb^{-1} of 13 TeV data [934]. The search has been optimised for $W' \to Bt$ and $W' \to Tb$, which both result in the tHb final state for the decays $B \to Hb$ and $T \to Ht$ [935]. The analysis targets high $m_{W'}$ and m_{VLQ}, such that the H and t are produced with large boost and can be reconstructed using large-R jets with $p_T > 300$ and 400 GeV, respectively. Even for the smallest mass differences considered in this search, $m_{W'} - m_{VLQ} = 200$ GeV, the b quark from the W' decay receives large enough momentum to be reconstructed with a b-tagged small-R jet with $p_T > 200$ GeV. The situation is different for the $W' \to Bt$ decay, where small mass differences lead to a t quark produced nearly at rest, and therefore not reconstructible with a single large-R jet. In this regime, the analysis loses sensitivity because the two decays $W' \to Tb$ and $W' \to Bt$ are assumed to happen with the same frequency. The H and t tagging algorithms select jets with a soft drop mass in the range 105–135 and 105–210 GeV, respectively. In addition, H jets have to pass a selection based on the discriminator from the double-b tagger and t jets have to have $\tau_{32} < 0.8$ and a subjet b tag. The signal region is defined by events with a H-, t- and b-tagged jet. The distribution in the reconstructed W' is used to search for a signal, obtained from the four-vector sum of the three identified jets in the event. Sideband and validation regions are used to estimate the multijet background. These are obtained by inverting the b-tagging, τ_{32} or jet mass requirements. A transfer function of the H tagger is derived as a function of p_T and η, describing the ratio of probabilities to pass the H- and inverted H-tagging selections. This transfer function is obtained from events with an inverted t tag. It is used to derive an event weight to construct a template for the reconstructed W' distribution in a control region with inverted H tag, which is used to predict the background in the signal region. This approach is validated in simulation and in a dedicated validation region in data. In both tests good agreement between the predicted and expected background distributions is observed. This method has the distinct advantage that it provides the background estimation for any distribution, such that the background model can be checked thoroughly. In the signal region, the multijet background constitutes about 70% of the total background, the remaining part originating from $t\bar{t}$ production. No significant deviation from the background prediction is observed in the data and cross section upper limits on W' production in the tHb decay mode are reported as a function of $m_{W'}$, for several m_{VLQ} hypotheses. This is the only search to date targeting the heavy-light decay of a W' boson.

No dedicated search for the decay modes $B \to Wt$, $T \to Wb$, $B \to Zb$ or $T \to Zt$ exists, which would result in $t\bar{t}W$, $b\bar{b}Z$ and tZb final states, respectively. While SM measurements in these final states have been carried out [936–943], these do not extend into the high energy tails of the production cross sections, which are relevant for BSM models. The best sensitivity to BSM effects in these final states is obtained by dedicated searches for resonant VLQ production, which will hopefully be carried out in the future.

5.4 Excited Third Generation Quarks

In models of compositeness, SM particles are built from strongly bound constituents, which obey more fundamental equations of motion. The strong force binding the constituents would induce contact interactions at scales lower than the compositeness scale Λ [944, 945]. For quarks, contact interactions change the distribution of the scattering angle in $qq \to qq$ scattering [946], which can be probed in pp collisions [947–956]. Another consequence of the internal structure of quarks is the existence of excited states, only possible for composite objects. While all quarks might be composite particles, the large masses of third generation quarks suggests that excited quarks couple primarily to b and t, which can be realised in Randall-Sundrum [957, 958] or composite Higgs models [898, 959]. Especially the large mass of the top quark has initiated studies of consequences from a possible top quark compositeness [960–963]. In some models of warped extra dimensions, the top and bottom quark excitations, t^* and b^*, can be spin-3/2 states, with dominant decays $t^* \to t + g$ [964–966] and $b^* \to b + g$ [967]. The phenomenology of excited third generation quarks is also related to VLQs, where the absence of mixing with SM quarks can lead to vanishing branching fractions into final states involving W, Z and H bosons. Instead, in models with an additional scalar S, loop-induced contributions can lead to dominant decays $T \to tg$, $T \to t\gamma$ and $T \to tS$ [968]. In these models, the dominant channel is the decay $T \to tg$ with a branching fraction of about 97% in a large part of the parameter space [969].

The dominant production and decay processes of excited quarks in pp collisions is through $qg \to q^* \to qg$, if the coupling to light quarks $q = \{u, d, s, c\}$ is sufficiently large. These would lead to resonant structures in the dijet mass spectrum at high p_{T} [970]. Recent searches by ATLAS and CMS with the total available 13 TeV data, collected in the years 2015–2018 and corresponding to an integrated luminosity of about 140 fb^{-1}, exclude excited quarks with masses below 6.7 TeV [971, 972]. The mass limits for a b^*, obtained for the process $bg \to b^* \to bg$, are reduced to 3.2 TeV because of b tagging efficiencies and the smaller production cross section due to the smallness of the b PDF. If the b^* has weak couplings in addition to the chromomagnetic bgb^* coupling, the decay $b^* \to Wt$ is possible with a branching fraction up to 40% [973]. The weak coupling is introduced in models where the heavy particle stabilises the Higgs boson mass at the electroweak scale [974, 975], or in so-called beautiful mirror models [976, 977], resolving the observed discrepancy of the forward-backward asymmetry in $Z \to b\bar{b}$ decays [107]. The most general Lagrange density describing the interaction of a heavy bottom quark with gluons is [970, 973, 978]

$$\mathcal{L} = \frac{g_s}{2\Lambda} G_{\mu\nu} \bar{b} \sigma^{\mu\nu} \left(\kappa_L P_L + \kappa_R P_R\right) b^* + \text{h.c.}, \tag{5.4}$$

where g_s is the strong coupling constant, Λ is the scale of compositeness which is usually set to the mass of the b^*, $G_{\mu\nu}$ is the gluon field strength tensor and $\sigma^{\mu\nu}$ are the relativistic spin matrices. The left- and right-handed coupling strengths are determined by the parameters κ_L and κ_R, with the corresponding projection operators P_L and P_R. Similarly to (5.3), the Lagrangian describing the weak interaction of the b^* is given by [978]

$$\mathcal{L} = \frac{g_W}{\sqrt{2}} W^+_\mu \bar{t} \gamma^\mu \left(g_L P_L + g_R P_R\right) b^* + \text{h.c.}, \tag{5.5}$$

where g_L and g_R determine the left- and right-handed b^* couplings to Wt. When both interactions are considered, $g_{L/R} \neq 0$ and $\kappa_{L/R} \neq 0$, the b^* can be produced through the strong interaction of (5.4) or the weak interaction of (5.5), where the latter results in an identical signature as single VLQ production. However, production through the weak interaction is largely suppressed because of the necessity of a top quark in the initial state, which requires a $g \to t\bar{t}$ splitting. The resulting production of a b^* in association with a t and a forward jet has a cross section more than an order of magnitude smaller than the cross section of $bg \to b^*$, such that a singly produced b^* is rarely produced with a forward jet. Besides this, the signature of $b^* \to Wt$ is similar to single B production with $B \to Wt$. Single VLQ searches targeting the Wt channel can be reinterpreted in b^* models, but the requirement of a forward jet, which is often used to define the signal region in single VLQ searches, reduces the sensitivity to a singly produced b^*. Vice versa, b^* searches not requiring a forward jet do not achieve the optimal sensitivity for single B production.

The first analysis searching for $b^* \to Wt$ has been performed by ATLAS using 4.7 fb^{-1} of 7 TeV data [979]. The analysis considers fully resolved b^* decays and selects events with either two leptons and exactly one small-R jet, or with one lepton and three small-R jets, one of which has to be b-tagged. The signal signature is similar to $t\bar{t}$ production, where one b is not reconstructed or produced outside the detector acceptance, such that this constitutes the largest background. The second largest background is the irreducible SM single top production in the Wt channel in the dilepton selection and W+jets production in the ℓ+jets selection. The b^* mass can be reconstructed in the ℓ+jets final state and serves as sensitive observable to discriminate between signal and background. In the dilepton final state S_T is used, which is defined as the scalar p_{T}-sum of the leptons, $p_{\mathrm{T}}^{\mathrm{miss}}$ and the small-R jet. For vector-like couplings, i.e. identical left- and right-handed b^* couplings $\kappa_L = \kappa_R = g_L = g_R = 1$, b^* masses below 1030 GeV can be excluded at 95% CL with this search. This search has been improved for the analysis of 8 TeV data, where also jet substructure information has been introduced in the ℓ+jets channel [905]. Large-R jets with a trimmed mass larger than 50 GeV are used to reconstruct either the W or t. No further requirements are imposed on large-R jets in order to be as inclusive as possible. The signal region is divided into a b^* and a B category, depending on the presence of a forward jet. Because of this categorisation, the analysis achieves sensitivity to both heavy bottom partners, produced through the strong and weak force. The main backgrounds in this search are estimated from simulation. The normalisation and modelling uncertainties are constrained with the help of control regions, which are obtained by requiring no b-tagged small-R jets for the W+jets control region and at least two b-tagged small-R jets for the $t\bar{t}$ control region. The signal and background contributions are fit to the data in the signal and control regions simultaneously. The combination of the ℓ+jets and dilepton final states results in a lower mass limit of 1.5 TeV for vector-like b^* couplings.

An analysis by CMS using 8 TeV data combines the all-hadronic, ℓ+jets and dilepton channels in the search for $b^* \rightarrow Wt$ [980]. While the ℓ+jets and dilepton channels target resolved topologies with exactly three and exactly one small-R jet, respectively, the all-hadronic selection is optimised for fully merged final states. This selection is based on two large-R jets with $p_{\mathrm{T}} > 425$ GeV, where one is W and the other one is t-tagged. The W tagging uses the pruned jet mass and τ_{21}; t tagging is achieved with the CMSTT, τ_{32} and a subjet b tag. The dominant backgrounds in the leptonic final states are estimated from simulation and validated in control regions. In the all-hadronic final state, the dominant multijet background is obtained by weighting events prior to the t tagging selection with the t-tagging misidentification rate. This misidentification rate is calculated as a function of jet p_{T} and η in a sample dominated by multijet events, obtained by inverting the W selection criteria. Because the t tagging algorithm has a misidentification rate of about 0.1% [243], the background obtained from the weighted pre-t-tagged sample can be considered statistically independent from the sample after the t tagging selection has been applied. The sensitivity to signals for b^* masses below 1 TeV is driven by the ℓ+jets and dilepton channels, but for higher masses the all-hadronic analysis provides better sensitivity. Similar to the 8 TeV analysis by ATLAS, lower mass limits of 1.5 TeV for vector-like b^* couplings are obtained.

A very recent analysis by CMS uses the full 13 TeV data with 137 fb^{-1} to search for $b^* \rightarrow Wt$ in the all-hadronic final state, targeting b^* masses >1.2 TeV [981]. The analysis uses two large-R jets with $p_{\mathrm{T}} > 400$ GeV and an angular separation in azimuth greater than $\pi/2$ to ensure a back-to-back topology of the two jets. The signal region is defined through a W- and a t-tagged jet. Both taggers use soft drop jet mass, where t jets in the signal region are required to have $105 < m_{\mathrm{jet}} < 220$ GeV and W jets $65 < m_{\mathrm{jet}} < 105$ GeV. In addition, τ_{32} and subjet b tagging is used for the definition of t jets, and τ_{21} for W jets. The analysis is performed in two dimensions, where the distribution in the plane (m_t, m_{Wt}) is analysed for a potential signal. Note that a region in m_t between 65 and 285 GeV is analysed, such that the signal region with $105 < m_{\mathrm{jet}} < 220$ GeV is enclosed by a low- and a high-mass sideband. In this analysis, m_t is the soft drop mass of the t jet and m_{Wt} is the mass of the dijet system. This allows for the use of a novel method to construct the multijet background template, which relies on a parametrisation of the pass-fail ratio as a function of m_t. The number of multijet events passing the t tagging requirement, n_{P}, in a given interval in the (m_t, m_{Wt}) plane is calculated as $n_{\mathrm{P}} = n_{\mathrm{F}} \cdot f(m_t, m_{Wt})$, where n_{F} is the number of events failing the t tagging requirement and $f(m_t, m_{Wt})$ is the two-dimensional pass-fail ratio. This ratio is obtained from data, with an initial estimate obtained from simulation in order to reduce the complexity of the function. It is found that a a surface parametrised by the product of a second-order polynomial in m_t and a first-order polynomial in m_{Wt} is sufficient to describe the data in the sideband regions. The advantage of this method is that it interpolates the pass-fail ratio into the signal region from the enclosing sidebands, such that the analysis can be fully tested and

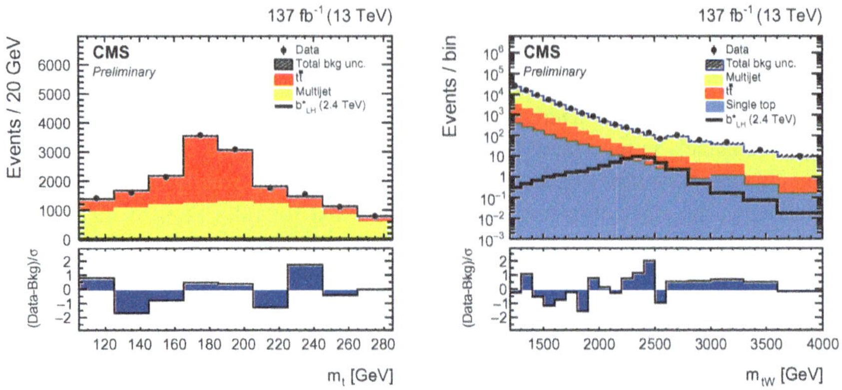

Fig. 5.19 Distribution in the soft drop jet mass of t-tagged jets in the $t\bar{t}$ control region (left), obtained in an all-hadronic b^* search by CMS. Two t-tagged large-R jets are required and the distribution is shown for dijet mass $1300 < m_{tt} < 1800\,\text{GeV}$. The distribution in m_{Wt} (right) is obtained for events with a W- and a t-tagged large-R jet, and is shown for the signal region with $105 < m_t < 220\,\text{GeV}$. Taken from [981]

verified before including the data in the signal region. Once the data in the signal region is examined, the predicted pass-fail ratio can be compared to the observed one to validate the multijet background estimation in the signal region. Besides the multijet background, $t\bar{t}$ is an important background in this search as well. In order to validate the modelling of this background by simulation, a dedicated control region is included. This region is obtained by changing the W tag into a t tag, such that multijet and $t\bar{t}$ constitute about 50% each to the total number of events in this region. The multijet background is estimated analogously to the signal region, where the dijet mass m_{Wt} becomes m_{tt}. In the final statistical evaluation, the signal and background models are fit to the data in the two (m_t, m_{Wt}) and (m_t, m_{tt}) planes. Two example distributions are shown in Fig. 5.19, where the background distributions are shown for their best-fit values. The large data sample allows for a precise validation of the $t\bar{t}$ modelling and an accurate estimation of the multijet background. In absence of a signal, a lower limit of 3 TeV on the b^* mass for vector-like couplings is reported, which improves the previous results by 1.5 TeV.

The production of a t^* is different from b^* because of the absence of top quarks in the initial state. Therefore, the QCD-induced pair production process $pp \to t^*\bar{t}^*$ dominates, with cross sections between 2 to 0.007 pb for a t^* with spin-3/2 and masses between 0.8 and 1.5 TeV. The cross section for spin-1/2 t^* pair production is about an order of magnitude smaller for masses of 0.8 TeV and a factor of about 5 smaller for masses of 1.5 TeV [969]. Because of the larger cross section, searches have focussed on spin-3/2 states so far. If mixing between spin-1/2 and spin-3/2 states is suppressed, the cross sections for $t^*\bar{t}$ and $\bar{t}^*t$ are smaller by about an order of magnitude than $t^*\bar{t}^*$, despite being kinematically favoured [964, 965]. Hence, existing t^* searches consider pair production only, and have been optimised for the $t\bar{t} + gg$ channel. The electroweak decay $t^* \to Wb$ is covered by VLQ searches.

The first search for the resonant production of $t\bar{t}$+2 jets at the LHC has been carried out by CMS using 19.5 fb^{-1} of 8 TeV data [982]. The analysis is carried out in the ℓ+jets final state, and has been optimised for t^* masses below 1 TeV with resolved t decays. Events are selected with one isolated lepton and at least six jets with $p_\text{T} > 30$ GeV, one of which has to be b tagged. The reconstructed t^* distribution is used as sensitive observable. It is obtained by all possible permutations of assignments of the six p_T-leading jets to the expected t^* decay partons. A kinematic fit is performed in order to improve the resolution in the reconstructed t^* mass. The fit has seven kinematic constraints from the W, t and t^* masses, and the conservation of p_T in the collision. The parameter estimation for this complex system is difficult and computationally expensive, such that a sampling of the allowed parameter space is performed before the actual numerical minimisation. However, only in about 11% of the simulated signal events a correct assignment is found. In most of the events with a wrong assignment, one of the actual jets from the t^* decay has failed the jet p_T requirement. However, it is found that lowering the jet p_T threshold increases the background without a gain in sensitivity. Despite this shortcoming, the t^* mass reconstruction results in a peaking distribution at the correct value for signal events, while the expected background is a smoothly falling distribution. The background is estimated by a parametric function with three free parameters, fit to the distribution in the reconstructed b^* mass. Lower mass limits of 800 GeV for a t^* with spin-3/2 are reported. Spin-1/2 particles can only be excluded between 470 and 510 GeV, but their existence at lower masses can be excluded by differential $t\bar{t}$ cross section measurements [627, 983] and $t\bar{t}$ production with additional jet activity [984, 985].

The CMS search for $t^*\bar{t}^*$ in the ℓ+jets final state has been updated using 35.9 fb^{-1} of 13 TeV data [986]. Small adjustments to the analysis have been made, for example two b-tagged jets instead of one are required in order to suppress contributions from non-$t\bar{t}$ backgrounds, and to reduce the complexity of finding the correct jet-parton assignment. In addition, this assignment is now based on an estimator using the sum of squares of differences between measured and expected W, t and t^* masses, without performing a kinematic fit. While this results in a worse t^* mass resolution for correct assignments, the overall sensitivity is nearly unchanged since only in 11% of the events all six jets are correctly assigned. No excess over the parametrised background is observed, resulting in lower mass limits of 1.2 TeV for a t^* with spin-3/2.

In future searches for t+g resonances, a significant improvement over previous results can only be achieved by the use of jet substructure information for the t identification. This will enable to probe higher masses through an increase in signal efficiency at high top quark p_T, and it will reduce the complexity of the t^* mass reconstruction. In addition, q/g discrimination applied on the two gluon jets could help to improve the signal-to-background ratio. With larger datasets it will become interesting to include the $t^* \to t\gamma$ decay, resulting in $t\bar{t}g\gamma$ final states. While this decay only has a branching fraction of about 3%, the presence of a photon can help to reduce the SM backgrounds considerably [969]. Last but not least, the single production channels $t^*\bar{t}$ and $\bar{t}^*t$ should be included [968] and t^* searches in the all-hadronic and dilepton final states should be envisioned.

5.5 Dark Matter and Mono-X

Dark matter has been a very intriguing subject in physics for decades. The history of dark matter can be traced back to the 1930s, but it was not until the 1970s that the abundance of non-luminous matter in the universe was recognised as a scientific problem. It took two more decades until the apparent discrepancy between the visible matter in galaxies and their gravitational mass inferred from rotation curves had been accepted to arise from an unknown particle species.[5] Further evidence for the existence of dark matter is provided by measurements of weak gravitational lensing [989]. While the nature of dark matter is still unknown and a subject of theoretical speculations, it is scientifically established that the unaccounted mass in our universe consists of one or more unknown particles. There are a number of viable candidates for dark matter. Axions have been postulated to solve the strong CP problem [990, 991] and have become a popular candidate for dark matter with masses of 10^{-6}–10^{-4} eV [992]. While SM neutrinos are not abundant enough to be viable dark matter particles [993], sterile neutrinos [994] are suggested to interact only gravitationally, apart from a small mixing with the three generations of SM neutrinos [995]. The most studied class of dark matter candidates are weakly interacting massive particles, WIMPs. These arise naturally from theories extending the SM, such as supersymmetry, which provides several dark matter candidates with masses from about 50 GeV to a few TeV. An excellent candidate is the neutralino, which has a self-coupling and interaction strength with ordinary matter just right that it can account for the observed astrophysical dark matter [996]. WIMPs can also appear in Little Higgs models [997] and in extra-dimension models, where the dark matter candidate is a stable Kaluza-Klein particle [998].

Several possibilities exist to search for traces of dark matter particles. The first method relies on the annihilation of dark matter particles, shown schematically in Fig. 5.20a. Because the annihilation rate of dark matter particles in the early universe is needed to be sufficiently large to ensure a stable equilibrium, the annihilation of relic dark matter particles to SM particles can lead to detectable signals from astrophysical sources. The signals include high-energy photons, neutrinos, positrons and anti-protons. Dedicated experiments have been designed for these hints of dark matter annihilation. Examples are the telescopes HESS [1000], MAGIC [1001] and VERITAS [1002] searching for high-energy cosmic rays, AMANDA [1003], Super-Kamiokande [1004] and IceCube [1005, 1006] have been designed to detect neutrinos, and the space-based experiments AMS [1007], PAMELA [1008, 1009], AMS-02 [1010] study the cosmic positron and anti-proton spectra. The second method are direct detection experiments, designed to detect the scattering of dark matter and SM particles, Fig. 5.20b. These experiments rely on the existence of relic dark matter in our solar system, with a density large enough to result in detectable signals. The signature is a nuclear recoil from the scattering, where different target materials and

[5] The interested reader is referred to [987, 988] for an exhaustive historical perspective on the scientific struggle, including the observational discoveries and the theoretical arguments, that preceded dark matter becoming part of the standard cosmological model.

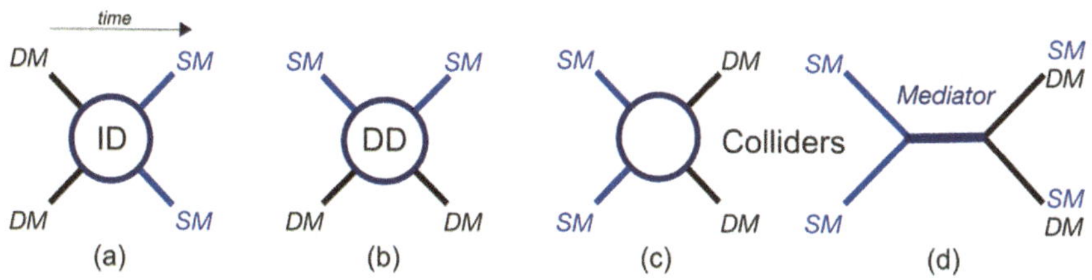

Fig. 5.20 Schematic diagrams showing dark matter (DM) interactions and their corresponding experimental detection techniques. Dark matter annihilation to SM particles is sought by indirect detection (ID) experiments (**a**). The scattering of dark matter and SM particles is targeted by direct detection (DD) experiments (**b**). At colliders, searches are designed to measure the production of dark matter particles from the interaction with SM particles (**c**), which can also occur through a mediator particle (**d**). Image by Doglioni and Boveia (ATLAS Collaboration), taken from [999]

detector technologies have been considered [1011]. Examples of recent experiments are XENON1T [1012], LUX [1013], PANDA-X [1014] and SuperCDMS [1015, 1016]. At the LHC, dark matter particles can be produced through their interaction with SM particles, either by the annihilation of a particle with its anti-particle or by radiation off a particle produced in the collision, Fig. 5.20c. The interaction of dark matter with SM particles can be described by an effective field theory (EFT) if the dark matter candidate is the only particle kinematically accessible at the LHC. In this case, the higher-dimensional operators allow to describe the interactions in a universal way [1017–1019]. This approach has been very successful in LHC searches of 7 and 8 TeV data, because the obtained bounds on the new physics scale Λ could be readily compared with results from direct and indirect detection experiments [1020]. However, the high energy of the LHC raises the question of the validity of EFTs, where the momentum expansion might break down [1021–1024]. A solution is offered by simplified dark matter models, which include a mediator particle, responsible for the interaction between dark matter and SM particles, Fig. 5.20d. The advantage is that diagrams involving s- and t-channel exchange of this mediator can be reliably included and thus the full kinematics of dark matter production are described [1025]. The price for this more complete description are additional parameters, describing the mass and couplings of the dark matter mediator. An extensive overview of simplified models and their parameter space is given by the report of the LHC Dark Matter Working Group [1026]. More complete models offer an even richer phenomenology and can provide guidance for unexplored signatures at the LHC [1027].

While many of the LHC searches have been designed with a WIMP dark matter particle in mind [1028], the results are usually applicable to a broad class of dark matter models. In what follows, the dark matter candidates will generically be referred to as χ with mass m_χ. In simplified models, the dark matter mediator ϕ has a mass m_ϕ with couplings g_χ and g_{SM} to dark matter and SM particles, respectively. Note that the mediator can either be a scalar or pseudo-scalar, usually denoted by ϕ or a, but this distinction is not relevant for the results presented here. The presence of one or more dark matter particles in the final state results in an imbalance in transverse momentum due to the undetected dark matter particles. The $\mathbf{p}_{\mathrm{T}}^{\mathrm{miss}}$ induced by the $\chi\bar{\chi}$ system is balanced by a SM particle, which seemingly has been produced singly. Hence the common notation Mono-X searches, where X can be any object identified in the detector, e.g. a jet, γ, W, Z, H, ...

This results in a special importance of $p_{\mathrm{T}}^{\mathrm{miss}}$ in dark matter searches, since any signal is expected to have a considerably different $p_{\mathrm{T}}^{\mathrm{miss}}$ spectrum than irreducible SM backgrounds, for example from $Z \to \nu\nu$ production. These backgrounds are difficult to model to a sufficiently high accuracy, such that these are usually validated and constrained by data in dedicated control regions. Care has to be taken that the control regions of a search in a given channel do not overlap with signal regions of other channels. If possible, control regions should be chosen with small $p_{\mathrm{T}}^{\mathrm{miss}}$, for example, a $Z \to \ell\ell$+jets selection can be used to estimate the background from $Z \to \nu\nu$+jets in mono-jet searches. More difficult are backgrounds where events are only partially reconstructed, for example $W \to \ell\nu$ with the lepton failing the selection requirements or produced outside the acceptance. In addition, there are events with spurious $p_{\mathrm{T}}^{\mathrm{miss}}$ from detector noise, beam halos or cosmic muons. These need to be rejected by dedicated algorithms, and their efficiency and false-positive rates have to be studied.

A large number of mono-X searches has been carried out by the ATLAS and CMS Collaborations since the start of the LHC. Not for all signatures explored jet substructure techniques are applicable or useful. However, in some channels these are indispensable tools, providing a large increase in sensitivity or even enabling these analyses. In general, boosted topologies are most relevant for $m_\chi \lesssim 100\,\mathrm{GeV}$, where the recoil of the $\chi\bar{\chi}$ system receives a large momentum. For masses greater than about 100 GeV, boosted final states loose sensitivity and should be complemented by resolved topologies where possible.

The first searches for dark matter have been mono-jet searches using 7 TeV data by ATLAS [1029, 1030] and CMS [1031]. In these searches, a jet from initial state radiation recoils against the $\chi\bar{\chi}$ system, resulting in a $p_{\mathrm{T}}^{\mathrm{miss}}$+jet signature. These searches have sensitivity to a large number of BSM models, where dark matter particles are predicted to couple to quarks and gluons. When considering the radiation of a photon instead of a jet from the initial-state quarks, the reduction of the signal cross section due to the electromagnetic interaction instead of the strong force is compensated by much smaller SM backgrounds from $Z\gamma$ and $W\gamma$ production. Analyses in the mono-γ channel by ATLAS [1032] and CMS [1033] using 7 TeV data achieve sensitivities to couplings and masses comparable to the mono-jet searches.

The four times larger data set and increased $\sqrt{s}$ results in improved constraints from mono-jet and mono-γ searches at 8 TeV [1034–1037]. For m_χ between 1 and 400 GeV, where the validity of the EFT approach is still guaranteed, upper limits on the χ-nucleon cross section could be improved by a factor of about three relative to the 7 TeV results. New channels have been investigated for the first time using 8 TeV data. Searches for mono-W and mono-Z production have been carried out. Naively, the radiation of a W boson from the initial state quark would not result in better sensitivity relative to gluon, quark or photon radiation. However, if there is constructive interference between the diagrams where the boson is radiated from the u or d

quark, the mono-W production can become the dominant process [1038]. Mono-Z production can probe direct couplings between the Z boson and the dark matter particles. In addition, mono-W and Z searches are sensitive to Higgs-mediated dark matter models through VH production with $H \rightarrow \chi\bar{\chi}$. Searches in the $Z(\ell\ell)$+$p_{\mathrm{T}}^{\mathrm{miss}}$ final states have been reported by ATLAS [1039] and CMS [1040]. The largest background for $p_{\mathrm{T}}^{\mathrm{miss}} > 100\,\mathrm{GeV}$ originates from ZZ production, which is an irreducible background in this channel and has to be obtained from simulation. CMS has also performed a search in the single-lepton channel, targeting $W(\ell\nu)$+$p_{\mathrm{T}}^{\mathrm{miss}}$ [1041]. This analysis has been optimised for $W' \rightarrow \ell\nu$, where the transverse mass between the lepton and $\mathbf{p}_{\mathrm{T}}^{\mathrm{miss}}$ is the optimal observable to distinguish signal from background events. In the search for dark matter, where the neutrino is accompanied by two dark matter particles escaping detection, the p_{T} of the W boson or $\chi\bar{\chi}$ system can not be reconstructed, resulting in a reduced sensitivity relative to mono-jet and mono-γ searches. The large hadronic branching fractions of the W and Z bosons are exploited in a mono-V search by ATLAS, where jet substructure is used for the first time in a dark matter search [1042]. Events are selected by the presence of one large-R jet with $p_{\mathrm{T}} > 250\,\mathrm{GeV}$ and a jet mass between 50 and 120 GeV after MDT grooming with $y_{\mathrm{cut}} = 0.16$ (see Sect. 2.4.3). Two signal regions are defined by $p_{\mathrm{T}}^{\mathrm{miss}} > 350\,\mathrm{GeV}$ and $p_{\mathrm{T}}^{\mathrm{miss}} > 500\,\mathrm{GeV}$. The m_{jet} distributions in these regions are analysed to search for a resonant signal over an approximately flat background, as shown in Fig. 5.21 (left). In the case of constructive interference, the mono-W production is enhanced by two orders of magnitude relative to destructive interference, resulting in much stronger bounds. For other cases, the sensitivity is comparable to the mono-jet and mono-γ searches, and about a factor of three better than for the leptonic decay channels. A CMS search with 8 TeV data [1043] could improve upon this result by taking into account fully merged and resolved V decays, as well as mono-jet events which do not fulfil the criteria of the V selection. The boosted selection requires one large-R jet with $p_{\mathrm{T}} > 200\,\mathrm{GeV}$, pruned jet mass in the range 60–110 GeV and $\tau_{21} < 0.5$. In the resolved selection, the dijet system of small-R jets has to have a mass between 60 and 110 GeV. In order to suppress backgrounds, a multi-variate discriminator is built from the jet pull angle, mass drop and the q/g discriminator. Events not falling into the boosted or resolved signal category, but with a small-R jet balanced with $p_{\mathrm{T}}^{\mathrm{miss}}$, are sorted into the mono-jet category. A total of nine control regions, three for each signal category, are used to determine the main backgrounds from $Z(\nu\nu)$+jets and $W(\ell\nu)$+jets. The control regions are obtained by selecting events with either a photon, or one or two muons. In the control regions, no $p_{\mathrm{T}}^{\mathrm{miss}}$ selection is applied. Instead, the p_{T} of the photon, the single muon or the dimuon system is removed and $p_{\mathrm{T}}^{\mathrm{miss}}$ is recalculated. This proxy for $p_{\mathrm{T}}^{\mathrm{miss}}$ is used to determine the $p_{\mathrm{T}}^{\mathrm{miss}}$ spectrum in the signal regions. The resulting $p_{\mathrm{T}}^{\mathrm{miss}}$ distribution in the fully merged signal region with a V-tagged jet is shown in Fig. 5.21 (right). By taking into account three signal regions and using the full information of the $p_{\mathrm{T}}^{\mathrm{miss}}$ spectra, this analysis improves over the previous CMS mono-jet search by 80% in sensitivity.

Searches for mono-H probe the direct H-χ interaction, because it is unlikely that the H is radiated from an initial state quark [1044, 1045]. Searches for H decays into invisible particles can probe the H-χ interaction as well if $m_H > 2m_\chi$. The

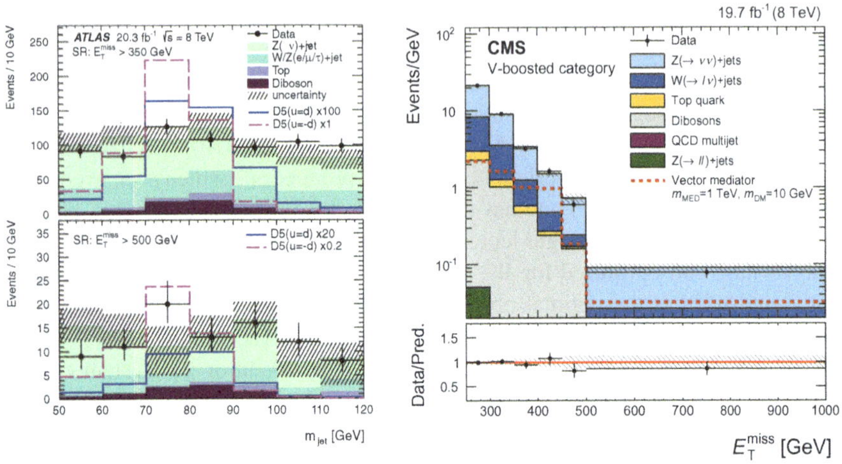

Fig. 5.21 Jet mass distribution for MDT groomed large-R jets in a mono-V search by ATLAS for two different intervals in $p_{\mathrm{T}}^{\mathrm{miss}}$ (left). Distribution in $p_{\mathrm{T}}^{\mathrm{miss}}$ in a CMS mono-V search, for events with V-tagged large-R jets (right). Taken from [1042] (left) and from [1043] (right)

current best upper limit at 95% CL on the branching fraction for invisible decays of the H boson is 0.13 [1046], which constitutes a very sensitive probe of the H-χ coupling. However, if the dark matter particle is too heavy for an on-shell decay of the H with $m_\chi > m_H/2$, mono-H searches are needed to probe this interaction. The first search of this kind has been carried out by ATLAS in the $H \to \gamma\gamma$ channel, using 8 TeV data. The requirements of two photons and $p_{\mathrm{T}}^{\mathrm{miss}}$ >90 GeV result in only 18 events in the signal region, with a best fit of 14.2 ± 4.0 from background processes, 1.1 ± 0.1 from SM H production and 2.7 ± 2.2 from BSM H production. The small numbers of events are due to the small value of the $H \to \gamma\gamma$ branching fraction. An opportunity to improve on this result is offered by mono-H searches in the $H \to b\bar{b}$ channel. The first search of this kind has been carried out by ATLAS on 8 TeV data [1047], where the $H \to b\bar{b}$ decays is reconstructed in resolved and boosted categories. This analysis differs from diboson resonance searches in the $VH \to \nu\nu b\bar{b}$ channel [723] by considering non-resonant production and measuring the $p_{\mathrm{T}}^{\mathrm{miss}}$ distribution as sensitive discriminant. In the resolved category, two b-tagged small-R jets with angular separation $\Delta R < 1.5$ and dijet mass in the range 90–150 GeV are required. The boosted category is defined by one large-R jet with $p_{\mathrm{T}} >$ 350 GeV, trimmed jet mass $90 < m_{\mathrm{jet}} < 150$ GeV and two matched b-tagged track-jets. The main background in this analysis originates from $Z(\nu\nu)$+jets production and is estimated similarly to the the mono-V search by CMS [1043], using a $Z \to \mu\mu$ selection for $p_{\mathrm{T}}^{\mathrm{miss}} < 200$ GeV and γ+jets for $p_{\mathrm{T}}^{\mathrm{miss}} > 200$ GeV. The boosted category is used to probe dark matter EFT operators, and the limits achieved are a few times stronger than the mono-H search in the $H \to \gamma\gamma$ channel. The resolved category is used to study a two-Higgs-doublet model with an additional Z', which predicts the decay cascade $Z' \to HA$ with $A \to \chi\bar{\chi}$ [679].

If dark matter production proceeds through the exchange of a scalar mediator, couplings to light quarks are suppressed and the sensitivity of mono-jet searches is limited. For this class of models, dark matter produced in association with third generation quarks is a promising search channel, including mono-b, $b\bar{b}$+p_T^{miss} and $t\bar{t}$+p_T^{miss} signatures [1048]. In addition, models with baryon number violating or flavour changing neutral current interactions predict the existence of mono-t signatures [1049], which can be probed at the LHC [1050, 1051]. ATLAS has reported a search with 8 TeV data in final states with large p_T^{miss} and one or two b-tagged small-R jets, optimised for $b+\chi\bar{\chi}$ and $b\bar{b}$+$\chi\bar{\chi}$ production [1052]. The same analysis considers final states with large p_T^{miss} and more than five jets, two of which have to be b tagged, or one lepton and more than four jets, where one has to be b tagged, to achieve sensitivity to $t\bar{t}$+$\chi\bar{\chi}$ production. Mono-t signatures are studied using 8 TeV data by ATLAS in the p_T^{miss}+$\ell\nu b$ channel [1053] and by CMS in the p_T^{miss}+$q\bar{q}'b$ channel, where the t decay is reconstructed with three small-R jets [1054]. These analyses are most sensitive for low values of m_χ, for which lower mass limits are derived of 1 TeV for a scalar coupling to b quarks and between 330 and 650 GeV on a scalar or vector mediator coupling to t quarks, respectively.

The use of jet substructure has proliferated in analyses of 13 TeV data and helped to improve the sensitivity of dark matter searches. Searches without jet substructure could also extend the coverage of the parameter space thanks to improved analysis and reconstruction methods compared to 8 TeV analyses, as well as the larger data set and increased $\sqrt{s}$. Mono-γ searches by ATLAS [779, 1055, 1056] and CMS [1057] using up to 139 fb^{-1} of 13 TeV data, report lower limits on the EFT scale Λ of 790 GeV at 95% CL. Dark matter particles with m_χ below 255–520 GeV and m_ϕ below 810–1310 GeV are excluded at 95% CL, where the ranges refer to different couplings. Mono-jet analyses by ATLAS, also using the full 13 TeV data recorded until 2018, exclude m_ϕ below 2.1 TeV for $m_\chi = 1$ GeV [1058–1060]. Mono-Z analysis in the p_T^{miss} +$\ell\ell$ final state have also been carried out by ATLAS [1061] and CMS [1062–1064] using up to 137 fb^{-1} of data. The results are interpreted in a number of models, including different dark matter mediators, two-Higgs-doublet models and models with large extra dimensions. These are complemented by mono-V searches in the all-hadronic final state, enabled by jet substructure methods. Two CMS analyses based on data taken in the year 2016 use pruned large-R jets with $p_T > 250$ GeV, $65 < m_{jet} < 105$ GeV and $\tau_{21} < 0.6$ to define the V-tag signal region [1065, 1066]. Events not passing this selection, but with a small-R jet with $p_T > 100$ GeV and $p_T^{miss} > 250$ GeV, are sorted into a mono-jet category. This has the advantage of a straight-forward combination of the mono-V and mono-jet channels. In ATLAS, resolved and fully merged V categories are considered [1067, 1068]. In the merged category, large-R jets have to fulfil p_T-dependent requirements on the trimmed jet mass and D_2. In addition, b-tagged track-jets are used for a further categorisation in order to gain sensitivity to $Z \to b\bar{b}$ decays. This analysis also considers a Z' produced in association with $\chi\bar{\chi}$ instead of the SM V, where $Z' \to q\bar{q}$ decays are assumed with a branching fraction of unity. To account for this possibility in the analysis, the jet mass window is shifted according to the mass of the Z' probed in the range from $0.85 m_{Z'}$ ($0.75 m_{Z'}$) to $m_{Z'} + 10$ GeV, for jets without (with) matched b-tagged small-

R jets. For Z' masses above 100 GeV, no fully merged selection is applied and the Z' decay is reconstructed in the resolved topology only. Constraints on several simplified models resulting in a mono-Z' signature are reported, with upper cross section limits in the range of 0.06–27 pb for Z' masses between 80 and 500 GeV [1068]. ATLAS and CMS exclude vector mediators with masses below 1.8 TeV for m_χ up to 700 GeV for couplings $g_{\rm SM} = 0.25$ and $g_\chi = 1$. These analyses are related to diboson searches in $\nu\nu J$ final states, optimised for resonant signals [711].

Mono-H searches at 13 TeV improved the sensitivity to dark matter models considerably, where the higher $\sqrt{s}$ and larger dataset are crucial in the $H \to \gamma\gamma$ channel [1069–1071]. The $H \to \tau\tau$ channel has been considered for the first time in dark matter searches in a CMS analysis of 35.9 fb^{-1} of data [1071], resulting in better sensitivity for $m_\phi > 700$ GeV relative to the $H \to \gamma\gamma$ channel, which is better at smaller masses. In addition to the simplified dark matter model, the results of this analysis are interpreted in a Z'-two-Higgs-doublet model with a Z' mediator, with the decay $Z' \to HA$ with $A \to \chi\bar{\chi}$. The combination of the two channels excludes Z' masses between 550 to 1265 GeV for $m_A = 300$ GeV and $m_\chi = 100$ GeV. The complementary channel with $H \to b\bar{b}$ has been analysed by ATLAS and CMS in several publications. Both collaborations have gradually improved the reconstruction algorithms and selection criteria in the development of these analyses. An example of the improvements in the ATLAS mono-$H(b\bar{b})$ searches using 3.2 fb^{-1} [1072], 36.1 fb^{-1} [1073] and 79.8 fb^{-1} [1074] is shown in Fig. 5.22. The trimmed jet mass of large-R jets is used as sensitive observable in the boosted signal region with $p_{\rm T}^{\rm miss} > 500$ GeV. Large-R jets are required to have two ghost-associated b-tagged track-jets. With respect to the analysis with 3.2 fb^{-1} of data, further selection requirements are implemented in the analysis with 36.1 fb^{-1} to suppress the $t\bar{t}$ background. These include a veto on events containing a reconstructed τ lepton, events with a b-tagged small-R jet not overlapping with the large-R jet are discarded, and the $p_{\rm T}$-sum of all small-R jets should be smaller than 0.57 times that sum added to $p_{\rm T}$ of the large-R jet. These criteria, together with an improved b-tagging algorithm, results in a large reduction of the SM backgrounds, as apparent when comparing the two distributions on top of Fig. 5.22. A further improvement is obtained by considering VR track-jets for subjet b tagging, as done in the updated analysis with 79.8 fb^{-1} of data. The signal efficiency is improved for high $p_{\rm T}$, where the track-jets merge when using a fixed-R algorithm. This leads to a large improvement in sensitivity for mediator >2.5 TeV, as shown in the bottom right of Fig. 5.22. In CMS, the mono-$H(b\bar{b})$ search with 2.3 fb^{-1} [1070] uses anti-$k_{\rm T}$ $R = 0.8$ jets with a pruned jet mass and subjet b-tagging for H tagging in the merged signal region, whereas small-R jets are used to define the resolved signal region. The later analysis with 35.9 fb^{-1} [1075] uses CA $R = 1.5$ jets to gain sensitivity for moderately boosted H bosons, making the resolved selection obsolete. A variant of the double-b tagger, trained on CA $R = 1.5$ jets, is used to identify the merged $H \to b\bar{b}$ decay. Further discrimination is achieved with a selection on the mass-decorrelated N_2, which is designed to be uncorrelated to the soft drop jet mass. The PUPPI algorithm enables the use of these substructure algorithms for CA $R = 1.5$ jets, where the contribution from pileup would result in a much worse resolution without a dedicated pileup mitigation. The

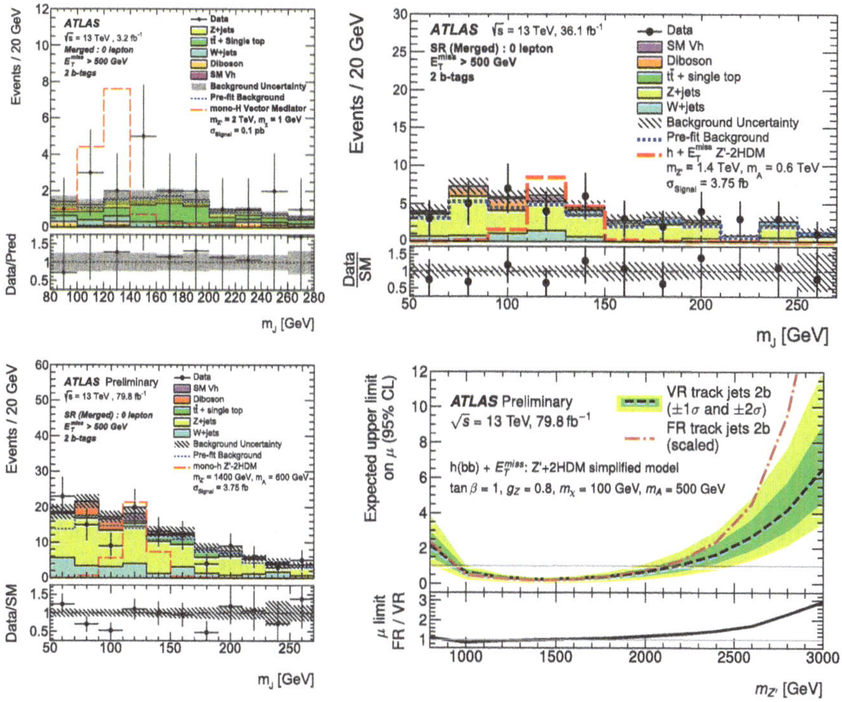

Fig. 5.22 Evolution of the ATLAS mono-$H(b\bar{b})$ search with 3.2 fb^{-1} (top left), 36.1 fb^{-1} (top right) and 79.8 fb^{-1} (bottom left) of 13 TeV data. Shown is the trimmed jet mass of large-R jets with two matched b-tagged track-jets. Note that the signal cross section is 100 fb in the first version of the analysis (top left) and about 25 times smaller (3.75 fb) in the later two versions with more data. Comparison of the expected upper limit using fixed-R track-jets for b tagging, as used in the analyses with up to 36.1 fb^{-1} of data, scaled to 79.8 fb^{-1}, and variable-R track-jets as used in the analysis with 79.8 fb^{-1} of data (bottom right). Taken from [1072] (top left), [1073] (top right) and [1074] (bottom)

analysis excludes dark matter particles up to $m_\chi = 430$ GeV for a mediator mass of 960 GeV in a simplified model. For the first time, also a two-Higgs-doublet model with an additional light pseudo-scalar a [1076] is studied, where A masses between 500 and 900 GeV are excluded for $m_a = 150$ GeV [1075]. The best constraints to date on models predicting mono-H signatures are obtained from a combination of the H decay channels WW, ZZ, $b\bar{b}$, $\gamma\gamma$ and $\tau\tau$ [1077]. Masses of the Z' between 500 to 3200 GeV are excluded in the Z'-two-Higgs-doublet model with $m_A = 300$ GeV and $m_\chi = 100$ GeV. For Z' masses above 800 GeV, the sensitivity of this combination is driven by the boosted $H \to b\bar{b}$ analysis.

A very recent analysis by ATLAS [1078], using the full 13 TeV data set, considers a "dark Higgs" boson s, which is responsible for the mass generation of χ [1079]. If s is the lightest state in the dark sector, the model shares similarities with the simplified models considered so far. Specifically, it predicts a mono-s signature, where the $s + \chi\bar{\chi}$ signal is produced through a heavy mediator, $q\bar{q} \to Z' \to s\chi\bar{\chi}$ [1080]. For s with masses larger than 160 and 180 GeV, the decays $s \to WW$ and $s \to ZZ$ become relevant, respectively [137]. The analysis considers the fully hadronic decay chain $s \to VV \to q\bar{q}^{(\prime)}q\bar{q}^{(\prime)}$, where the four quarks are reconstructed in a single large-R jet. This is one of the few analyses at the LHC targeting four-prong decays in single jets. If the reconstruction of the fully merged decay fails, a partially merged reconstruction is attempted, where a large-R jet and one or two accompanying small-R jets are combined to form the $s \to VV$ decay. Discrimination between signal and background jets is improved by the use of track-assisted reclustered jets [430], see Sect. 3.2. The improvement in the resolution of N-subjettiness ratios used for $VV \to 4q$ tagging results in a stronger expected discovery significance by a factor of up to 2.5. Besides the gain in resolution, another advantage of using track-assisted reclustered jets is the possibility to propagate experimental uncertainties consistently. This results in a realistic estimation of the signal efficiency without a dedicated measurement, which is not possible for this channel because of a lack of SM processes with sufficiently large cross sections. In the fully merged category, signal jets are required to have $p_{\mathrm{T}} > 300$ GeV, a trimmed mass in the range 100–400 GeV and satisfy the combination of N-subjettiness ratios $\tau_{42} < 0.3$ and $\tau_{43} < 0.6$. In the partially merged category, the large-R jet is combined with two small-R jets if its mass is between 60 and 100 GeV and with one small-R jet if it has a mass larger than 100 GeV. The considered small-R jets have to be within $\Delta R < 2.5$ of the large-R jet. This recovers some of the signal events, but with the drawback of larger backgrounds in the partially merged category. The two distributions in the reconstructed s mass, measured in the fully merged and partially merged categories, are used to derive upper cross section limits on the signal model. The limits range between between 0.32 and 0.03 pb for s with mass between 160 and 360 GeV, respectively, and $m_\chi = 200$ GeV and $m_{Z'} = 1$ TeV.

Dark matter searches targeting models with mediators coupling to third generation quarks have also profited from developments in jet substructure reconstruction. In an ATLAS search for $\chi\bar{\chi}$ produced in association with b, $b\bar{b}$ or $t\bar{t}$, using 36.1 fb^{-1} of 13 TeV data [1081], the jet mass of reclustered jets is used to enhance the sensitivity. Two signal regions in the jets+$p_{\mathrm{T}}^{\mathrm{miss}}$ final state have been optimised for $t\bar{t}$ +$\chi\bar{\chi}$ production. Mediators with $100 < m_\phi < 350$ GeV result in a moderately boosted $t\bar{t}$ topology, which is identified by requiring the jet masses of the two p_{T}-leading $R = 1.2$ reclustered jets to be >140 and 80 GeV. For lower masses, $m_\phi < 100$ GeV, the on average smaller boost results in semi-merged topologies, which are selected by requiring the jet masses to be greater than 80 GeV for both p_{T}-leading $R = 0.8$ reclustered jets. Other topological selection criteria are applied to suppress SM backgrounds, resulting in a negligible contribution from multijet production. Two further signal regions in the jets+$p_{\mathrm{T}}^{\mathrm{miss}}$ final state are optimised for $b\bar{b} + \chi\bar{\chi}$ and $b + \chi\bar{\chi}$ production, and a fifth signal region considers dilepton $t\bar{t}$ events. Mediator masses between 10 and 50 GeV for scalar mediators coupling to top quarks are excluded for $m_\chi = 1$ GeV. While the sensitivity of the search is driven by the dilepton category for $m_\phi < 150$ GeV, for larger mediator masses the jets+$p_{\mathrm{T}}^{\mathrm{miss}}$ final state is better. The dilepton final state has been updated using the full 13 TeV data set with

139 fb^{-1}, resulting in an exclusion of $m_\phi < 250$ GeV [1082]. The ℓ+jets final state has been analysed in a dedicated ATLAS analysis using 36.1 fb^{-1} of 13 TeV data [528]. This analysis considers 16 individual signal regions, each optimised for a particular decay topology. Hadronic decays of boosted t and W are selected by reclustering small-R jets with an iterative approach. Large-R jets are obtained with an initial value $R = 3.0$, which is then reduced to $R = 2m/p_\mathrm{T}$, where m is either m_t or m_W. If a large-R jet loses a large fraction of its p_T in this iteration, it is discarded. The masses of the reclustered jets help to select events with t and W jets. The obtained upper cross section limits from a combination of all signal regions are compatible with the limits obtained in the jets+$p_\mathrm{T}^\mathrm{miss}$ [1081] and dilepton final states [1082], but no exclusions can be derived from this analysis alone. In models with b quark couplings, the upper cross section limits are about two orders of magnitude above the predicted signal cross sections [1081], showing the challenge of this final state at the LHC.

An analysis by CMS in search for $t\bar{t} + \chi\bar{\chi}$ and $b\bar{b} + \chi\bar{\chi}$ is based on 2.3 fb^{-1} of 13 TeV data [1083], and combines dilepton, ℓ+jets and jets+$p_\mathrm{T}^\mathrm{miss}$ final states. In order to suppress the large $t\bar{t}$ and $Z(\rightarrow \nu\bar{\nu})$+jets backgrounds in the jets+$p_\mathrm{T}^\mathrm{miss}$ final state, a resolved top tagger is employed. In this algorithm, triplets of small-R jets are built from which a discriminant is constructed using a BDT. The inputs are the q/g [411] and b tagging discriminants for each jet, the opening angle between the candidate b jet and the two jets from the W boson decay, and the output value of a kinematic fit to m_t and m_W. The working point chosen in this analysis corresponds to an efficiency of 94% for correctly identifying the $t\bar{t}$ decay with two resolved top tags, with a misidentification rate of 48%. In the dilepton and ℓ+jets final states, backgrounds are suppressed with angular selections between the leptons, jets and $\mathbf{p}_\mathrm{T}^\mathrm{miss}$. In the final combination, the analyses in the ℓ+jets and all-hadronic final states show the best sensitivity to $t\bar{t} + \chi\bar{\chi}$ production. The analysis has been updated using 35.9 fb^{-1} [1084], resulting in excluded masses of scalar mediators coupling to top quarks below 160 GeV for $m_\phi = 2m_\chi$ and unity couplings, $g_\mathrm{SM} = 1$ and $g_\chi = 1$.

At 13 TeV, also mono-t searches have profited from jet substructure techniques. A CMS analysis with 36 fb^{-1} of data [1085] has been optimised for a scalar mediator with flavour-changing couplings, resulting in $t + \chi\bar{\chi}$ production [1049, 1050]. The same experimental signature of $t + p_\mathrm{T}^\mathrm{miss}$ is obtained by a coloured mediator, decaying to a top quark and a dark fermion, $t + \psi$ [1086], which is also considered in this analysis. The analysis is carried out in the $J + p_\mathrm{T}^\mathrm{miss}$ final state, where J denotes a t-tagged large-R jet. For t tagging, CA $R = 1.5$ jets are used with $p_\mathrm{T} > 250$ GeV. The jets are corrected for pileup effects using the PUPPI algorithm and groomed using soft drop with $z_\mathrm{cut} = 0.15$ and $\beta = 1$. Jets are selected if their soft drop mass is in the range 110–210 GeV and a b-tagged subjet is found. A BDT is used to combine substructure information to achieve better background rejection. The input to the BDT are τ_{32} and eleven ratios of generalised ECFs ${}_v e_N^{(\beta)}$ (see Sect. 2.4.2), with varying parameters β, N and v. In addition, it was found that the variable f_rec, which describes the deviation of the ratio of the reconstructed values of m_W and m_t from its expectation and has originally been defined for the HTTv2 [234], helps to improve the performance at low p_T. It is not clear what the additional information carried by f_{rec} is, compared to the other inputs τ_{32} and the ECF ratios. However, f_{rec} carries some additional discrimination power since adding more ECF ratios and N-subjettiness observables could not compensate for dropping f_{rec} from the BDT [526]. Note that this t tagger is special in the sense that all substructure observables except for f_{rec} are groomed using soft drop, which improves the performance at low p_T. For the calculation of f_{rec}, mass drop and filtering are used. The gain from the BDT compared to τ_{32} alone is a reduction of the background efficiency from 6.9 to 4.7% for a constant signal efficiency of 50%. Two signal regions are defined by a selection based on large p_T^{miss} and two intervals of the BDT discriminator, resulting in a loose and a tight t-tagged jet category. The main backgrounds are obtained from data in control regions and extrapolated into the signal region with transfer functions. A statistical interpretation of the measured distribution in p_T^{miss} results in excluded masses between 0.2 and 1.75 TeV for flavour-changing vector and axial-vector mediators and $m_\chi \lesssim 100$ GeV, $g_{SM} = 0.25$ and $g_\chi = 1$. This exclusion is higher by 820 GeV compared to the 8 TeV result [1054], which is a remarkable result. In ATLAS, the mono-t signature is analysed together with the single production of a VLQ T in the $T \to Zt$ channel, with $Z \to \nu\bar{\nu}$, resulting in the same signature. The analysis, based on 36.1 fb^{-1}, considers $J + p_T^{miss}$ and ℓ+jets final states [908]. The dark matter signal region is similar to the VLQ signal region, except for the absence of a forward jet (see Sect. 5.3.2). The sensitivity is comparable with the CMS analysis, with excluded vector mediator masses below 2 TeV for $m_\chi = 1$ GeV, $g_{SM} = 0.5$ and $g_\chi = 1$.

Searches for $t\bar{t} + \chi\bar{\chi}$ so far discussed have neglected the possibility of dark matter production in association with a single top quark, where the interaction is mediated by a scalar particle [1087–1089]. The same interaction giving rise to the $t\bar{t} + \chi\bar{\chi}$ signature leads to diagrams involving the t-channel exchange of a W boson, resulting in $t + \chi\bar{\chi}$ in association with a b quark and a forward jet, as well as dark matter produced in association with tW. These are different than mono-t signatures, where a vector or pseudo-vector interaction gives rise to $t + p_T^{miss}$ final states without additional particles. Since $tbj + \chi\bar{\chi}$ and $tW + \chi\bar{\chi}$ can be kinematically favoured over $t\bar{t} + \chi\bar{\chi}$, the inclusion of the corresponding final states in dark matter searches is expected to improve the sensitivity to scalar mediators with couplings to top quarks by 30% up to a factor of two [1087]. This has been exploited in recent searches by ATLAS [1090] and CMS [1091], with 139 and 36 fb^{-1}, respectively. While ATLAS analyses ℓ+jets and dilepton final states, the CMS analysis is optimised for ℓ+jets and jets+p_T^{miss} final states. The CMS analysis results in lower mass limits on scalar and pseudoscalar mediators of 290 and 300 GeV, respectively, thereby improving the previous limits [1084] based on the same amount of data by 130 GeV. The ATLAS analysis interprets the results in the context of a two-Higgs-doublet model [1088], such that the results can not be compared directly. In this model, masses of the pseudoscalar mediator can be excluded below 200 GeV for charged Higgs boson masses $m_{H^\pm}$ in the range 0.4–1.4 TeV and up to 330 GeV for $m_{H^\pm} = 900$ GeV. These searches are sensitive to single VLQ production in the $T \to tZ$ channel with $Z \to \nu\bar{\nu}$, but these interpretations have not been done so far.

The best sensitivity to dark matter models is achieved by combining experimental results from a number of different analyses, such as searches for dijet and dilepton resonances, $t\bar{t}$ resonances or mono-X signatures. The resonance searches cover the mediator interactions with SM particles, while the mono-X searches constrain the parameter space where the coupling to dark matter particles is larger. A recent comprehensive combination has been performed by ATLAS [1092], considering a number of mediator-based simplified dark matter models. A result of this combination is shown in Fig. 5.23, which shows the limits obtained from dijet, $t\bar{t}$ and di-b-jet resonance searches and mono-X searches in the mediator versus dark matter mass plane for a vector mediator. Since the mediator-quark coupling is rather large, the strongest limits are obtained by dijet searches. The $t\bar{t}$ and di-b-jet resonance searches provide coverage for cases where the mediator couples predominantly to third generation quarks. The mono-X searches have unique sensitivity to mediator masses below 500 GeV. In leptophilic models, where the mediator-lepton coupling is increased, the dijet searches lose sensitivity for mediator masses larger than $2m_\chi$. In this case, the parameter space for mediator masses below 1 TeV is covered by mono-X searches alone. Under certain assumptions, the results from collider searches can be compared to direct and indirect detection experiments [1093]. The interpretation in spin-dependent and spin-independent χ-nucleon interaction depends on the mediator mass and assumptions on couplings. For the spin-independent case, direct detection experiments dominate the sensitivity for m_χ above 10 GeV, but for small

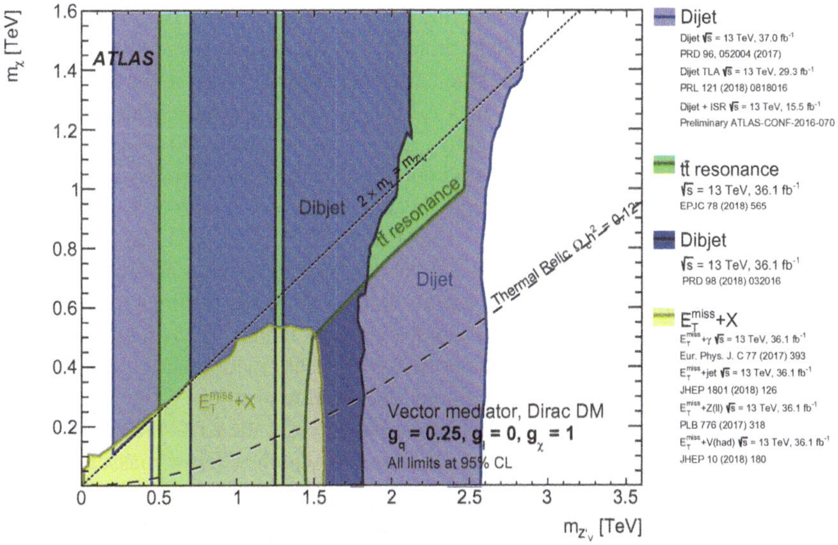

Fig. 5.23 Regions in the mediator-dark matter mass plane excluded by different searches for a simplified model with a vector mediator. The dashed curve corresponds to combinations of dark matter and mediator mass values that are consistent with a dark matter density of 0.12; above the curve annihilation processes deplete the density to below 0.12. Taken from [1092]

dark matter masses the collider results complement direct detection results. In this region, direct detection experiments have little sensitivity because of the small recoil energy such dark matter particles would induce [1094].

Searches for mono-X signatures are complementary to resonance searches at the LHC, where the dark matter mediator can contribute to the s-channel production of pairs of SM particles. An analysis considering a simplified model of spin-2 mediators [1095] finds that the strongest constraints are obtained from $\gamma\gamma$ and $\ell\ell$ resonance searches. However, if these modes are suppressed, mono-X and resonance searches provide stringent constraints on the mass and couplings of the dark matter mediator. Pseudo-scalar mediators can be constrained by searches for $t\bar{t}+\chi\bar{\chi}$ and $t\bar{t}$ and dijet resonances, as well as final states with τ lepton pairs [1096]. Only a complete search programme at the LHC in all conceivable mono-X and di-SM-particle resonance channels can shed light on dark matter in the future.

5.6 Light Resonances Coupling to Quarks or Gluons

Searches for resonances decaying to pairs of quarks or gluons, generically denoted by Z', have always been among the first analyses performed when collisions at unexplored centre-of-mass energies have become available at hadron-hadron colliders. These searches are typically performed as a bump-hunt [706] in the dijet mass spectrum, or by analysing the scattering angle in dijet events [946]. There are many well-motivated extensions of the SM predicting these resonances [700, 1097–1101], some of those have been proposed already more than 30 years ago [797, 858, 945, 970, 1102]. Searches for dijet resonances have been performed already at the CERN Sp$\bar{\text{p}}$S collider experiments UA1 [1103] and UA2 [1104, 1105], and have been extended in mass reach by the Tevatron experiments CDF [1106–1110] and D0 [1111–1113]. At the LHC, the reach in resonance mass could be extended considerably every time a new record centre-of-mass energies have been reached, at 7 TeV [948, 949, 952, 953, 1114–1120], 8 TeV [950, 954, 1121–1124] and 13 TeV [801, 951, 955, 956, 971, 972, 1125–1129]. These searches focus on heavy particles with resonance masses above 1 TeV and place stringent constraints on their allowed couplings to quarks and gluons. With the advent of simplified models for dark matter, where a mediator coupling to quarks is responsible for the production of dark matter particles, a renewed interest in low mass mediators has emerged. In addition to mono-X signatures, these mediators can be probed in dijet signatures. However, at the LHC the high multijet rates initiated by the strong force do not allow for searches below masses of about 1 TeV.

Data scouting [1130, 1131] is a way to circumvent the bandwidth limitations of the trigger, event processing and data storage. It has been introduced by CMS to store events with trigger-level information only, after passing the level-1 trigger and being processed by the high-level trigger. To maintain the high rate of 1 kHz of the high-level trigger, only the four-momenta of small-R calorimeter jets reconstructed online are stored. The corresponding events are then processed with the standard software and analysis tools. A dijet search performed on these scouting events with 18.8 fb^{-1} of 8 TeV data [1132] results in sensitivities for resonance masses in the range of 500–700 GeV better by a factor of two and more than obtained from prescaled triggers [1121], where only a fraction of events is recorded at low p_{T}. Analyses based on data scouting by ATLAS [1133] and CMS [1128] using 13 TeV data could improve the bounds from the 8 TeV analysis in the mass region 450 GeV to 1.5 TeV by another factor of two. A recent analysis by CMS based on the same 13 TeV scouting data with 18.3 fb^{-1}, could extend the mass range down to 350 GeV, by reconstructing the resonance mass using three jets instead of two [1134].

The region below 350 GeV in resonance mass can be probed in signatures where the Z' recoils against a high-p_{T} photon. The photon produced by ISR reduces the acceptance, but presents a way to trigger these events and reduce the multijet background. A search by ATLAS based on 79.8 fb^{-1} of 13 TeV data [1135] uses events recorded by a single-photon trigger with $p_{\mathrm{T}} > 140$ GeV, or a trigger requiring a photon with $p_{\mathrm{T}} >$ 75–85 GeV, depending on the year of data acquisition, and two small-R jets with $p_{\mathrm{T}} > 50$ GeV. The single photon trigger allows for a lower jet p_{T} threshold, such that invariant dijet masses down to 169 GeV can be probed. Constraints on resonance masses in the range from 225 to 1100 GeV are placed. Instead of a photon, the recoiling system can be W decaying leptonically, resulting in a lepton+dijet signature. This signature probes also models with more than the minimal particle content [1026], for example where a heavy resonance X decays to a W and a Z' boson [1136, 1137]. Recently, ATLAS has performed a search for dijet resonances recoiling against a charged lepton using 139 fb^{-1} of 13 TeV data [1138]. Events are required to have one isolated lepton with $p_{\mathrm{T}} > 60$ GeV and two small-R jets with $p_{\mathrm{T}} > 20$ GeV. Events with dijet mass larger than 220 GeV are selected. Below this value the dijet mass distribution is not monotonically falling due to a kinematic bias from the lepton p_{T} selection, such that the parametric background fit employed in this search fails. Upper limits on the signal cross section times acceptance and branching ratio are reported ranging from 100 to 0.1 fb for a Z' mass between 0.25 and 6 TeV. A dark matter mediator between 0.25 and 1.2 TeV is excluded for $g_{\mathrm{SM}} = 0.25$ and $g_{\chi} = 1$.

Resonance masses lower than about 250 GeV can not be probed in traditional dijet resonance searches at the LHC, but jet substructure techniques enable the exploration of Z' masses down to about 10 GeV. For low mass resonances, the recoiling ISR particle results in a boosted Z', which can be reconstructed by a single large-R jet [1139, 1140]. The first search of this kind has been performed by CMS on 2.6 fb^{-1} of 13 TeV data [1141], where a high-p_{T} ISR jet provides enough energy to satisfy the trigger requirements. The ISR jet recoils against another high-p_{T} signal jet, such that the analysis strategy is very similar to the measurement of H+jet production in the $H \to b\bar{b}$ channel (see Sect. 4.2.2). The highest-p_{T} large-R jet in the event is assumed to originate from the boosted Z' decay and is analysed further. The jet is corrected for pileup effects with PUPPI and the soft drop mass with $z_{\mathrm{cut}} = 0.1$ and $\beta = 0$ is chosen as sensitive observable to distinguish signal from background events. The signal is enriched using a selection on τ_{21}, transformed using the DDT technique to $\tau_{21}^{\mathrm{DDT}} = \tau_{21} + 0.063\rho'$, where $\rho' = \ln[m_{\mathrm{SD}}^2/(p_{\mathrm{T}}\mu)]$ and

$\mu = 1\,\text{GeV}$ [251]. This definition of ρ' is different from the usual dimensionless scaling variable $\rho = \ln[m_{\text{SD}}^2/p_{\text{T}}^2]$ [237], and has been chosen in order to decorrelate τ_{21}^{DDT} from the jet mass. The range in ρ' is restricted to 0–4, corresponding to a soft drop jet mass between 30 and 300 GeV for jets with $p_{\text{T}} > 500\,\text{GeV}$. For $\rho' < 0$ non-perturbative effects become important and for $\rho' > 4$ hard wide-angle radiation leads to out-of-cone effects, both effects resulting in a deviation of the average of the τ_{21}^{DDT} distribution from a constant. Signal events are classified by $\tau_{21}^{\text{DDT}} < 0.38$. The dominant background from multijet production is obtained using a pass-fail ratio technique, where the events failing the substructure selection with $\tau_{21}^{\text{DDT}} > 0.38$ are weighted by the pass-fail ratio. This ratio is obtained from a fit of polynomial functions to data as a function of τ_{21}^{DDT} and ρ', which corrects for any residual correlations between τ_{21}^{DDT} and ρ'. In order to avoid a bias due to a potential signal, a region in the soft drop mass around the probed Z' mass is removed from the fit. Resonant backgrounds from simulated V+jets production are subtracted from the data prior to the determination of the pass-fail ratio. By analysing the soft drop jet mass, the analysis places limits on a Z' in the mass range from 100 to 300 GeV. In the region 140–300 GeV, these limits are better by a factor of up to four than the previous limits by the UA2 experiment. These have been the first experimental results for Z' masses in the range 100–140 GeV.

An update of this analysis by CMS, based on 35.9 fb^{-1} [489], uses N_2 for the substructure selection and ρ for the decorrelation. For similar reasons as above, only jets with $-5.5 < \rho < -2.0$ are considered, with different numerical values because of the change to ρ instead of ρ'. This translates into jet masses in the range from 25 to 185 GeV for $p_{\text{T}} = 500\,\text{GeV}$. The definition of N_2^{DDT} is improved in relation to the previous analysis, where an exact decorrelation is performed on simulated multijet events. This requires a map in the plane of ρ and p_{T}, defining a constant 5% background efficiency for all values of ρ and p_{T} considered. Residual differences between data and simulation, which can affect this decorrelation, are accounted for by allowing for deviations in the pass-fail from a constant. Since the pass-fail ratio is determined simultaneously to the signal and background fit to data, and because it is only slowly varying as function of ρ and p_{T}, it is not necessary any more to remove a window around the Z' mass from its determination. The soft drop jet mass is measured in five intervals of jet p_{T}, where the distribution for $500 < p_{\text{T}} < 600\,\text{GeV}$ is shown in Fig. 5.24. The multijet background, together with smaller contributions from V+jets and $t\bar{t}$ production, describe the data well. The improved background estimation allows to extend the reach in Z' masses down to 50 GeV, with an upper reach of 300 GeV. The upper reach of this search can be extended by considering larger jet distance parameters, as done in a search by CMS using data recorded in the year 2017 with an integrated luminosity of 41.1 fb^{-1} [1142]. The analysis strategy is unchanged in relation to the search based on 2016 data [489], but the analysis is extended to use CA $R = 1.5$ jets in addition to anti-k_{T} $R = 0.8$ jets. The CA $R = 1.5$ jets are required to have $p_{\text{T}} > 575\,\text{GeV}$ and $-4.7 < \rho < -1.0$, which translates into a jet mass range from 55 to 350 GeV for $p_{\text{T}} = 575\,\text{GeV}$ and from 81 to 500 GeV for $p_{\text{T}} = 850\,\text{GeV}$. The analysis based on $R = 0.8$ jets is used to probe Z' masses in

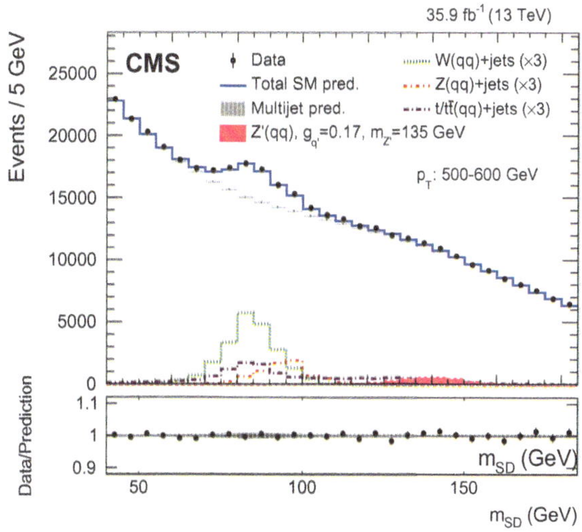

Fig. 5.24 Soft drop mass distribution of large-R jets with $500 < p_{\mathrm{T}} < 600\,\mathrm{GeV}$, selected by a requirement on N_2^{DDT}, in a search for a boosted light hadronic resonance. The multijet background distribution is obtained from a control region with events failing the N_2^{DDT} selection. Taken from [489]

the range 50–175 GeV and $R = 1.5$ jets are used to probe the range 175–500 GeV. For the mass range 50–225 GeV, this analysis places the most stringent constraints on Z'-quark couplings from direct searches to date.

Another possibility to select boosted Z' resonances is offered by an ISR photon, which results in smaller multijet backgrounds and lower p_{T} thresholds compared to an ISR jet. ATLAS has exploited this in an analysis of 13 TeV data corresponding to $36.1\,\mathrm{fb}^{-1}$ [1144], using separate signal regions for ISR jet and ISR photon events. In this case, the trimmed $R = 1.0$ jet mass with $R_{\mathrm{sub}} = 0.2$ and $f_{\mathrm{cut}} = 0.05$ is used to separate signal from background jets. The analysis uses τ_{21}^{DDT}, which is decorrelated from the jet mass using ρ', similar to the method applied in the first boosted dijet analysis by CMS [1141]. The multijet background is estimated using the pass-fail ratio method and validated in V+jets and $V+\gamma$ validation regions. The measured large-R jet mass distributions in the photon and jet channels result in constraints on Z' masses between 100 and 220 GeV. The combined results are comparable to the results by CMS using the same amount of data [489], but the sensitivity is worse by a factor of about 1.5, which can be attributed to the use of trimming and τ_{21}^{DDT} compared to soft drop and N_2^{DDT} by CMS. CMS has exploited the lower trigger thresholds of single-photon triggers in an analysis based on $35.9\,\mathrm{fb}^{-1}$ of data [1143], which allows to select large-R jets with $p_{\mathrm{T}} > 200\,\mathrm{GeV}$. Similar to ISR jet analyses, the soft drop jet mass and N_2^{DDT} are used to separate signal and background jets. In this analysis, the optimal sensitivity is obtained for a constant background efficiency of 10%, which is used to calculate the map in ρ and p_{T}, defining N_2^{DDT}. The jet mass distribution is measured in the range 10–200 GeV and shown in Fig. 5.25. The multijet background is obtained using the pass-fail ratio method and resonant backgrounds are estimated using simulation. Distributions for three Z' signals with masses of 10, 25 and 50 GeV

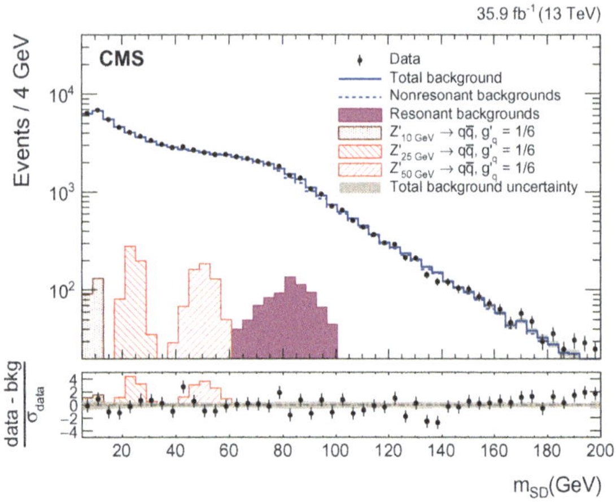

Fig. 5.25 Soft drop mass distribution of large-R jets with $p_T > 200\,\text{GeV}$, selected by a requirement on N_2^{DDT} γ+jet events. The multijet background distribution is obtained from a control region with events failing the N_2^{DDT} selection. Taken from [1143]

are shown as well, highlighting the sensitivity of this search for Z' masses as low as 10 GeV. This is the first search probing such low dijet resonance masses at a hadron collider.

The results from dijet searches at different colliders with various centre-of-mass energies and different analysis strategies can be compared in the mass-coupling plane of a leptophobic Z' benchmark model [1145]. The two relevant parameters are the Z' mass and its coupling to quarks, g_q', which can be associated with g_{SM} in simplified dark matter models. A comparison of the observed limits in the mass-coupling plane obtained by analyses from ATLAS, CMS, CDF and UA2 is shown in Fig. 5.26. Analyses by ATLAS and CMS nowadays cover the range from 10 GeV to 5 TeV, where this large coverage has been achieved by innovative analysis strategies. Coverage for masses of 10–225 GeV is given by boosted dijet searches; the region 225–700 GeV is probed by analyses using ISR radiation and with a resolved dijet system; Z' masses between 700 GeV and 1.5 TeV are analysed using data scouting techniques, and high-p_T dijet searches cover the region 1.5–5 TeV. The LHC analyses achieve sensitivities better by factors between two and three compared to results from other experiments, where available. The results described in this section place important constraints on a number of models of new physics, such as dark matter models and extensions of the SM gauge groups. Note that many of these results have been obtained for resonances coupling to quarks, but also coloured resonances with gluon couplings result in the same final states. In resolved resonance decays, this difference will not be observable and the results can readily be reinterpreted for resonances decaying to gg or qg. In jet substructure analyses, the richer radiation pattern of gluons will lead to a decrease in signal efficiency, where the signal distribution in τ_{21}^{DDT} and N_2^{DDT} will become more background-like. Estimating this effect from the difference in the distributions of $H \to c\bar{c}$ and $H \to gg$ decays, a decrease in signal efficiency of about 10% is expected. This should be taken into account when re-interpreting these results in models with gluon couplings, but a more detailed study of the effect would be desirable.

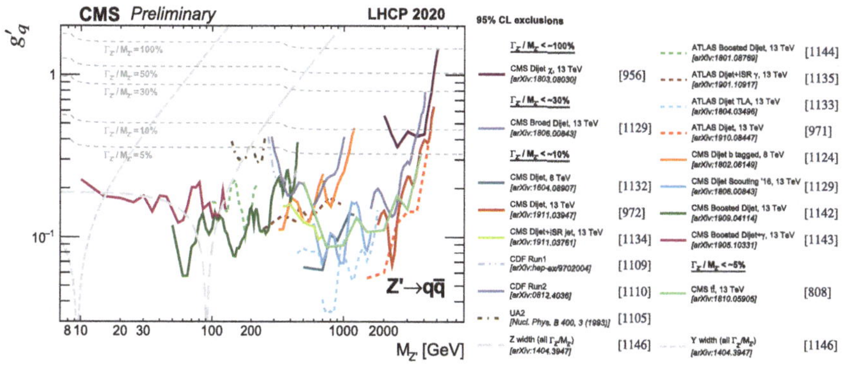

Fig. 5.26 Limits on the universal coupling g'_q between a leptophobic Z' boson and quarks from dijet analyses by ATLAS, CMS, CDF and UA2. The observed limits are shown by coloured lines, with the excluded area above the lines. The grey dashed lines show the g'_q values for fixed values of relative Z' widths. Also shown are indirect constraints on g'_q from the Υ meson and Z boson widths [1146]. Taken from [1147]

5.7 Supersymmetry

Supersymmetry (SUSY) is an elegant concept, offering solutions to many of the open questions in particle physics. SUSY introduces a partner for each SM particle, which differs from its SM counterpart in its spin by half a unit, thus transforming fermions into bosons and vice versa. It provides a framework for the unification of particle physics and gravity, an explanation of the large hierarchy between the energy scale that characterises electroweak symmetry breaking and the Planck scale, a solution to the matter-antimatter asymmetry of the universe, and a weakly interacting massive particle which is a viable candidate for dark matter. A variant of SUSY, including right-handed neutrinos, can also solve the problem of explaining non-vanishing neutrino masses [1148]. Extensive reviews on SUSY exist in the literature and the interested reader is referred to these and references therein [1149–1151]. Due to its compelling theoretical nature, SUSY has been one of the primary targets for BSM searches at every generation of particle colliders. At the LHC, more than 200 papers have already been published by ATLAS and CMS on searches for SUSY, see the article by Canepa for a recent review [1152]. Here, a focus is on searches where jet substructure methods have been applied to improve the sensitivity, or even made the analysis feasible in the first place. SUSY searches are an example of LHC analyses where jet substructure methods have traditionally not been relevant, but which have started to profit from the developments in this field. Previous SUSY searches have pushed the mass limits of supersymmetric particles to values above 1–2 TeV, making collimated particle decays an important consideration in the design of new analyses. In addition, there are specific SUSY scenarios which can only be tested with jet substructure methods. Once large regions of the parameter space of minimal supersymmetric models are excluded, searches for these more complex scenarios become imperative.

A prime example where jet substructure methods help to improve the sensitivity are searches for top squark production, where squarks denote superpartners of the SM quarks. The two supersymmetric partners of the t_L and t_R SM chiral states are denoted by $\tilde{t}_1$ and $\tilde{t}_2$, where the lighter state will be denoted simply by $\tilde{t}$ in the following. The lightest neutralino $\tilde{\chi}_1^0$ is in many models the stable, lightest supersymmetric particle (LSP). In simplified models [1153–1155], the $\tilde{t}$ can decay directly through a two-body decay, $\tilde{t} \to t\tilde{\chi}_1^0$, or it can decay via the three-body decay $\tilde{t} \to bW\tilde{\chi}_1^0$. The direct production of $\tilde{t}$ pairs with these subsequent decays are shown in Fig. 5.27 (left, middle). For large $\tilde{t}$ masses and large mass differences with the $\tilde{\chi}_1^0$, a large boost for the t is obtained in the two-body decay. The three body decay is predicted to happen only for mass differences Δm between the $\tilde{t}$ and $\tilde{\chi}_1^0$ smaller than m_t, such that the W is produced with small p_T. The relevant signature for boosted $\tilde{t}$ searches is $t\bar{t} + p_T^{\text{miss}}$, where the p_T^{miss} arises from the undetectable $\tilde{\chi}_1^0$s, which is similar to signatures of dark matter searches. Another scenario is motivated by compressed models [1157–1159], where Δm is comparable to m_t. In this case, the signature of $\tilde{t}$ pair production is similar to that of $t\bar{t}$ production, and therefore difficult to discriminate from the SM background. However, in scenarios where the superpartner of the gluon, the gluino $\tilde{g}$, is heavy and decays to a light $\tilde{t}$, the resulting signature $\tilde{g} \to \tilde{t}t$ with $\tilde{t} \to t\tilde{\chi}_1^0$, results in a boosted top quark from the gluino decay and an accompanying top quark with lower p_T, which is often described by an effective interaction as shown in Fig. 5.27 (right).

The $t\bar{t}+p_T^{\text{miss}}$ signature has been targeted by an ATLAS $\tilde{t}$ search using 3.2 fb^{-1} of 13 TeV data [457]. Events with a lepton, p_T^{miss} and jets are selected. Large-R jets are built from a reclustering of small-R jets, and the trimmed mass is used to select events with semi-merged and fully merged t decays. Three signal regions with increasing values of p_T^{miss} are used to search for a signal. Updates of this analysis, based on 36.1 and 139 fb^{-1} of data [528, 1160], consider more signal regions, optimised for various $\tilde{t}$ decays and mass splittings, and for $t\bar{t} + \chi\bar{\chi}$ production. The t tagging has been refined using an iterative reclustering of small-R jets, where the reclustering is repeated with decreasing R until an optimal distance parameter of $R = 2m_t/p_T$ is obtained. This iterative reclustering can be seen as an approximation to the VR algorithm, and is related to the optmimal-R parameter of the HTTv2 [234]. No further

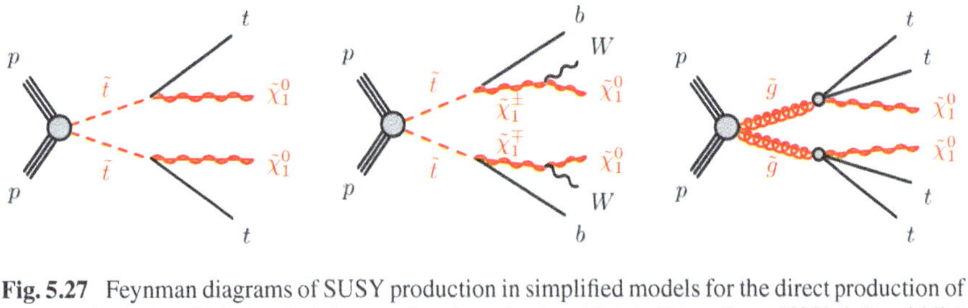

Fig. 5.27 Feynman diagrams of SUSY production in simplified models for the direct production of top squarks (left, middle) and gluino-mediated production (right). Taken from [528] (left, middle) and [1156] (right)

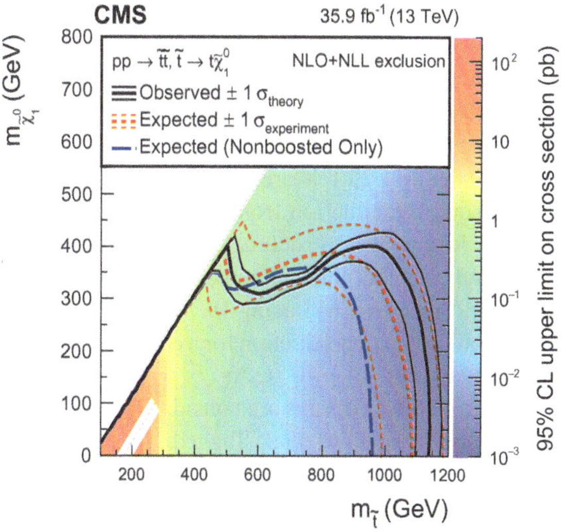

Fig. 5.28 Upper cross section limits and mass exclusion limits at 95% CL in the plane of $\tilde{t}$ and $\tilde{\chi}_1^0$ masses, as obtained by a CMS search for pair production of $\tilde{t}$ with the decay $\tilde{t} \to t\tilde{\chi}_1^0$. The expected mass exclusion, derived using categories without considering boosted W and t jets, is shown as well (nonboosted only). Taken from [1161]

jet substructure techniques are introduced; the signal is enriched using kinematic properties of the expected signal events. Top-squark masses below 1200 GeV are excluded for a massless $\tilde{\chi}_1^0$ in two body decays. A CMS search for the $t\bar{t}+p_T^{miss}$ signature on 35.9 fb^{-1} of 13 TeV data [1161] uses so-called razor variables [1162, 1163] to discriminate signal from background. Leptons are identified using mini-isolation to retain efficiency for boosted t decays in the ℓ+jets channel. Signal regions are defined by W and t tagged jets, where the taggers use the soft drop jet mass, N-subjettiness ratios and subjet-b tagging. The analysis uses signal categories defined by small-R jet, b-, W- and t-tagged jet multiplicities and analyses distributions in the razor variables to take advantage of the varying signal-to-background ratio. The improvement in sensitivity due to the inclusion of the W- and t-tagged jet categories is shown in Fig. 5.28 for the two body decay $\tilde{t} \to t\tilde{\chi}_1^0$, exemplary for all production and decay modes studied. For high $\tilde{t}$ and small $\tilde{\chi}_1^0$ masses, the improvement in the expected exclusion limit is significant, where the boosted categories improve the limit by about 120 GeV. In the regime of large $\tilde{\chi}_1^0$ masses, the improvement is smaller because the t receives smaller boosts in the $\tilde{t}$ decay. A search by CMS using 137 fb^{-1} of 13 TeV data [1164] uses bins in p_T^{miss} instead of razor variables to search for a signal. These bins are measured in 16 categories, where t-tagged categories are defined by either resolved or merged t decays. The t-tagged categories rely on improvements of existing taggers using deep neural networks. The resolved top tagger improves upon an earlier version [1083], where small-R jet triplets are built and examined for their compatibility with a resolved t decay. Efficiencies of 45% for misidentification rates of 10% are achieved. The DeepAK8 tagger is used to identify merged t decays, where a working point corresponding to an efficiency of 40% with a misidentification rate of 5% is chosen. The analysis sets stringent limits on the mass of $\tilde{t}$ for its direct production through the strong force. Masses below 1.2 TeV are excluded for a light $\tilde{\chi}_1^0$. For a $\tilde{t}$ mass of 1 TeV, $\tilde{\chi}_1^0$ masses below 600 GeV are excluded. An ATLAS search in the all-hadronic final state, using 139 fb^{-1} [1165] and targeting the $t\bar{t}+p_{\mathrm{T}}^{\mathrm{miss}}$ signature, uses the mass of reclustered $R = 0.8$ and $R = 1.2$ jets to select events with fully-merged and partially-merged t decays. The sensitivity of the analysis is comparable to the CMS search in the ℓ+jets channel [1164], where $\tilde{t}$ masses below 1.25 TeV are excluded for $\tilde{\chi}_1^0$ masses below 200 GeV.

Top squark production from the decays of gluinos is another example of SUSY searches where jet substructure techniques help to improve the sensitivity. One of the first searches in this channel in the all-hadronic final state considering W tagging has been performed by CMS using 19.7 fb^{-1} of 8 TeV data [1166]. This search has been optimised for $\tilde{g}$ pair production for a compressed spectrum with $\Delta m \approx m_t$ or smaller, in the decay channel $\tilde{g} \to \tilde{t}t$. At $\sqrt{s} = 8$ TeV, the $\tilde{g}$ mass reach is below 1 TeV, such that semi-merged t decays are considered and identified with a large-R W jet and an accompanying small-R b jet. Tagged W jets serve as a means to select events, where small-R jets are used to build razor variables to discriminate signal from background. The SM backgrounds are obtained from control regions, defined by inverting the τ_{21} selection used for W tagging and other kinematic requirements. The analysis excludes $\tilde{g}$ masses below 700 GeV and $\tilde{t}$ masses below 300 GeV for $\Delta m = 175$ GeV. A CMS search using 2.3 fb^{-1} of 13 TeV data [1167] considers triplets of small-R jets to construct resolved t decays [1168], as well as doublets and mono-jet signatures to account for semi-merged and fully-merged t decays. While the tagging efficiency of the algorithm is high, increasing from 30% at $p_{\mathrm{T}} = 200$ GeV to 80% at $p_{\mathrm{T}} = 1$ TeV, the misidentification rate is also high, between 30–40% on average. The reason for the large misidentification rate lies in the use of ungroomed small-R jets, where the mass increases for background jets with increasing p_{T} following the behaviour of the Sudakov peak. Nevertheless, at high $p_{\mathrm{T}}^{\mathrm{miss}}$ it is beneficial to have high signal efficiency, where the SM backgrounds are small. For models with gluino-mediated $\tilde{t}$ production $pp \to \tilde{g}\tilde{g}$, with $\tilde{g} \to t\bar{t}\tilde{\chi}_1^0$, gluino masses up to 1550 GeV are excluded. The deficiencies of the t identification have been addressed in a later CMS analysis based on 35.9 fb^{-1} [529]. The fully-merged and semi-merged t decays are identified using large-R jets with soft drop grooming and N-subjettiness ratios. These changes result in a reduction of a factor of about two in the average misidentification rate, while retaining the signal efficiency. Consequently, the $\tilde{g}$ mass limits could be improved by nearly 500 GeV, masses below 2040 GeV are excluded for a light $\tilde{\chi}_1^0$. This also corresponds to an improvement by about 100 GeV relative to searches in the all-hadronic final state based on the same amount of data, but without top tagging or jet substructure [1169, 1170].

In most SUSY decay cascades, a large number of jets is produced. These may arise from gluino-mediated $\tilde{t}$ production with a $t\bar{t}t\bar{t}+2\tilde{\chi}_1^0$ final state as discussed above, from chargino and neutralino decays or gauge-mediated SUSY symmetry breaking scenarios [1171–1173] with W, Z, H bosons in the final state, or from R-parity violating (RPV) [1174, 1175] decays with SUSY particle decays to SM quarks. Boosted heavy SM particles are produced in scenarios with large mass splittings between the SUSY particles produced in the primary interaction, the LSP, and the SM particles. A way to discriminate between these signals and SM background processes is the scalar sum of large-R jet masses M_J^Σ [1176, 1177], which is large for signal events and small for SM production of light quark and gluon jets. Even if the signal does not result in boosted heavy particles, M_J^Σ can still be a good discriminator in scenarios where the jet multiplicity is very large, such that on average jets accidentally overlap more often than in SM processes [1178]. The advantage of M_J^Σ over other discriminating observables is that it is largely uncorrelated to kinematic quantities such as H_T or p_T^{miss}, which are frequently used in SUSY searches. It is an excellent quantity to define signal and control regions. An ATLAS search using 36.1 fb^{-1} of 13 TeV data [1179] is optimised for RPV SUSY scenarios with six to ten quarks produced at tree level in the SUSY decay cascade. Trimmed large-R jets are used to select events, where four or five jets with $p_T > 200$ GeV are required in the signal regions. Depending on the decay scenario, the value of M_J^Σ has to be larger than 0.6, 0.8 or 1 TeV. The background is obtained from deriving templates in the jet mass distribution of large-R jets in signal-depleted control regions. For each jet, a probability density function is derived, which gives the relative probability for a jet with given p_T and η to have a certain mass [1180]. For each jet in the signal region, a randomised value for the jet mass is generated from the probability density function, in order to to derive the background estimate. The procedure is verified in validation regions and the corresponding uncertainty is obtained from dedicated side bands with smaller jet multiplicities. The analysis excludes gluino masses between 1000 and 1875 GeV in gluino cascade decays with ten quarks, and improves upon previous results based on 8 TeV data by more than 800 GeV [1180]. A recent analysis by ATLAS in the all-hadronic final state, based on 139 fb^{-1} of 13 TeV data [1156], targets cascade decays involving W and Z bosons, gluino-mediated four-top production and RPV cascades involving top quarks. The analysis uses PF jets and the corresponding p_T^{miss} significance, which improves the separation between events where p_T^{miss} originates from detector effects or from weakly interacting particles with high p_T. Large-R jets are reclustered from small-R jets, where small-R jets originating from pileup interactions are removed using the jet-vertex-tagger [442] prior to the reclustering. Ungroomed large-R jets with $p_T > 100$ GeV are then used to calculate M_J^Σ. The final results are based on analysing the measured distributions in p_T^{miss} significance. The gluino mass limits for top squark-mediated gluino decays with RPV couplings are 1.5 TeV for $\tilde{t}$ masses between 400 GeV and 1.1 TeV. The variable M_J^Σ is useful in ℓ+jets final states as well, as demonstrated in a CMS search using 137 fb^{-1} of 13 TeV data [1181]. Similar as in ATLAS, a reclustering of small-R jets is employed, but with $R = 1.4$. Because of the large value of R, ISR radiation can lead to large tails in M_J^Σ, such that a reliable background model needs to be derived from control regions in data. This is achieved by a method exploiting that M_J^Σ is uncorrelated from the transverse mass m_T, calculated between the lepton and $\mathbf{p}_T^{\text{miss}}$. The plane in M_J^Σ versus m_T is used to define three background-dominated regions, from which the background is predicted in the signal region. The analysis results in excluded gluino masses below 2150 GeV for $\tilde{\chi}_1^0$ masses up to 700 GeV, in gluino-mediated $\tilde{t}$ production.

Another example of SUSY scenarios where jet substructure plays an important role are models where the next-to-lightest SUSY particle is a neutralino $\tilde{\chi}_2^0$, with large mass splitting to the $\tilde{\chi}_1^0$ LSP. In addition, the mass splitting $\Delta m(\tilde{g}, \tilde{\chi}_2^0)$ is small, such that pair-produced gluinos decay into a soft quark, anti-quark and $\tilde{\chi}_2^0$. The $\tilde{\chi}_2^0$ decays into a $\tilde{\chi}_1^0$ and a Z or H boson. For light $\tilde{\chi}_1^0$ with masses $O(1\,\text{GeV})$, boosted Z and H bosons are predicted to be produced in association with large p_T^miss and additional soft quarks. This signature arises typically in models which preserve naturalness despite large $\tilde{g}$ and $\tilde{t}$ masses [1182], as suggested by LHC searches in other decay channels. In these scenarios, jet substructure techniques are paramount to suppress SM backgrounds. A search by CMS based on 35.9 fb^{-1} of 13 TeV data [1183] targets final states with two H jets and large p_T^miss. Two large-R jets are required to have $p_\text{T} > 300\,\text{GeV}$ and identified using the jet mass and the discriminator of the double b tagger. Events are classified by the number of identified H jets, where events with jets failing the tagging requirements are used to predict the SM background. The p_T^miss distribution is analysed in search for a signal in events with one and two identified H jets. Gluino masses up to 2010 GeV are excluded for a $\tilde{\chi}_1^0$ mass of 1 GeV and $\Delta m(\tilde{g}, \tilde{\chi}_2^0) = 50\,\text{GeV}$, assuming $\mathcal{B}(\tilde{\chi}_2^0 \rightarrow H\tilde{\chi}_1^0) = 1$. A search by CMS using 137 fb^{-1} of 13 TeV data [1184] targets the same model, but with $\mathcal{B}(\tilde{\chi}_2^0 \rightarrow Z\tilde{\chi}_1^0) = 1$. Large-$R$ jets are required to have $p_\text{T} > 200\,\text{GeV}$ and are groomed with the soft drop algorithm. The use of soft drop facilitates a linear fit to the jet mass distribution in jet mass side bands, which can be interpolated into the signal region with $70 < m_\text{jet} < 100\,\text{GeV}$ in order to predict the background from non-resonant SM production. The distribution of background events in bins of p_T^miss can then be predicted under the assumption of a minimal correlation between m_jet and p_T^miss, which is verified in simulation. The p_T^miss distribution in events with two identified Z jets is used to interpret the results in terms of exclusion limits. The search excludes $\tilde{g}$ masses up to 1920 GeV in this model.

There are SUSY scenarios predicting multijet final states with very little p_T^miss and non-isolated photons or leptons. The predicted final states are very difficult to disentangle from the large SM backgrounds, such that these scenarios were named stealth SUSY [1185, 1186]. In these models, a number of superpartners have minimal coupling to the SUSY breaking mechanism and are therefore mass degenerate. In a simplified model of stealth SUSY, the hidden sector can be represented by a fermionic state $\tilde{S}$ and a scalar S [1187]. The largest cross section at the LHC comes from $\tilde{g}$ pair production, followed by a long decay cascade. The first decay $\tilde{g} \rightarrow q\bar{q}\tilde{\chi}_1^0$ is followed by the neutralino decay $\tilde{\chi}_1^0 \rightarrow \gamma\tilde{S}$ and the $\tilde{S}$ decays to a massless gravitino $\tilde{G}$ and the scalar S. Because of the mass degeneracy of $\tilde{S}$ and S, the $\tilde{G}$ is produced nearly at rest and leads to no detectable signature. The S decays to two gluons. If the mass difference $\Delta m(\tilde{g}, \tilde{\chi}_1^0)$ is large, the $\tilde{\chi}_1^0$ is produced with large boost and its decay products merge into a single jet. This results in jets containing the photon from the $\tilde{\chi}_1^0$ decay and the fragmentation products from the two gluons from the S decay. Naturally, these signatures are indistinguishable from the large multijet background without using jet substructure information. A dedicated analysis by CMS using 35.9 fb^{-1} of 13 TeV data [1188] targets this signature. Large-R anti-k_T jets with $p_\text{T} > 200\,\text{GeV}$ are examined for the presence of a photon, i.e. a particle candidate with an ECAL energy deposit consistent with the shower shape of a photon and no associated hits in the pixel detector.[6] For each jet with a photon candidate, the jet is reclustered with the k_{T} algorithm using the photon and all other jet constituents as input. Reversing the clustering history and looking at the last clustering step, the less massive of the two pseudojets is the first subjet. In the second step the more massive pseudojet is declustered, resulting in the second and third subjets. Signal jets are required to have three subjets with $p_{\mathrm{T}} > 10\,\mathrm{GeV}$ and $\tau_{31} < 0.4$. The photon subjet energy fraction f_γ is defined as the fraction of the photon p_{T} relative to the p_{T} of the subjet containing the photon. This is a measure of the additional activity in proximity to the photon and serves as a way to discriminate between signal photons and photons from the decay of hadrons, which are abundant in SM multijet production. The distribution of f_γ is shown in Fig. 5.29 for data, multijet background and signal jets. The distributions for signal jets show a more pronounced peak at $f_\gamma \approx 1$ for increasing $\tilde{\chi}_1^0$ mass, because of a larger relative photon p_{T} with respect to the jet axis, resulting in more isolated photons. Signal jets are classified into tight and loose photon jets by $f_\gamma > 0.9$ and $f_\gamma < 0.9$, respectively. Six signal regions are defined for events with three or four and more large-R jets, where at least one jet has to be identified as tight photon jet. In each of these signal regions, the H_T distribution is measured and serves as a discriminator between signal and background. The background is obtained from data by weighting events with at most one loose photon jet with the misidentification probability per jet, obtained as a function of p_{T} and η. The largest uncertainty in this search originates from the measurement of the signal efficiency, which is performed in a control region enriched with dileptonic $t\bar{t}$ decays. Fully-merged t decays of the kind $t \to e\nu b$ with an additional gluon from ISR, reconstructed in the same jet, are utilised as a proxy for the $gg\gamma$ signal jets. The pixel veto is reversed to allow for an efficient selection of electrons instead of photons. Differences between samples where the hadronisation is simulated with PYTHIA and HERWIG result in uncertainties in the signal efficiency between 24 and 85% for p_{T} between 200 and 500 GeV. This analysis would benefit from a better understanding of the differences in the parton shower and hadronisation in these simulations. The analysis excludes $\tilde{g}$ masses up to 1.7 TeV for a $\tilde{\chi}_1^0$ mass of 200 GeV. This is the first result exploiting jet substructure methods to identify single jets originating from two gluons and a photon.

In non-minimal SUSY scenarios [1189, 1190], the extended scalar sector contains additional CP-even and CP-odd scalar particles. These are the result of the symmetry breaking of a scalar superfield, which acquires a vacuum expectation value and dynamically generates the interactions of the Higgs doublet superfields. The lightest CP-odd scalar, the axion a, is considerably lighter than the other scalars for a wide range of the parameter space [1191, 1192]. For axion masses $m_a < m_H/2$, the decay $H \to aa$ is possible and is an intriguing channel to discover these new states [1193,

[6] Photon conversions $\gamma \to e^+e^-$ in the tracking detector may lead to tracks pointing in the direction of the photon candidate, such that a higher selection efficiency is obtained by a pixel veto, instead of discarding all candidates with an associated track.

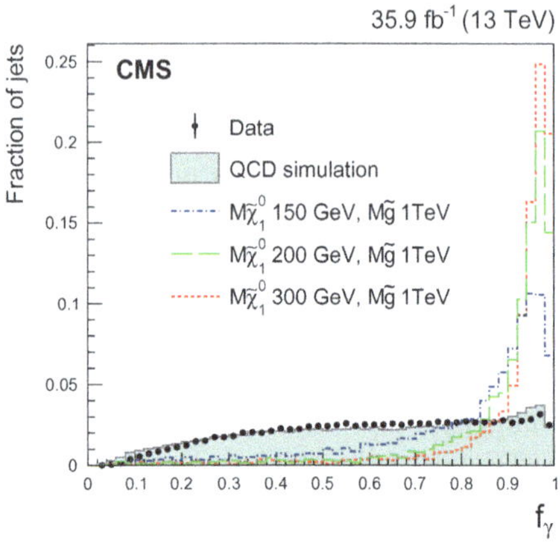

Fig. 5.29 Distribution in the photon subjet energy fraction f_γ for large-R jets with $p_T > 200$ GeV, three reconstructed subjets and $\tau_{31} < 0.4$. Taken from [1188]

1194].[7] Current experimental constraints on non-SM branching fractions of the H boson result in upper bounds of 19 at 95% CL [1195, 1196], which can be taken as the upper bound for $\mathcal{B}(H \to aa)$. Since the a boson mixes with the H boson, the hierarchy of predicted branching fractions into SM particles is similar, with $a \to b\bar{b}$, $a \to \tau\tau$ and $a \to \mu\mu$ the largest, if kinematically accessible [1197]. In parameter regions resulting in small fermionic couplings, the loop induced decays $a \to \gamma\gamma$ and $a \to gg$ can be dominant [1198]. Searches at the LHC for the decays $H \to aa$ are facilitated by constraints from the fact that the a bosons have the same mass and their combination gives m_H. The experimental challenges include soft decay products of the a bosons, as well as merged a decays for small m_a. The Jacobian peak of the $H \to aa$ decay implies that most of the a bosons have p_T of about 60 GeV or smaller, which is shared between the a decay products, making their reconstruction challenging. For masses $m_a \lesssim 15$ GeV, these are separated in ΔR by 0.5 and smaller, resulting in insufficient selection efficiencies when using isolation requirements or considering resolved decays only. Searches for $H \to aa$ have been conducted by ATLAS and CMS using 8 TeV data, analysing the channels 4τ [1199, 1200], 4μ [1201], $2\mu 2\tau$ [1200, 1202] and $2\mu 2b$ [1200]. A combination of 8 TeV searches by CMS excludes $\mathcal{B}(H \to aa)$ between 4 and 17%, depending on the specific model and the choice of parameters, for m_a between 5 and 65 GeV [1200]. At 13 TeV, leptonic final states have been analysed by ATLAS in search for 4μ decays [1203], and by CMS in search for $2\mu 2\tau$ decays [1204, 1205], where collimated signatures are reconstructed with dedicated lepton reconstruction techniques. An ATLAS analysis targets loop-induced decays in the $2\gamma 2g$ channel [1206]. The analysis is carried out in the VBF production mode since it has a higher cross section than VH production, but allows to suppress SM backgrounds through the presence of forward jets. In order to reduce the contamination of pileup jets in the forward region, forward jet vertex tagging is used [445]. This significantly improves the VBF

[7] Note that this signature also arises in other models of new physics, for example two-Higgs-doublet models with an additional pseudoscalar [1076].

signal purity up to a factor of four for events with an average pileup of 35. Searches including $a \to b\bar{b}$ decays have been reported by ATLAS [1207] and CMS [1208] in $2\mu 2b$ final states. Resolved a decays are considered in these analyses, which have been optimised for $20 < m_a < 62.5$ GeV. Smaller masses are not accessible in these searches, which would need boosted reconstruction techniques. The first search in the $2\tau 2b$ channel has been reported by CMS [1209], considering $e\tau_\mathrm{h}$, $\mu\tau_\mathrm{h}$ and $e\mu$ final states and resolved b decays. Final states with $\tau_\mathrm{h}\tau_\mathrm{h}$ are not considered because of high τ_h trigger thresholds. The large branching fractions for $a \to b\bar{b}$ and $a \to \tau\tau$ result in upper limits as low as 6% on $\mathcal{B}(H \to aa)$ in certain BSM scenarios from this analysis alone. The $4b$ final state is not accessible in gluon-gluon fusion production at the LHC, because of the overwhelming background from multijet production. However, it can be analysed in VH production, with leptonic V decays. The first search of this kind has been performed by ATLAS with 36.1 fb^{-1} of 13 TeV data [1210]. Resolved a decays are targeted, reconstructed using small-R jets with $p_\mathrm{T} > 20$ GeV. The signal is enriched using BDTs in signal regions categorised by the lepton, jet and b-jet multiplicity. The analysis is sensitive to m_a in the range 20–60 GeV, but the best sensitivity is achieved for $m_a \approx 30$ GeV. Below this value, the acceptance decreases because of overlapping b jets. This is remedied in a recent analysis by ATLAS, using the same dataset, but considering collimated $a \to b\bar{b}$ decays, optimised for m_a in the range 15–30 GeV [1211]. This signature needs a new identification method for boosted $a \to b\bar{b}$ jets, because of their much smaller p_T compared to typical boosted signatures. Large-R jets with $R = 0.8$ are obtained from a reclustering of small-R jets. The corresponding ghost-associated tracks are clustered using the exclusive k_T algorithm, returning either two or three track-jets. These serve as proxies for the flight directions of the two b quarks, where three track-jets are considered to capture cases with significant additional radiation. The use of exclusive k_T clustering instead of inclusive k_T clustering, as used for the identification of $H \to b\bar{b}$ jets, results in an improved matching of b quarks to track-jets. While the inclusive clustering finds correct matches only in 46% of the cases for $m_a = 20$ GeV, the exclusive clustering has a success rate of nearly 100%. A BDT is trained to identify $a \to b\bar{b}$ jets, using b-tagging information calculated from the track-jets, their angular separation ΔR, and their p_T asymmetry. The BDT achieves signal efficiencies of 25% and 35% for background rejections of 1% and 2.6%, respectively, where the latter is derived for b jets from $t\bar{t}$ decays. Efficiencies and corrections for simulated events are measured in a multijet sample enriched in $g \to b\bar{b}$ splittings. Since the distributions of the BDT input variables can change due to the non-zero a mass, the measurement is repeated using input distributions as expected from the decay $a \to b\bar{b}$ instead of distributions from a massless gluon. Within the statistical uncertainties of the measurement, the correction factors are found to be independent of m_a. Two signal regions are constructed to test for $H \to aa$ decays, defined by high purity and low purity $a \to b\bar{b}$ jets. The sensitivity achieved by this analysis improves the sensitivity at $m_a = 20$ GeV by a factor of 2.5 compared to the resolved $4b$ analysis [1210]. This highlights the potential gain from jet substructure methods in future analyses in $a \to b\bar{b}$ decays.

5.8 Leptoquarks

Leptoquarks (LQs)[8] are hypothetical particles that can decay to SM quarks and leptons. They are triplets with respect to the strong interaction, have fractional electric charge, and can be either scalar (spin 0) or vector (spin 1) particles. Many extensions to the SM, among them grand unification [1213–1215], technicolour [1216, 1217], and compositeness models [1218, 1219], predict the existence of these particles. The effective Buchmüller–Rückl–Wyler model [1220] incorporates the assumption that LQ interactions with SM fermions are renormalisable and gauge invariant, leading to restrictions on the allowed quantum numbers of LQs [1221]. Depending on its quantum numbers and the coupling structure, a given LQ can decay to any one of a number of different combinations of SM fermions. The couplings of LQs to leptons and quarks of different generations introduce flavour changing neutral currents that may be observable in precision measurements [1222]. Therefore, most searches at the Tevatron [1223], HERA [1224] and the LHC have focussed on LQs with couplings to quarks and leptons of the same generation. While simultaneous couplings to the first and second generations are tightly constrained by experimental data [1225], the bounds are weaker for couplings to the second and third generation, thus allowing the existence of LQs with non-diagonal couplings in the generation matrix [1226–1228]. Leptoquarks have received considerable theoretical and experimental attention recently, because of significant deviations from the predictions of the SM in measurements of decays of B mesons. In particular, deviations have been seen in the values of the ratio $R_{D^{(*)}}$, defined as the ratio of the $B \to D^{(*)}\tau\nu$ branching fraction to the $B \to D^{(*)}\mu\nu$ branching fraction. These deviations from the SM were first reported by BaBar [1229, 1230] and Belle [1231–1234] and have been confirmed by LHCb [1235, 1236] with a combined significance of about three standard deviations [1237, 1238]. The ratios of the branching fractions of $B \to K^{(*)}\mu\mu$ to $B \to K^{(*)}ee$, R_K and $R_{K^{(*)}}$, as measured by LHCb [1239–1243], show departures from lepton universality by 2.6 and 2.4 standard deviations, respectively. The measurement of the muon anomalous magnetic moment a_μ, one of the most precisely measured quantities in particle physics [1244], also deviates from the SM prediction by 3.5 standard deviations [1245]. These anomalies are among the most significant deviations from the SM observed so far. The existence of LQs with masses at the TeV scale and large couplings to third-generation quarks [1226, 1227, 1246–1256] has been proposed as a possible explanation for one, two, or all of these deviations. At the LHC, pair production of LQs is possible via gluon-gluon fusion or quark-antiquark annihilation, allowing direct searches to be performed. Single LQ production via quark-gluon scattering is subdominant for LQs coupled to heavy quarks, as it requires a heavy quark in the initial state. The pair production cross section depends on the mass of the scalar LQ and is known at NLO precision [1257]. The pair production cross section for vector LQs has been calculated at LO [1258] and is much larger than the scalar LQ cross section. The cross section for vector LQs depends

---

[8]The text in the first paragraph of this subsection has been taken from [1212] and has been written by the author. It has been adjusted to fit this book.

on an additional parameter κ, which is a dimensionless coupling and takes a value of $\kappa = 1$ in the Yang-Mills case and $\kappa = 0$ in the minimal coupling case [1259]. In experimental searches for LQ pair production, the acceptance is unchanged when considering vector instead of scalar LQs, such that the upper cross sections limits apply to both cases. The mass exclusion is stronger for vector LQs due to the larger production cross section.

Collider searches for the pair production of LQs with decays to third-generation quarks have been performed in the decay channels LQ $\rightarrow b\tau$, and LQ $\rightarrow b\nu$ at $\sqrt{s} = 8$ TeV [1168, 1260–1265] and at $\sqrt{s} = 13$ TeV [1266, 1267]. The first search targeting top-quark couplings in the decay channel LQ $\rightarrow t\tau$, has been performed by CMS at 8 TeV [1268]. The search is performed in $\ell\tau_h$+jets final states. One of the difficulties in this search is the estimation of $t\bar{t}$ and W+jets backgrounds from misidentified τ_h at high p_T, where dedicated measurements of the τ_h misidentification rate have been essential. With 35.9 fb^{-1} of 13 TeV data [1269], this search has been improved by adding the $\ell\tau_h\tau_h$ final state, made possible by the larger data set and higher $\sqrt{s}$. Additionally, a background estimation from data has been developed, making measurements of the τ_h misidentification rate unnecessary. The background from misidentified τ_h candidates is obtained from control regions with inverted τ_h isolation criteria. The measurement uses the reconstructed top quark p_T as sensitive observable, which is predicted to have a very different shape for LQ $\rightarrow t\tau$ production than from SM processes. In contrast to p_T of the τ_h candidate, this observable is not biased from the inversion of the τ_h isolation. The measured distributions of top quark p_T in two signal regions are shown in Fig. 5.30, where a different background composition is observed for events with opposite-sign and same-sign lepton charges. This search excludes scalar LQ masses below 900 GeV for $\mathcal{B}$(LQ $\rightarrow t\tau) = 1$. Very recently, ATLAS has performed an analysis in search for LQ $\rightarrow t\tau$ using 139 fb^{-1} of 13 TeV data [1270]. The search considers $\ell\tau_h$, $\ell\tau_h\tau_h$, $\ell\ell\tau_h$, $\ell\ell\tau_h\tau_h$ and $\ell\ell\ell\tau_h$ final states, where the categories with additional light leptons could be included thanks to the larger dataset compared to the CMS search at 13 TeV. The search defines seven signal categories, defined by the lepton charges, and measures the distribution in the effective LQ mass in each category. The sensitivity is dominated by the $\ell\tau_h$ category, but the $\ell\ell\tau_h$ and $\ell\ell\ell\tau_h$ categories bring a significant improvement. The search excludes scalar LQs below 1430 GeV for $\mathcal{B}$(LQ $\rightarrow t\tau) = 1$.

Constraints on LQ decays to neutrinos can be obtained from a reinterpretation of SUSY searches for direct squark production, in the limit of a massless neutralino. Since squarks are scalar particles and the neutralino is a spin-1/2 fermion, the same phenomenological signature is obtained for $\tilde{q} \rightarrow q\tilde{\chi}_1^0$ as expected from LQ $\rightarrow q\nu$ [1228]. This has been exploited for the first time by CMS in a reinterpretation of a SUSY search for the direct production of third-generation squarks [1170], which results in excluded scalar LQ masses below 1100 and 1020 GeV for $\mathcal{B}$(LQ $\rightarrow b\nu) = 1$ and $\mathcal{B}$(LQ $\rightarrow t\nu) = 1$, respectively [1271]. ATLAS has combined a search for LQ $\rightarrow b\tau$ using 36.1 fb^{-1} of 13 TeV data [1272], with the reinterpretation of four SUSY searches [528, 1273–1275], to cover all possible final states with third-generation couplings, LQ $\rightarrow (t/b)(\tau/\nu)$. Masses of scalar LQs below 800 GeV are excluded for any combination of branching fractions. The highest mass limit of

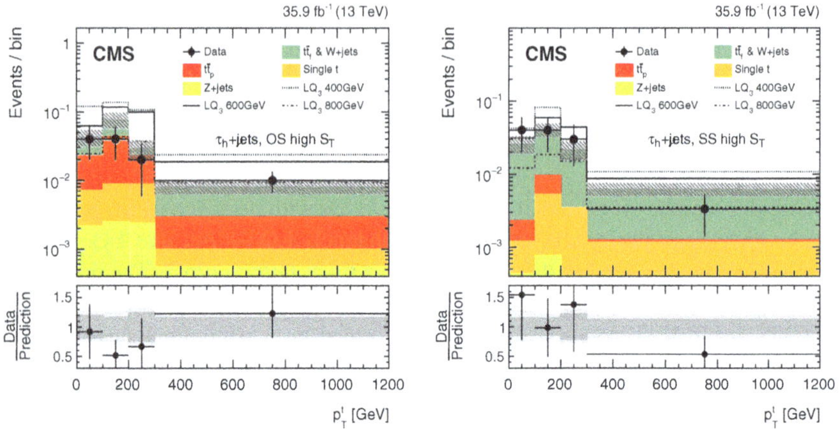

Fig. 5.30 Distributions in reconstructed top quark p_T for events with a muon and a τ_h candidate, for opposite-sign (left) and same-sign (right) lepton charges. Taken from [1269]

1050 GeV is obtained for $\mathcal{B}(\mathrm{LQ} \to b\tau) = 1$, for which this search has been optimised. A recent ATLAS search for top squarks in the $t\bar{t}+p_T^{\mathrm{miss}}$ channel in all-hadronic final states, using 139 fb^{-1} of 13 TeV data [1165], can be reinterpreted as a LQ search as well. The search is described above and relies on top tagging for the identification of t decays. The obtained mass limit is 1240 GeV for $\mathcal{B}(\mathrm{LQ} \to t\nu) = 1$.

The first search considering non-diagonal couplings in the quark and lepton generation matrix is a search for LQ $\to t\mu$, carried out by CMS with 35.9 fb^{-1} of 13 TeV data [1212]. The search considers $\mu\mu$+jets and $\mu\mu\ell$+jets final states, where the mass of any two muons has to be greater than 111 GeV to suppress the Z+jets and ZV backgrounds. This search considers only small-R jets, but large boosts of the t quarks are taken into account by considering events where the hadronic t quark decay can be reconstructed with only one or two small-R jets. For each event, the $t \to \ell\nu b$ decay is constructed from permutations of one or more of the seven p_T-leading small-R jets, one of the three p_T-leading muons or the p_T-leading electron, and p_T^{miss}. The hadronic top quark candidates are constructed using all permutations of small-R jets not assigned to the $t \to \ell\nu b$ decay. The LQ candidates are assembled from top quark candidates and the two p_T-leading muons that have not been associated to the leptonic top quark. A χ^2 variable that takes into account the mass of each top quark candidate and the relative mass difference between the two LQ candidates is used to select the best pair of LQ candidates for each event. The reconstructed LQ mass is measured for events with three leptons, and the S_T distribution is measured for events with two leptons. While the former category has higher sensitivity for LQ masses below 400 GeV, the latter category allows to probe high LQ masses. The analysis excludes scalar LQ masses below 1420 GeV for $\mathcal{B}(\mathrm{LQ} \to t\mu) = 1$, and a combination with searches for LQ $\to t\tau$ [1269] and LQ $\to b\nu$ [1271] allows to exclude LQ masses below 900 GeV for any combination of branching fractions fulfilling $\mathcal{B}(\mathrm{LQ} \to t\mu) + \mathcal{B}(\mathrm{LQ} \to t\tau) = 1$ or $\mathcal{B}(\mathrm{LQ} \to t\mu) + \mathcal{B}(\mathrm{LQ} \to b\nu) = 1$.

The excluded mass range for LQs coupling to top quarks goes beyond 1 TeV in many cases, suggesting that searches considering boosted top quarks and jet substructure can improve the sensitivity at the LHC. In particular, when combining signatures from pair and single production, the sensitivity can be drastically improved by the use of top tagging [1276, 1277]. The reason for the improvement lies in the mechanisms of single production, which results in the same $t\ell t\ell$ final state for LQs coupling to top quarks as pair production, but with a low-p_T t produced in association with the $\ell + \mathrm{LQ}(\to t\ell)$ system from the production vertex. The corresponding signature can be captured by selecting events with an $\ell\ell$ pair and a t-tagged large-R jet, capturing events from single and pair production simultaneously. This strategy has been followed by a CMS analysis using 137 fb^{-1} of 13 TeV data [1278], which has been optimised for the scenario of $\mathcal{B}(\mathrm{LQ} \to t\tau) = \mathcal{B}(\mathrm{LQ} \to b\nu) = 0.5$, resulting in $pp \to t\tau\nu b$ production. For single production, the b can be soft or very forward, outside of the detector acceptance. Events are selected with $p_\mathrm{T}^\mathrm{miss} > 200$ GeV, one τ_h candidate, at least one b-tagged small-R jet and a reconstructed t candidate. The analysis considers fully merged, partially merged and resolved t decays. Fully merged t decays are reconstructed using the soft drop jet mass and τ_{32}. Partially merged decays are reconstructed with a W jet, identified using the pruned jet mass and τ_{21}, and a small-R jet with a combined mass close to m_t. Resolved decays are reconstructed using three small-R jets. Signal events are categorised by a successful boosted or resolved t reconstruction and the b jet multiplicity. The distribution in S_T is measured in the resulting four signal regions, where S_T is the sum of $p_\mathrm{T}^\mathrm{miss}$ and p_T of the reconstructed t and τ_h. The analysis excludes scalar LQ masses up to 950 GeV, considering pair production only, which constitutes an improvement by about 140 GeV compared to the previous CMS search for $\mathrm{LQ} \to t\tau$, which obtained a mass limit of 810 GeV for $\mathcal{B}(\mathrm{LQ} \to t\tau) = 0.5$ [1269]. Taking into account single and pair production, the limit improves to 980 and 1020 GeV for LQ-quark-lepton couplings of 1.5 and 2.5, respectively.

Very recently, ATLAS has completed a search for cross-generation couplings in the $\mathrm{LQ} \to te$ and $\mathrm{LQ} \to t\mu$ channels using 139 fb^{-1} of 13 TeV data [1279]. The analysis aims at LQ masses above 1 TeV with decays of boosted t quarks. Events are selected in e^+e^- and $\mu^+\mu^-$ final states with dilepton masses greater than 120 GeV and two trimmed large-R jets with $p_\mathrm{T} > 200$ GeV and $m_\mathrm{jet} > 50$ GeV. The analysis treats the e^+e^- and $\mu^+\mu^-$ channels separately, because simultaneous LQ couplings to electrons and muons are tightly constrained by the absence of flavour-changing neutral currents [1225]. The signal is enhanced over SM backgrounds with the use of a BDT with 29 inputs in the electron channel and 32 inputs in the muon channel. The inputs include the discriminating variables from a LQ reconstruction using the recursive jigsaw technique [1280], kinematic variables derived from the leptons, jets and $p_\mathrm{T}^\mathrm{miss}$, the large-$R$ jet masses, and three jet substructure variables: the k_T-splitting scale $\sqrt{d_{32}}$, τ_{32} and the mass of the two pseudojets in the second-to-last clustering step of the k_T algorithm, Q_W [486]. Interestingly, the jet substructure variables are only included in the muon channel, because of an overlap removal of electrons and large-R jets in the electron channel. The most important input variables are the dilepton invariant mass, the scalar sum of the lepton p_{T} and the jet masses of the two leading-p_{T} large-R jets. Three regions in the BDT output are used to define the signal region. Because the BDT is adjusted to include the signal mass as input [1281], the measured distributions change for each signal mass probed. The analysis reports a lower limit on the mass of scalar LQs of 1480 GeV for $\mathcal{B}(\mathrm{LQ} \rightarrow te) = 1$ and and 1470 GeV for $\mathcal{B}(\mathrm{LQ} \rightarrow t\mu) = 1$. The LQ $\rightarrow te$ result comprises the first experimental result from a direct search in this channel. Comparing the expected limits with the previous analysis by CMS [1212], the sensitivity in the LQ $\rightarrow t\mu$ channel improved by 120 GeV. The obtained mass limits are weaker by about 300 GeV than the ones obtained in an ATLAS search for LQs decaying to electrons and muons together with first- and second-generation quarks using the same dataset [1282]. This is a general difficulty of LQ searches involving top quarks, where the different t decay modes and the difficulties in reconstructing and identifying the t decay reduce the sensitivity compared to light-quark initiated high-p_{T} jets. However, this can also be encouraging because it shows the potential left for future analyses in search for LQs decaying to top quarks.

Chapter 6
Summary

Abstract While the field of jet substructure is still relatively young, it has led to a myriad of studies and measurements at the LHC. These probe regions which are inaccessible with other methods, emphasising the importance of jet substructure techniques. These techniques will be refined and improved in the years to come, resulting in indispensable tools for experimental studies.

Jet substructure has permeated a multitude of analyses at the LHC. These range from searches for new physical phenomena, a better understanding of the strong force and top quark production, to studies of the Higgs boson. Jet substructure analyses of LHC data have been facilitated by dedicated experimental studies of the detector response and elaborate theoretical calculations, made possible by the precise formulation of jet substructure algorithms. Many of the results published by the ATLAS and CMS Collaborations have profited from these developments, either improving the precision or enabling the execution of analyses in the first place.

Numerous theoretical and experimental challenges in the field of jet substructure make the successful application of these techniques a remarkable achievement. On the theory side, either fixed-order perturbative calculations with virtual corrections and matched parton showers are needed to achieve sufficient accuracy for a reliable prediction of jet substructure observables; or all-order resummations of the leading and subleading effects need to be performed. Insights into the advantages and limitations of one or the other approach can be gained by comparing results from different calculations. An additional complication arises from non-perturbative effects, such as hadronisation and particle decays, which can lead to sizeable corrections that are difficult to estimate from first principles. These can be approximated through corresponding models implemented in event generators or non-perturbative parameters in analytic calculations, leading to distortions of distributions. In some models, these corrections are intricately connected with the underlying event, which also has a perturbative and a non-perturbative part. Only by simulating all of these effects, realistic estimates of the performance of jet substructure algorithms can be obtained, and the estimation of fundamental SM parameters from jet substructure measurements becomes feasible. On the experimental side, an excellent understanding of the reconstruction efficiencies of all particles produced in the collision is needed

R. Kogler, *Advances in Jet Substructure at the LHC*, Springer Tracts in Modern Physics 284, https://doi.org/10.1007/978-3-030-72858-8_6

for jet substructure analyses. These include hit efficiencies in the tracking detectors, the reconstruction of tracks in dense environments, the clustering and calibration of energy deposits in granular calorimeters and the possibility to combine tracking and calorimetric measurements. The influence of in-time and out-of-time pileup is an important aspect that needs to be corrected for. Calibrations of the jet energy scale and resolution are as important as calibrations of the jet mass scale and resolution. In experimental analyses, jet collections with different distance parameters are frequently used, such that corrections for different types of jets need to be derived. The performance of jet substructure taggers needs to be measured in dedicated samples of real collision data in order to validate their expected performance from simulation. A reliable prediction of jet substructure observables by event generators is key to precise measurements, such that the tuning of free parameters to experimental data is another important aspect.

Measurements that have tested our understanding of jet substructure constitute unfolded measurements of the groomed and ungroomed jet mass in light quark and gluon jets. These help to understand the regions of validity in the approximations made in event generators for describing jet substructure observables. These measurements also show the efficacy of grooming algorithms. Measurements of the jet mass in soft-drop groomed jets show good agreement with analytic calculations in their region of validity, establishing confidence in these calculations. This opens the possibility for a precise determination of the strong coupling constant from jet substructure measurements. Differential measurements of the radiation pattern in light quark and gluon jets show an overall good description of the data in regions dominated by both, perturbative and non-perturbative effects. Together with measurements of (charged) particle multiplicities, these measurements will help to improve the modelling of jet substructure observables in future tunes of event generators, where gluon-initiated jets are less well constrained than quark initiated jets. Unfolded jet mass measurements of W and t jets have shown a reliable calibration and modelling of fully merged two- and three-prong decays. The measurements of t jets have enabled the determination of the top quark mass from the production of highly boosted top quarks. Unfolded measurements of jet substructure observables in resolved, partially and fully merged t decays resulted in comprehensive data that can be used for improvements in the description of additional radiation and the modelling of colour flow and other non-perturbative effects in $t\bar{t}$ production.

Developments of jet substructure techniques have enabled a number of SM measurements at high p_{T}. Measurements of electroweak vector boson production in V+jets processes have to cope with large backgrounds from multijet production. Conversely, studies of diboson production have to reduce V+jets processes for reliable measurements. Efficient substructure taggers, in combination with methods for estimating the non-resonant backgrounds, enabled first studies of these processes in hadronic channels and allowed to constrain anomalous production mechanisms. Jet substructure techniques are key to measurements of high-p_{T} Higgs boson production in $b\bar{b}$ and $c\bar{c}$ decays. Developments of b and c flavour tagging in dense environments have been crucial for first results in these channels. In studies of $t\bar{t}$ production, top quark tagging has been indispensable for measurements at high p_{T}. A combination of b flavour tagging on subjets and identification of the three-prong structure of jets has extended the reach of measurements in the lepton+jets channel and enabled measurements in the all-hadronic channel. These measurements allow for comparisons with precision calculations in regions where electroweak and higher-order QCD corrections are important. Additionally, these measurements have sensitivity to the gluon PDF at high longitudinal momentum fraction, making unbiased simultaneous extractions of the top quark mass, the strong coupling constant and proton PDFs possible in the future.

Data analyses in search for new physical phenomena have relied on jet substructure techniques since the first pp collisions have been recorded at the LHC. Algorithms for electroweak boson tagging have been crucial in searches for resonant diboson production. Similarly, Higgs boson taggers made searches for HH and VH resonances feasible. Top quark tagging has been central in searches for resonant $t\bar{t}$ production. Heavy partners of third-generation quarks have been searched for using the full suite of V, H and t taggers, in some cases combined in multiclass taggers. Jet substructure techniques have also been instrumental in searches for dark matter, probing parameter regions of large mediator and small dark matter masses. In addition, searches for very light mediators have been made possible with jet substructure methods, by analysing the jet mass distribution in events where a light mediator is expected to recoil against a highly energetic jet from initial state radiation. In searches for supersymmetry, t tagging has been employed in searches for top squarks, probing the region of high squark and gluino masses. In supersymmetry models with R-parity violation, dedicated substructure taggers have allowed to probe a parameter space inaccessible before. Last but not least, t taggers have been employed in searches for leptoquarks with large couplings to top quarks, improving the sensitivity at high masses. In all of these searches, dedicated methods have been employed to estimate the backgrounds from SM processes, in some cases ameliorated by jet substructure methods. A very important ingredient is the knowledge of the signal efficiency, directly related to the efficiency of jet substructure taggers, and consequently affecting the analysis sensitivity. Measurements in control regions are performed to assess the efficiencies, and estimate the corresponding uncertainties. The background estimation and the measurement of signal efficiencies are the most crucial aspects in searches for new physics, and often represent the most arduous parts. Despite the numerous results from ATLAS and CMS in search for new phenomena, there are still a number of final states uncovered. An example are cascade decays of beyond-the-SM resonances, which can be classified by the SM particles produced [1283, 1284]. In some cases, these decays can be theoretically motivated [1285], in other cases, the final states are signal agnostic, but could lead to an unexpected discovery.

Besides analyses where jet substructure methods have been used explicitly, there are many analyses profiting from the developments in this field. Improvements of particle flow algorithms, progress in novel reconstruction methods such as TCCs and UFOs, together with dedicated calibrations, have led to performance improvements for jet measurements. The development and commissioning of pileup mitigation methods such as PUPPI, SoftKiller and Constituent Subtraction have led to a stable jet reconstruction performance up to very high instantaneous luminosities. Lepton and photon identification algorithms also profit from these developments, as their performance is influenced by contributions from pileup. Virtually every analysis performed by the ATLAS and CMS Collaborations uses some aspect of jet substructure methods, such that a rudimentary understanding of these is beneficial for everyone working in this field.

Jet substructure studies in heavy ion collisions constitute an aspect that has not been discussed in this book. In these dense environments, grooming methods allow to access the partonic structure of jets, while mitigating the effects of the underlying event and hadronisation. The partonic energy loss and p_T-broadening of jets induced by gluon radiation from traversing a large nucleus can be enhanced significantly in hot matter, as produced in heavy-ion collisions [1286]. These can be accessed through measurements of the groomed momentum sharing of the two-prong jet substructure in pp and heavy ion collisions [1287]. Insights into the interactions of quarks and gluons with the hot medium can also be obtained from groomed [1288] and ungroomed [1289] jet mass measurements. Future measurements can benefit from jet substructure algorithms to separate the partonic energy loss and gluon-induced radiation from non-perturbative effects in the scattering with the medium [1290]. Jet grooming may also help to develop a factorisation theorem in SCET, allowing to resum the large logarithms arising from final state measurements when summing over multiple interactions of the jet with the medium [1291].

In the future, jet substructure methods will continue to gain in importance. This is in part due to the upgrade to the HL-LHC and the resulting larger data sample, together with an increase in centre-of-mass energy to 14 TeV, allowing for more precise studies of high p_T processes. Even the high-energy tails of rare electroweak processes will become accessible, and jet substructure taggers will play an important role in their measurements. The drawback of the high instantaneous luminosities aimed at with the HL-LHC are higher levels of pileup. These will require advanced mitigation strategies, where jet substructure methods will be indispensable. The upgrades of the ATLAS and CMS detectors will enable the use of particle flow and pileup mitigation algorithms already at the first trigger level, allowing for online jet substructure selections at a rate of 40 MHz. Upgrades of the inner tracking detectors will extend the tracking coverage to pseudorapidities of about 4, and higher granularities of the calorimeters will ensure an excellent angular resolution in the reconstruction of particles. Overall, the detector capabilities for jet substructure analyses will be enhanced with respect to the present detectors. Another reason for the future importance of jet substructure algorithms are studies using advanced machine learning techniques. These can be used to design powerful jet substructure taggers, but they can also help to identify variables carrying additional information, which have not been considered so far. For example, machine learning can help to construct a single observable, which is the product of a number of jet substructure observables, carrying most of the discrimination power of a complex artificial neural network [1292, 1293]. Similar techniques can also help to identify jet substructure observables for the identification of electrons inside jets [1294] or additional variables for jet tagging [1295]. Classifying the information learned by algorithms of artificial intelligence, and assessing the uncertainty on the corresponding predictions, is an active field of study with the potential to reveal new insights in the physics of jet substructure.

The capabilities for jet substructure measurements will be a key consideration in the design of new detectors. Future collider concepts [1296] are being discussed, which could raise the centre-of-mass energy in *pp* collisions from 14 to 27 TeV [1297], even up to 100 TeV [1298]. These will result in jets with energies in the range of several TeV, where high granularity calorimeters and excellent tracking capabilities are needed for an optimal performance of jet substructure taggers [1299, 1300]. It will be an exciting scientific endeavour to understand the dynamics of jets at these energies.

Whatever the future will bring, jet substructure techniques will play an important role in all aspects of particle physics at the highest energy collisions produced by particle accelerators.

References

1. ATLAS Collaboration, Observation of a new particle in the search for the Standard Model Higgs boson with the ATLAS detector at the LHC. Phys. Lett. B **716**, 1 (2012). https://doi.org/10.1016/j.physletb.2012.08.020. arXiv:1207.7214
2. C.M.S. Collaboration, Observation of a new boson at a mass of 125 GeV with the CMS experiment at the LHC. Phys. Lett. B **716**, 30 (2012). https://doi.org/10.1016/j.physletb.2012.08.021. arXiv:1207.7235
3. F. Englert, R. Brout, Broken symmetry and the mass of gauge vector mesons. Phys. Rev. Lett. **13**, 321 (1964). https://doi.org/10.1103/PhysRevLett.13.321
4. P.W. Higgs, Broken symmetries and the masses of gauge bosons. Phys Rev Lett. **13**, 508 (1964). https://doi.org/10.1103/PhysRevLett.13.508
5. G. Guralnik, C. Hagen, T. Kibble, Global conservation laws and massless particles. Phys. Rev. Lett. **13**, 585 (1964). https://doi.org/10.1103/PhysRevLett.13.585
6. ATLAS Collaboration, Measurements of the Higgs boson production and decay rates and coupling strengths using *pp* collision data at $\sqrt{s} = 7$ and 8 TeV in the ATLAS experiment. Eur. Phys. J. C **76**, 6 (2016). https://doi.org/10.1140/epjc/s10052-015-3769-y. arXiv:1507.04548
7. ATLAS, CMS Collaborations, Measurements of the Higgs boson production and decay rates and constraints on its couplings from a combined ATLAS and CMS analysis of the LHC pp collision data at $\sqrt{s} = 7$ and 8 TeV. JHEP **08**, 45 (2016). https://doi.org/10.1007/JHEP08(2016)045. arXiv:1606.02266
8. ATLAS Collaboration, Combined measurements of Higgs boson production and decay using up to 80 fb^{-1} of proton-proton collision data at $\sqrt{s} = 13$ TeV collected with the ATLAS experiment. Phys. Rev. D **101**, 012002 (2020). https://doi.org/10.1103/PhysRevD.101.012002. arXiv:1909.02845
9. C.M.S. Collaboration, Precise determination of the mass of the Higgs boson and tests of compatibility of its couplings with the standard model predictions using proton collisions at 7 and 8 TeV. Eur. Phys. J. C **75**, 212 (2015). https://doi.org/10.1140/epjc/s10052-015-3351-7. arXiv:1412.8662
10. CMS Collaboration, Combined measurements of Higgs boson couplings in proton-proton collisions at $\sqrt{s} = 13$ TeV. Eur. Phys. J. C **79**, 421 (2019). https://doi.org/10.1140/epjc/s10052-019-6909-y. arXiv:1809.10733

R. Kogler, *Advances in Jet Substructure at the LHC*, Springer Tracts in Modern Physics 284, https://doi.org/10.1007/978-3-030-72858-8

11. S. Rappoccio, The experimental status of direct searches for exotic physics beyond the standard model at the Large Hadron Collider. Rev. Phys. **4**, 100027 (2019). https://doi.org/10.1016/j.revip.2018.100027. arXiv:1810.10579
12. Gfitter Group, H. Flächer, M. Goebel, J. Haller, A. Hoecker, K. Mönig, J. Stelzer, Revisiting the global electroweak fit of the standard model and beyond with Gfitter. Eur. Phys. J. C **60**, 543 (2009). https://doi.org/10.1140/epjc/s10052-009-0966-6. arXiv:0811.0009. Erratum: Eur. Phys. J. C **71**, 1718 (2011)
13. Gfitter Group, M. Baak et al., Updated status of the global electroweak fit and constraints on new physics. Eur. Phys. J. C **72**, 2003 (2012). https://doi.org/10.1140/epjc/s10052-012-2003-4. arXiv:1107.0975
14. Gfitter Group, M. Baak et al., The electroweak fit of the standard model after the discovery of a new boson at the LHC. Eur. Phys. J. C **72**, 2205 (2012). https://doi.org/10.1140/epjc/s10052-012-2205-9. arXiv:1209.2716
15. Gfitter Group, M. Baak, J. Cúth, J. Haller, A. Hoecker, R. Kogler, K. Mönig et al., The global electroweak fit at NNLO and prospects for the LHC and ILC. Eur. Phys. J. C **74**, 3046 (2014). https://doi.org/10.1140/epjc/s10052-014-3046-5. arXiv:1407.3792
16. Gfitter Group, J. Haller, A. Hoecker, R. Kogler, K. Mönig, T. Peiffer, J. Stelzer, Update of the global electroweak fit and constraints on two-Higgs-doublet models, Eur. Phys. J. C **78**, 675 (2018). https://doi.org/10.1140/epjc/s10052-018-6131-3. arXiv:1803.01853
17. K. Babu, C.F. Kolda, Higgs mediated $B^0 \to \mu^+\mu^-$ in minimal supersymmetry. Phys. Rev. Lett. **84**, 228 (2000). https://doi.org/10.1103/PhysRevLett.84.228. arXiv:hep-ph/9909476
18. G. D'Ambrosio, G. Giudice, G. Isidori, A. Strumia, Minimal flavor violation: an effective field theory approach. Nucl. Phys. B **645**, 155 (2002). https://doi.org/10.1016/S0550-3213(02)00836-2. arXiv:hep-ph/0207036
19. G. Bevilacqua, M. Worek, Constraining BSM physics at the LHC: Four top final states with NLO accuracy in perturbative QCD. JHEP **07**, 111 (2012). https://doi.org/10.1007/JHEP07(2012)111. arXiv:1206.3064
20. J. Baglio, A. Djouadi, R. Gröber, M. Mühlleitner, J. Quevillon, M. Spira, The measurement of the Higgs self-coupling at the LHC: Theoretical status. JHEP **04**, 151 (2013). https://doi.org/10.1007/JHEP04(2013)151. arXiv:1212.5581
21. A. Altheimer et al., Jet substructure at the Tevatron and LHC: new results, new tools, new benchmarks. J. Phys. G **39**, 063001 (2012). https://doi.org/10.1088/0954-3899/39/6/063001. arXiv:1201.0008
22. A. Altheimer et al., Boosted objects and jet substructure at the LHC. Report of BOOST2012, held at IFIC Valencia, 23rd–27th of July 2012. Eur. Phys. J. C **74**, 2792 (2014). https://doi.org/10.1140/epjc/s10052-014-2792-8. arXiv:1311.2708
23. D. Adams et al., Towards an understanding of the correlations in jet substructure. Eur. Phys. J. C **75**, 409 (2015). https://doi.org/10.1140/epjc/s10052-015-3587-2. arXiv:1504.00679
24. A.J. Larkoski, I. Moult, B. Nachman, Jet substructure at the Large Hadron Collider: a review of recent advances in theory and machine learning. Phys. Rept. **841**, 1 (2020). https://doi.org/10.1016/j.physrep.2019.11.001. arXiv:1709.04464
25. S. Marzani, G. Soyez, M. Spannowsky, Looking inside jets: an introduction to jet substructure and boosted-object phenomenology. Springer Lect. Notes Phys. **958** (2019). https://doi.org/10.1007/978-3-030-15709-8
26. R. Kogler, B. Nachman, A. Schmidt et al., (eds.), Jet substructure at the Large Hadron Collider: experimental review. Rev. Mod. Phys. **91**, 045003 (2019). https://doi.org/10.1103/RevModPhys.91.045003. arXiv:1803.06991
27. J.A.D.E. Collaboration, Observation of planar three jet events in e^+e^- annihilation and evidence for gluon bremsstrahlung. Phys. Lett. B **91**, 142 (1980). https://doi.org/10.1016/0370-2693(80)90680-2
28. PLUTO Collaboration, Evidence for gluon bremsstrahlung in e^+e^- annihilations at high energies. Phys. Lett. B **86**, 418 (1979). https://doi.org/10.1016/0370-2693(79)90869-4
29. MARK-J Collaboration, Discovery of three jet events and a test of quantum chromodynamics at PETRA energies. Phys. Rev. Lett. **43**, 830 (1979). https://doi.org/10.1103/PhysRevLett.43.830

30. TASSO Collaboration, Evidence for planar events in e^+e^- annihilation at high energies. Phys. Lett. B **86**, 243 (1979). https://doi.org/10.1016/0370-2693(79)90830-X
31. D0 Collaboration, Observation of the top quark. Phys. Rev. Lett. **74**, 2632 (1995). https://doi.org/10.1103/PhysRevLett.74.2632. arXiv:hep-ex/9503003
32. C.D.F. Collaboration, Observation of top quark production in $\bar{p}p$ collisions with the Collider Detector at Fermilab. Phys. Rev. Lett. **74**, 2626 (1995). https://doi.org/10.1103/PhysRevLett.74.2626. arXiv:hep-ex/9503002
33. ATLAS Collaboration, Observation of Higgs boson production in association with a top quark pair at the LHC with the ATLAS detector. Phys. Lett. B **784**, 173 (2018). https://doi.org/10.1016/j.physletb.2018.07.035. arXiv:1806.00425
34. CMS Collaboration, Observation of $t\bar{t}$ H production. Phys. Rev. Lett. **120**, 231801 (2018). https://doi.org/10.1103/PhysRevLett.120.231801. arXiv:1804.02610
35. ATLAS Collaboration, Observation of $H \rightarrow b\bar{b}$ decays and VH production with the ATLAS detector. Phys. Lett. B **786**, 59 (2018). https://doi.org/10.1016/j.physletb.2018.09.013. arXiv:1808.08238
36. CMS Collaboration, Observation of Higgs boson decay to bottom quarks. Phys. Rev. Lett. **121**, 121801 (2018). https://doi.org/10.1103/PhysRevLett.121.121801. arXiv:1808.08242
37. M.H. Seymour, Searches for new particles using cone and cluster jet algorithms: a comparative study. Z. Phys. C **62**, 127 (1994). https://doi.org/10.1007/BF01559532
38. J.M. Butterworth, B.E. Cox, J.R. Forshaw, WW scattering at the CERN LHC. Phys. Rev. D **65**, 096014 (2002). https://doi.org/10.1103/PhysRevD.65.096014. arXiv:hep-ph/0201098
39. K. Agashe, A. Belyaev, T. Krupovnickas, G. Perez, J. Virzi, LHC signals from warped extra dimensions. Phys. Rev. D **77**, 015003 (2008). https://doi.org/10.1103/PhysRevD.77.015003. arXiv:hep-ph/0612015
40. J.M. Butterworth, A.R. Davison, M. Rubin, G.P. Salam, Jet substructure as a new Higgs search channel at the LHC. Phys. Rev. Lett. **100**, 242001 (2008). https://doi.org/10.1103/PhysRevLett.100.242001. arXiv:0802.2470
41. D.E. Kaplan, K. Rehermann, M.D. Schwartz, B. Tweedie, Top tagging: a method for identifying boosted hadronically decaying top quarks. Phys. Rev. Lett. **101**, 142001 (2008). https://doi.org/10.1103/PhysRevLett.101.142001. arXiv:0806.0848
42. G. Hanson, G.S. Abrams, A.M. Boyarski, M. Breidenbach, F. Bulos, W. Chinowsky et al., Evidence for jet structure in hadron production by e^+e^- annihilation. Phys. Rev. Lett. **35**, 1609 (1975). https://doi.org/10.1103/PhysRevLett.35.1609
43. J.D. Bjorken, S.J. Brodsky, Statistical model for electron-positron annihilation into hadrons. Phys. Rev. D **1**, 1416 (1970). https://doi.org/10.1103/PhysRevD.1.1416
44. S. Brandt, C. Peyrou, R. Sosnowski, A. Wroblewski, The principal axis of jets-an attempt to analyse high-energy collisions as two-body processes. Phys. Lett. **12**, 57 (1964). https://doi.org/10.1016/0031-9163(64)91176-X
45. E. Farhi, Quantum chromodynamics test for jets. Phys. Rev. Lett. **39**, 1587 (1977). https://doi.org/10.1103/PhysRevLett.39.1587
46. E. Farhi, Quantum chromodynamics test for jets. Phys. Rev. Lett. **39**, 1587–1588 (1977). https://doi.org/10.1103/PhysRevLett.39.1587
47. G. Sterman, S. Weinberg, Jets from quantum chromodynamics. Phys. Rev. Lett. **39**, 1436 (1977). https://doi.org/10.1103/PhysRevLett.39.1436
48. G.P. Salam, G. Soyez, A Practical Seedless Infrared-Safe Cone jet algorithm. JHEP **05**, 86 (2007). https://doi.org/10.1088/1126-6708/2007/05/086. arXiv:0704.0292
49. J.A.D.E. Collaboration, Experimental studies on multijet production in e^+e^- annihilation at PETRA energies. Z. Phys. C **33**, 23 (1986). https://doi.org/10.1007/BF01410449
50. J.A.D.E. Collaboration, Experimental investigation of the energy dependence of the strong coupling strength. Phys. Lett. B **213**, 235 (1988). https://doi.org/10.1016/0370-2693(88)91032-5
51. S. Catani, Y. Dokshitzer, M. Olsson, G. Turnock, B. Webber, New clustering algorithm for multijet cross sections in e^+e^- annihilation. Phys. Lett. B **269**, 432 (1991). https://doi.org/10.1016/0370-2693(91)90196-W

52. S. Catani, Y. Dokshitzer, B. Webber, The $k_{\perp}$-clustering algorithm for jets in deep inelastic scattering and hadron collisions. Phys. Lett. B **285**, 291 (1992). https://doi.org/10.1016/0370-2693(92)91467-N
53. S.D. Ellis, D.E. Soper, Successive combination jet algorithm for hadron collisions. Phys. Rev. D **48**, 3160–3166 (1993). https://doi.org/10.1103/PhysRevD.48.3160. arXiv:hep-ph/9305266
54. S. Catani, Y. Dokshitzer, M. Seymour, B. Webber, Longitudinally-invariant $k_{\perp}$-clustering algorithms for hadron-hadron collisions. Nucl. Phys. B **406**, 187 (1993). https://doi.org/10.1016/0550-3213(93)90166-M
55. Y.L. Dokshitzer, G.D. Leder, S. Moretti, B.R. Webber, Better jet clustering algorithms. JHEP **08**, 001 (1997). https://doi.org/10.1088/1126-6708/1997/08/001. arXiv:hep-ph/9707323
56. M. Wobisch, T. Wengler, Hadronization corrections to jet cross-sections in deep inelastic scattering, in *Monte Carlo generators for HERA physics. Proceedings, Workshop*, Hamburg, Germany, 1998–1999 (1998), arXiv:hep-ph/9907280
57. M. Cacciari, G.P. Salam, G. Soyez, The anti-k_t jet clustering algorithm. JHEP **04**, 063 (2008). https://doi.org/10.1088/1126-6708/2008/04/063. arXiv:0802.1189
58. R.K. Ellis, W.J. Stirling, B.R. Webber, QCD and collider physics. Camb. Monogr. Part. Phys. Nucl. Phys. Cosmol. **8** (1996)
59. G.P. Salam, Towards jetography. Eur. Phys. J. C **67**, 637 (2010). https://doi.org/10.1140/epjc/s10052-010-1314-6. arXiv:0906.1833
60. S.D. Ellis, Z. Kunszt, D.E. Soper, Jets at hadron colliders at order α_s^3: a look inside. Phys. Rev. Lett. **69**, 3615 (1992). https://doi.org/10.1103/PhysRevLett.69.3615. arXiv:hep-ph/9208249
61. O.P.A.L. Collaboration, A direct observation of quark-gluon jet differences at LEP. Phys. Lett. B **265**, 462 (1991). https://doi.org/10.1016/0370-2693(91)90082-2
62. ALEPH Collaboration, Study of the subjet structure of quark and gluon jets. Phys. Lett. B **346**, 389 (1995). https://doi.org/10.1016/0370-2693(95)00037-L
63. DELPHI Collaboration, Energy dependence of the differences between the quark and gluon jet fragmentation. Z. Phys. **C70**, 179 (1996). https://doi.org/10.1007/s002880050095
64. L3 Collaboration, K_s^0 and Λ production in quark and gluon jets at LEP. Phys. Lett. B **407**, 389 (1997). https://doi.org/10.1016/S0370-2693(97)00864-2. Erratum: Phys. Lett. B **427**, 409 (1998)
65. CDF Collaboration, A measurement of jet shapes in $p\bar{p}$ collisions at $\sqrt{s} = 1.8$ TeV. Phys. Rev. Lett. **70**, 713 (1993). https://doi.org/10.1103/PhysRevLett.70.713
66. D0 Collaboration, Transverse energy distributions within jets in $p\bar{p}$ collisions at $\sqrt{s} = 1.8$ TeV. Phys. Lett. B **357**, 500 (1995). https://doi.org/10.1016/0370-2693(95)00889-S
67. Z.E.U.S. Collaboration, Measurement of jet shapes in photoproduction at HERA. Eur. Phys. J. C **2**, 61 (1998). https://doi.org/10.1007/s100520050124. arXiv:hep-ex/9710002
68. Z.E.U.S. Collaboration, Measurement of jet shapes in high Q^2 deep inelastic scattering at HERA. Eur. Phys. J. C **8**, 367 (1999). https://doi.org/10.1007/s100520050471. arXiv:hep-ex/9804001
69. H1 Collaboration, Measurement of internal jet structure in dijet production in deep inelastic scattering at HERA. Nucl. Phys. B **545**, 3 (1999). https://doi.org/10.1016/S0550-3213(99)00118-2. arXiv:hep-ex/9901010
70. CMS Collaboration, Jet energy scale and resolution in the CMS experiment in pp collisions at 8 TeV. JINST **12**, P02014 (2017). https://doi.org/10.1088/1748-0221/12/02/P02014. arXiv:1607.03663
71. ALEPH Collaboration, Performance of the ALEPH detector at LEP. Nucl. Instrum. Meth. A **360**, 481 (1995). https://doi.org/10.1016/0168-9002(95)00138-7
72. DELPHI Collaboration, Performance of the DELPHI detector. Nucl. Instrum. Meth. A **378**, 57 (1996). https://doi.org/10.1016/0168-9002(96)00463-9
73. M. Thomson, Particle flow calorimetry and the PandoraPFA algorithm. Nucl. Instrum. Meth. A **611**, 25 (2009). https://doi.org/10.1016/j.nima.2009.09.009. arXiv:0907.3577
74. CMS Collaboration, Particle-flow event reconstruction in CMS and performance for jets, taus, and MET. CMS Physics Analysis Summary (2009), CMS-PAS-PFT-09-001, https://cds.cern.ch/record/1194487

75. CMS Collaboration, Commissioning of the particle-flow reconstruction in minimum-bias and jet events from pp collisions at 7 TeV. CMS Physics Analysis Summary (2010), CMS-PAS-PFT-10-002, https://cds.cern.ch/record/1279341
76. G. Arnison et al., Experimental observation of isolated large transverse energy electrons with associated missing energy at $\sqrt{s} = 540$ GeV. Phys. Lett. B **122**, 103 (1983). https://doi.org/10.1016/0370-2693(83)91177-2
77. M. Banner et al., Observation of single isolated electrons of high transverse momentum in events with missing transverse energy at the CERN pp collider. Phys. Lett. B **122**, 476 (1983). https://doi.org/10.1016/0370-2693(83)91605-2
78. Particle Data Group, K.A. Olive et al., Review of particle physics. Chin. Phys. C **38**, 090001 (2014). https://doi.org/10.1088/1674-1137/38/9/090001
79. The ALEPH, DELPHI, L3, OPAL Collaborations, the LEP Electroweak Working Group, Electroweak measurements in electron-positron collisions at W-boson-pair energies at LEP, Phys. Rept. **532**, 119 (2013), arXiv:1302.3415
80. CDF, D0 Collaborations, T.E.W. Group, 2012 update of the combination of CDF and D0 results for the mass of the W Boson (2012), arXiv:1204.0042
81. W.J. Marciano, A. Sirlin, Deviations from electron-muon universality in the leptonic decays of the intermediate bosons. Phys. Rev. D **8**, 3612 (1973). https://doi.org/10.1103/PhysRevD.8.3612
82. D. Albert, W.J. Marciano, D. Wyler, Z. Parsa, Decays of intermediate vector bosons, radiative corrections and QCD jets. Nucl. Phys. B **166**, 460 (1980). https://doi.org/10.1016/0550-3213(80)90208-4
83. K. Inoue, A. Kakuto, H. Komatsu, S. Takeshita, Radiative corrections for $W \rightarrow e\bar{\nu}_e$ decay in the Weinberg-Salam Model with arbitrary number of generations. Prog. Theor. Phys. **64**, 1008 (1980)
84. M. Consoli, S.L. Presti, L. Maiani, Higher-order effects and the vector boson physical parameters. Nucl. Phys. B **223**, 474 (1983). https://doi.org/10.1016/0550-3213(83)90066-4
85. D.Y. Bardin, S. Riemann, T. Riemann, Electroweak one-loop corrections to the decay of the charged vector boson. Z. Phys. C **32**, 121 (1986). https://doi.org/10.1007/BF01441360
86. F. Jegerlehner, Vector boson parameters: Scheme dependence and theoretical uncertainties. Z. Phys. C **32**, 425 (1986). https://doi.org/10.1007/BF01551840
87. A. Denner, T. Sack, The W-boson width. Z. Phys. **C46**, 653 (1990). https://doi.org/10.1007/BF01560267
88. B.A. Kniehl, F. Madricardo, M. Steinhauser, Gauge independent W boson partial decay widths. Phys. Rev. D **62**, 073010 (2000). https://doi.org/10.1103/PhysRevD.62.073010. arXiv:hep-ph/0005060
89. T. Chang, K. Gaemers, W.V. Neerven, QCD corrections to the mass and width of the intermediate vector bosons. Nucl. Phys. B **202**, 407 (1982). https://doi.org/10.1016/0550-3213(82)90407-2
90. T. Alvarez, A. Leites, J. Terron, Top mass corrections to weak boson decays. Nucl. Phys. B **301**, 1 (1988). https://doi.org/10.1016/0550-3213(88)90159-9
91. S. Gorishny, A. Kataev, S.A. Larin, The $\mathcal{O}(\alpha_s^3)$ corrections to $\sigma_{tot}(e^+e^- \rightarrow \text{Hadrons})$ and $\Gamma(\tau \rightarrow \nu_\tau + \text{Hadrons})$ in QCD. Phys. Lett. B **259**, 144 (1991). https://doi.org/10.1016/0370-2693(91)90149-K
92. L.R. Surguladze, M.A. Samuel, Total hadronic cross section in $e^+ e^-$ annihilation at the four-loop level of perturbative QCD. Phys. Rev. Lett. **66**, 560 (1991). https://doi.org/10.1103/PhysRevLett.66.560
93. K.G. Chetyrkin, Corrections of $\mathcal{O}(\alpha_s^3)$ to R (had) in pQCD with light gluinos. Phys. Lett. B **391**, 402 (1997). https://doi.org/10.1016/S0370-2693(96)01478-5. arXiv:hep-ph/9608480
94. K.G. Chetyrkin, J.H. Kuhn, Quadratic mass corrections of $\mathcal{O}(\alpha_s^3 m_q^2/s)$ to the decay rate of Z and W bosons. Phys. Lett. B **406**, 102–109 (1997). https://doi.org/10.1016/S0370-2693(97)00631-X. arXiv:hep-ph/9609202
95. D. Kara, Corrections of order $\alpha\alpha_s$ to W Boson decays. Nucl. Phys. B **877**, 683 (2013). https://doi.org/10.1016/j.nuclphysb.2013.10.024. arXiv:1307.7190

96. W.J. Stirling, E. Vryonidou, Electroweak gauge boson polarisation at the LHC. JHEP **07**, 124 (2012). https://doi.org/10.1007/JHEP07(2012)124. arXiv:1204.6427
97. S. De, V. Rentala, W. Shepherd, Measuring the polarization of boosted, hadronic W bosons with jet substructure observables, arXiv:2008.04318
98. J.C. Collins, D.E. Soper, Angular distribution of dileptons in high-energy hadron collisions. Phys. Rev. D **16**, 2219 (1977). https://doi.org/10.1103/PhysRevD.16.2219
99. E. Mirkes, Angular decay distribution of leptons from w-bosons at NLO in hadronic collisions. Nucl. Phys. B **387**, 3 (1992). https://doi.org/10.1016/0550-3213(92)90046-E
100. E. Mirkes, J. Ohnemus, W and Z polarization effects in hadronic collisions. Phys. Rev. D **50**, 5692 (1994). https://doi.org/10.1103/PhysRevD.50.5692. arXiv:hep-ph/9406381
101. E. Mirkes, J. Ohnemus, Angular distributions of Drell-Yan lepton pairs at the Tevatron: order $\alpha - s^2$ corrections and Monte Carlo studies. Phys. Rev. D **51**, 4891 (1995). https://doi.org/10.1103/PhysRevD.51.4891. arXiv:hep-ph/9412289
102. Z. Bern et al., Left-handed W bosons at the LHC. Phys. Rev. D **84** (2011). https://doi.org/10.1103/PhysRevD.84.034008. arXiv:1103.5445
103. C.M.S. Collaboration, Measurement of the polarization of W bosons with large transverse momenta in W+jets events at the LHC. Phys. Rev. Lett. **107**, 021802 (2011). https://doi.org/10.1103/PhysRevLett.107.021802. arXiv:1104.3829
104. ATLAS Collaboration, Measurement of the polarisation of W bosons produced with large transverse momentum in pp collisions at $\sqrt{s} = 7$ TeV with the ATLAS experiment. Eur. Phys. J. C **72**, 2001 (2012). https://doi.org/10.1140/epjc/s10052-012-2001-6. arXiv:1203.2165
105. UA1 Collaboration, G. Arnison et al., Experimental observation of lepton pairs of invariant mass around 95 GeV/c^2 at the CERN SPS collider. Phys. Lett. B **126**, 398 (1983). https://doi.org/10.1016/0370-2693(83)90188-0
106. UA2 Collaboration, P. Bagnaia et al., Evidence for $Z^0 \rightarrow e^+e^-$ at the CERN $\bar{p}p$ collider. Phys. Lett. B **129**, 130 (1983). https://doi.org/10.1016/0370-2693(83)90744-X
107. SLD Electroweak Group, DELPHI, ALEPH, SLD, SLD Heavy Flavour Group, OPAL, LEP Electroweak Working Group, L3, Precision electroweak measurements on the Z resonance. Phys. Rept. **427**, 257 (2006). https://doi.org/10.1016/j.physrep.2005.12.006. arXiv:hep-ex/0509008
108. G. Degrassi, P. Gambino, A. Vicini, Two loop heavy top effects on the $m_Z - m_W$ interdependence. Phys. Lett. B **383**, 219 (1996). https://doi.org/10.1016/0370-2693(96)00720-4. arXiv:hep-ph/9603374
109. G. Degrassi, P. Gambino, A. Sirlin, Precise calculation of M_W, $sin^2\theta_W(M_Z)$, and $sin^2\theta_{\text{eff}}^{\text{lept}}$. Phys. Lett. B **394**, 188 (1997). https://doi.org/10.1016/S0370-2693(96)01677-2. arXiv:hep-ph/9611363
110. G. Degrassi, P. Gambino, Two loop heavy top corrections to the Z^0 boson partial widths. Nucl. Phys. B **567**, 3 (2000). https://doi.org/10.1016/S0550-3213(99)00729-4. arXiv:hep-ph/9905472
111. A. Freitas, Higher-order electroweak corrections to the partial widths and branching ratios of the Z boson. JHEP **04**, 070 (2014). https://doi.org/10.1007/JHEP04(2014)070. arXiv:1401.2447
112. A. Djouadi, C. Verzegnassi, Virtual very heavy top effects in LEP/SLC precision measurements. Phys. Lett. B **195**, 265 (1987). https://doi.org/10.1016/0370-2693(87)91206-8
113. A. Djouadi, $\mathcal{O}(\alpha\alpha_s)$ vacuum polarization functions of the standard model gauge bosons. Nuovo Cim. A **100**, 357 (1988). https://doi.org/10.1007/BF02812964
114. B.A. Kniehl, Two loop corrections to the vacuum polarizations in perturbative QCD. Nucl. Phys. B **347**, 86 (1990). https://doi.org/10.1016/0550-3213(90)90552-O
115. B.A. Kniehl, A. Sirlin, Dispersion relations for vacuum polarization functions in electroweak physics. Phys. B **371**, 141 (1992). https://doi.org/10.1016/0550-3213(92)90232-Z
116. A. Djouadi, P. Gambino, Electroweak gauge bosons selfenergies: complete QCD corrections. Phys. Rev. D **49**, 3499 (1994), https://doi.org/10.1103/PhysRevD.49.3499, https://doi.org/10.1103/PhysRevD.53.4111. arXiv:hep-ph/9309298. Erratum: Phys. Rev. D **53**, 4111 (1996)

117. L. Avdeev, J. Fleischer, S. Mikhailov, O. Tarasov, $\mathcal{O}(\alpha\alpha_s^2)$ correction to the electroweak ρ parameter. Phys. Lett. B **336**, 560 (1994). https://doi.org/10.1016/0370-2693(94)90573-8, https://doi.org/10.1016/0370-2693(95)00269-Q. arXiv:hep-ph/9406363. Erratum: Phys. Lett. B **349**, 597 (1995)
118. K.G. Chetyrkin, J.H. Kühn, M. Steinhauser, Corrections of order $\mathcal{O}(G_F M_t^2 \alpha_s^2)$ to the ρ parameter. Phys. Lett. B **351**, 331 (1995). https://doi.org/10.1016/0370-2693(95)00380-4. arXiv:hep-ph/9502291
119. K.G. Chetyrkin, J.H. Kühn, M. Steinhauser, QCD corrections from top quark to relations between electroweak parameters to order α_s^2. Phys. Rev. Lett. **75**, 3394 (1995). https://doi.org/10.1103/PhysRevLett.75.3394. arXiv:hep-ph/9504413
120. J.J. van der Bij, K.G. Chetyrkin, M. Faisst, G. Jikia, T. Seidensticker, Three loop leading top mass contributions to the ρ parameter. Phys. Lett. B **498**, 156 (2001). https://doi.org/10.1016/S0370-2693(01)00002-8. arXiv:hep-ph/0011373
121. M. Faisst, J.H. Kühn, T. Seidensticker, O. Veretin, Three loop top quark contributions to the ρ parameter. Nucl. Phys. B **665**, 649 (2003). https://doi.org/10.1016/S0550-3213(03)00450-4. arXiv:hep-ph/0302275
122. Y. Schröder, M. Steinhauser, Four-loop singlet contribution to the ρ parameter. Phys. Lett. B **622**, 124 (2005). https://doi.org/10.1016/j.physletb.2005.06.085. arXiv:hep-ph/0504055
123. K.G. Chetyrkin, M. Faisst, J.H. Kühn, P. Maierhofer, C. Sturm, Four-loop QCD corrections to the ρ parameter. Phys. Rev. Lett. **97**, 102003 (2006). https://doi.org/10.1103/PhysRevLett.97.102003. arXiv:hep-ph/0605201
124. R. Boughezal, M. Czakon, Single scale tadpoles and $\mathcal{O}(G_F m_t^2 \alpha_s^3)$ corrections to the ρ parameter. Nucl. Phys. B **755**, 221 (2006). https://doi.org/10.1016/j.nuclphysb.2006.08.007. arXiv:hep-ph/0606232
125. K.G. Chetyrkin, J.H. Kühn, A. Kwiatkowski, QCD corrections to the e^+e^- cross-section and the Z boson decay rate. Phys. Rept. **277**, 189 (1996). https://doi.org/10.1016/S0370-1573(96)00012-9. arXiv:hep-ph/9503396
126. P.A. Baikov, K.G. Chetyrkin, J.H. Kühn, $\mathcal{O}(\alpha_s^4)$ QCD corrections to Z and τ decays. Phys. Rev. Lett. **101**, 012002 (2008). https://doi.org/10.1103/PhysRevLett.101.012002. arXiv:0801.1821
127. P. Baikov, K. Chetyrkin, J.H. Kühn, J. Rittinger, Complete $\mathcal{O}(\alpha_s^4)$ QCD corrections to hadronic Z decays. Phys. Rev. Lett. **108**, 222003 (2012). https://doi.org/10.1103/PhysRevLett.108.222003. arXiv:1201.5804
128. A. Czarnecki, J.H. Kühn, Nonfactorizable QCD and electroweak corrections to the hadronic Z boson decay rate. Phys. Rev. Lett. **77**, 3955 (1996). https://doi.org/10.1103/PhysRevLett.77.3955. arXiv:hep-ph/9608366
129. R. Harlander, T. Seidensticker, M. Steinhauser, Complete corrections of order $\alpha\alpha_s$ to the decay of the Z boson into bottom quarks. Phys. Lett. B **426**, 125 (1998). https://doi.org/10.1016/S0370-2693(98)00220-2. arXiv:hep-ph/9712228
130. J. Fleischer, O.V. Tarasov, F. Jegerlehner, P. Raczka, Two loop $\mathcal{O}(alpha_s G_\mu m_t^2)$ corrections to the partial decay width of the Z^0 into $b\bar{b}$ final states in the large top mass limit. Phys. Lett. B **293**, 437 (1992). https://doi.org/10.1016/0370-2693(92)90909-N
131. G. Buchalla, A.J. Buras, QCD corrections to the $\bar{s}dZ$ vertex for arbitrary top quark mass. Nucl. Phys. B **398**, 285 (1993). https://doi.org/10.1016/0550-3213(93)90110-B
132. G. Degrassi, Current algebra approach to heavy top effects in $Z \to b + \bar{b}$. Nucl. Phys. B **407**, 271–289 (1993). https://doi.org/10.1016/0550-3213(93)90058-W. arXiv:hep-ph/9302288
133. K.G. Chetyrkin, A. Kwiatkowski, M. Steinhauser, Leading top mass corrections of order $\mathcal{O}(\alpha\alpha_s m_t^2/M_W^2)$ to partial decay rate $\Gamma(Z \to b\bar{b})$. Mod. Phys. Lett. A **8**, 2785, https://doi.org/10.1142/S0217732393003172
134. J. Stirling, E. Vryonidou, Polarisation of electroweak gauge bosons at the LHC. EPJ Web Conf. **49**, 14005 (2013). https://doi.org/10.1051/epjconf/20134914005. arXiv:1302.1365
135. ATLAS, CMS Collaborations, Combined measurement of the Higgs boson mass in pp collisions at $\sqrt{s} = 7$ and 8 TeV with the ATLAS and CMS experiments. Phys. Rev. Lett. **114**, 191803 (2015). https://doi.org/10.1103/PhysRevLett.114.191803. arXiv:1503.07589

136. LHC Higgs Cross Section Working Group, S. Dittmaier et al., Handbook of LHC Higgs cross sections: 2. Differential distributions, arXiv:1201.3084
137. LHC Higgs Cross Section Working Group, J.R. Andersen et al., Handbook of LHC Higgs cross sections: 3. Higgs properties, arXiv:1307.1347
138. LHC Higgs Cross Section Working Group, D. de Florian et al., Handbook of LHC Higgs cross sections: 4. Deciphering the nature of the Higgs sector. CYRM-2017-002 (2017). https://doi.org/10.23731/CYRM-2017-002. arXiv:1610.07922
139. A. Hoecker, Physics at the LHC Run 2 and beyond, in Proceedings, 2016 European School of High-Energy Physics (ESHEP2016): Skeikampen, Norway (2016), arXiv:1611.07864
140. ATLAS, CDF, CMS, D0 Collaborations, First combination of Tevatron and LHC measurements of the top-quark mass, arXiv:1403.4427
141. M. Jezabek, J.H. Kuhn, QCD corrections to semileptonic decays of heavy quarks. Nucl. Phys. B **314**, 1 (1989). https://doi.org/10.1016/0550-3213(89)90108-9
142. G. Calderon, G. Lopez Castro, Convolution formula and finite W boson width effects in top quark width. Int. J. Mod. Phys. **A23**, 3525 (2008). https://doi.org/10.1142/S0217751X08041578. arXiv:hep-ph/0108088
143. A. Denner, T. Sack, The top width. Nucl. Phys. **B358**, 46 (1991). https://doi.org/10.1016/0550-3213(91)90530-B
144. G. Eilam, R.R. Mendel, R. Migneron, A. Soni, Radiative corrections to top quark decay. Phys. Rev. Lett. **66**, 3105 (1991). https://doi.org/10.1103/PhysRevLett.66.3105
145. I.R. Blokland, A. Czarnecki, M. Slusarczyk, F. Tkachov, Heavy to light decays with a two loop accuracy. Phys. Rev. Lett. **93**, 062001 (2004). https://doi.org/10.1103/PhysRevLett.93.062001. arXiv:hep-ph/0403221
146. I.R. Blokland, A. Czarnecki, M. Slusarczyk, F. Tkachov, Next-to-next-to-leading order calculations for heavy-to-light decays. Phys. Rev. D **71**, 054004 (2005). https://doi.org/10.1103/PhysRevD.79.019901, https://doi.org/10.1103/PhysRevD.71.054004. arXiv:hep-ph/0503039. Erratum: Phys. Rev. D **79**, 019901 (2009)
147. I. Bigi, Y. Dokshitzer, V. Khoze, J. Kühn, P. Zerwas, Production and decay properties of ultra-heavy quarks. Phys. Lett. B **181**, 157 (1986). https://doi.org/10.1016/0370-2693(86)91275-X
148. J.H. Kühn, Theory of top quark production and decay, in *The Top Quark and the Electroweak Interaction*, in *Proceedings, 23rd Summer Institute on Particle Physics*, Stanford, USA, 10–21 July 1995 (1996), p. 1, arXiv:hep-ph/9707321, http://www.slac.stanford.edu/pubs/confproc/ssi95/ssi95-001.html
149. W. Bernreuther, Z.-G. Si, Top quark spin correlations and polarization at the LHC: standard model predictions and effects of anomalous top chromo moments. Phys. Lett. B **725**, 115 (2013). https://doi.org/10.1016/j.physletb.2013.06.051, https://doi.org/10.1016/j.physletb.2015.03.035. arXiv:1305.2066. Erratum: Phys. Lett. B **744**, 413 (2015)
150. ATLAS Collaboration, Measurement of top quark polarization in top-antitop events from proton-proton collisions at $\sqrt{s} = 7$ TeV using the ATLAS detector. Phys. Rev. Lett. **111**, 232002 (2013). https://doi.org/10.1103/PhysRevLett.111.232002. arXiv:1307.6511
151. C.M.S. Collaboration, S. Chatrchyan et al., Measurements of $t\bar{t}$ spin correlations and top-quark polarization using dilepton final states in pp collisions at $\sqrt{s} = 7$ TeV. Phys. Rev. Lett. **112**, 182001 (2014). https://doi.org/10.1103/PhysRevLett.112.182001. arXiv:1311.3924
152. A. Kadeer, J.G. Korner, Radiative corrections to top quark decays, in *Heavy Quark Physics, Proceedings, Helmholtz International School, HQP08*, Dubna, Russia, 11–21 August 2008 (2009), p. 89, arXiv:0906.3474
153. A. Czarnecki, J.G. Korner, J.H. Piclum, Helicity fractions of W bosons from top quark decays at NNLO in QCD. Phys. Rev. D **81**, 111503 (2010). https://doi.org/10.1103/PhysRevD.81.111503. arXiv:1005.2625
154. ATLAS Collaboration, Measurement of the W boson polarization in top quark decays with the ATLAS detector. JHEP **06**, 088 (2012). https://doi.org/10.1007/JHEP06(2012)088. arXiv:1205.2484

155. C.M.S. Collaboration, S. Chatrchyan et al., Measurement of the W-boson helicity in top-quark decays from $t\bar{t}$ production in lepton+jets events in pp collisions at $\sqrt{s} = 7$ TeV. JHEP **10**, 167 (2013). https://doi.org/10.1007/JHEP10(2013)167. arXiv:1308.3879
156. C.M.S. Collaboration, V. Khachatryan et al., Measurement of the W boson helicity in events with a single reconstructed top quark in pp collisions at $\sqrt{s} = 8$ TeV. JHEP **01**, 053 (2015). https://doi.org/10.1007/JHEP01(2015)053. arXiv:1410.1154
157. CMS Collaboration, V. Khachatryan et al., Measurement of the W boson helicity fractions in the decays of top quark pairs to lepton+jets final states produced in pp collisions at $\sqrt{s} = 8$ TeV. Phys. Lett. B **762**, 512 (2016). https://doi.org/10.1016/j.physletb.2016.10.007. arXiv:1605.09047
158. J. E. Huth et al., Toward a standardization of jet definitions, in *DPF Summer Study on High-energy Physics: Research Directions for the Decade (Snowmass 90), Snowmass*, Colorado, 25 June–13 July 1990 (1990), p. 134, http://lss.fnal.gov/cgi-bin/find_paper.pl?conf-90-249
159. CTEQ Collaboration, R. Brock et al., Handbook of perturbative QCD: Version 1.0. Rev. Mod. Phys. **67**, 157–248 (1995). https://doi.org/10.1103/RevModPhys.67.157
160. M. Cacciari, G.P. Salam, Dispelling the N^3 myth for the k_t jet-finder. Phys. Lett. B **641**, 57–61 (2006). https://doi.org/10.1016/j.physletb.2006.08.037. arXiv:hep-ph/0512210
161. S.D. Ellis, J. Huston, K. Hatakeyama, P. Loch, M. Tonnesmann, Jets in hadron-hadron collisions. Prog. Part. Nucl. Phys. **60**, 484–551 (2008). https://doi.org/10.1016/j.ppnp.2007.12.002. arXiv:0712.2447
162. C. Buttar et al., Standard model handles and candles working group: tools and jets summary report, in *Physics at TeV Colliders, La physique du TeV aux Collisionneurs*, Les Houches, France, 11–29 June 2007 (2008), p. 121, arXiv:0803.0678
163. S. Sapeta, QCD and jets at hadron colliders. Prog. Part. Nucl. Phys. **89**, 1 (2016). https://doi.org/10.1016/j.ppnp.2016.02.002. arXiv:1511.09336
164. M. Cacciari, G.P. Salam, G. Soyez, FastJet user manual. Eur. Phys. J. C **72**, 1896 (2012). https://doi.org/10.1140/epjc/s10052-012-1896-2. arXiv:1111.6097
165. M. Cacciari, Phenomenological and theoretical developments in jet physics at the LHC. Int. J. Mod. Phys. **A30**, 1546001 (2015). https://doi.org/10.1142/S0217751X1546001X. arXiv:1509.02272
166. A. Schwartzman, Jet energy calibration at the LHC. Int. J. Mod. Phys. **A30**, 1546002 (2015). https://doi.org/10.1142/S0217751X15460021. arXiv:1509.05459
167. P. Francavilla, Measurements of inclusive jet and dijet cross sections at the Large Hadron Collider. Int. J. Mod. Phys. **A30**, 1546003 (2015). https://doi.org/10.1142/S0217751X15460033. arXiv:1510.01943
168. P. Kokkas, Measurements of jet-related observables at the LHC. Int. J. Mod. Phys. **A30**, 1546004 (2015). https://doi.org/10.1142/S0217751X15460045. arXiv:1509.02144
169. J. Rojo, Constraints on parton distributions and the strong coupling from LHC jet data. Int. J. Mod. Phys. A **30**, 1546005 (2015). https://doi.org/10.1142/S0217751X15460057. arXiv:1410.7728
170. M. Campanelli, Forward jets and large rapidity gaps. Int. J. Mod. Phys. **A30**, 1546006 (2015). https://doi.org/10.1142/S0217751X15460069. arXiv:1508.04631
171. U. Blumenschein, Jet production in association with vector bosons or top quarks. Int. J. Mod. Phys. A **30**, 1546007 (2015). https://doi.org/10.1142/S0217751X15460070. arXiv:1509.04885
172. M. Voutilainen, Heavy quark jets at the LHC. Int. J. Mod. Phys. **A30**, 1546008 (2015). https://doi.org/10.1142/S0217751X15460082. arXiv:1509.05026
173. K. Kousouris, Search for physics beyond the Standard Model using jet observables. Int. J. Mod. Phys. A **30**, 1546009 (2015). https://doi.org/10.1142/S0217751X15460094
174. C. Roland, Jets in heavy ion collisions at the LHC. Int. J. Mod. Phys. A **30**, 1546010 (2015). https://doi.org/10.1142/S0217751X15460100
175. CDF Collaboration, F. Abe et al., Topology of three-jet events in $\overline{p}p$ collisions at $\sqrt{s} = 1.8$ TeV. Phys. Rev. D **45**, 1448 (1992). https://doi.org/10.1103/PhysRevD.45.1448

176. G.C. Blazey et al., Run II jet physics, in QCD and weak boson physics in Run II, in *Proceedings*, Batavia, USA, March 4–6, June 3–4, November 4–6, 1999 (2000), p. 47, arXiv:hep-ex/0005012
177. S.D. Ellis, J. Huston, M. Tonnesmann, On building better cone jet algorithms. eConf **C010630**, 513 (2001), arXiv:hep-ph/0111434
178. M.H. Seymour, C. Tevlin, A comparison of two different jet algorithms for the top mass reconstruction at the LHC. JHEP **11**, 052 (2006). https://doi.org/10.1088/1126-6708/2006/11/052. arXiv:hep-ph/0609100
179. I.W. Stewart, F.J. Tackmann, J. Thaler, C.K. Vermilion, T.F. Wilkason, XCone: N-jettiness as an exclusive cone jet algorithm. JHEP **11**, 072 (2015). https://doi.org/10.1007/JHEP11(2015)072. arXiv:1508.01516
180. J. Thaler, T.F. Wilkason, Resolving boosted jets with XCone. JHEP **12**, 051 (2015). https://doi.org/10.1007/JHEP12(2015)051. arXiv:1508.01518
181. H. Georgi, A simple alternative to jet-clustering algorithms, arXiv:1408.1161
182. Y. Bai, Z. Han, R. Lu, J_{E_T}: a global jet finding algorithm. JHEP **03**, 102 (2015). https://doi.org/10.1007/JHEP03(2015)102. arXiv:1411.3705
183. Y. Bai, Z. Han, R. Lu, $J_{E_T}^{II}$: a two-prong jet finding algorithm, arXiv:1509.07522
184. M. Cacciari, G.P. Salam, Pileup subtraction using jet areas. Phys. Lett. B **659**, 119–126 (2008). https://doi.org/10.1016/j.physletb.2007.09.077. arXiv:0707.1378
185. M. Cacciari, G.P. Salam, G. Soyez, The catchment area of jets. JHEP **04**, 005 (2008). https://doi.org/10.1088/1126-6708/2008/04/005. arXiv:0802.1188
186. M. Dasgupta, L. Magnea, G.P. Salam, Non-perturbative QCD effects in jets at hadron colliders. JHEP **02**, 055 (2008). https://doi.org/10.1088/1126-6708/2008/02/055. arXiv:0712.3014
187. D. Krohn, J. Thaler, L.-T. Wang, Jets with variable R. JHEP **06**, 059 (2009). https://doi.org/10.1088/1126-6708/2009/06/059. arXiv:0903.0392
188. I.W. Stewart, F.J. Tackmann, W.J. Waalewijn, N-Jettiness: an inclusive event shape to veto jets. Phys. Rev. Lett. **105**, 092002 (2010). https://doi.org/10.1103/PhysRevLett.105.092002. arXiv:1004.2489
189. T.T. Jouttenus, I.W. Stewart, F.J. Tackmann, W.J. Waalewijn, The soft function for exclusive N-jet production at hadron colliders. Phys. Rev. D **83**, 114030 (2011). https://doi.org/10.1103/PhysRevD.83.114030. arXiv:1102.4344
190. C.W. Bauer, F.J. Tackmann, J.R. Walsh, S. Zuberi, Factorization and resummation for dijet invariant mass spectra. Phys. Rev. D **85**, 074006 (2012). https://doi.org/10.1103/PhysRevD.85.074006. arXiv:1106.6047
191. S. Alioli, C.W. Bauer, C. Berggren, F.J. Tackmann, J.R. Walsh, S. Zuberi, Matching fully differential NNLO calculations and parton showers. JHEP **06**, 089 (2014). https://doi.org/10.1007/JHEP06(2014)089. arXiv:1311.0286
192. J. Gaunt, M. Stahlhofen, F.J. Tackmann, J.R. Walsh, N-jettiness subtractions for NNLO QCD calculations. JHEP **09**, 058 (2015). https://doi.org/10.1007/JHEP09(2015)058. arXiv:1505.04794
193. R. Boughezal, X. Liu, F. Petriello, N-jettiness soft function at next-to-next-to-leading order. Phys. Rev. D **91**, (2015). https://doi.org/10.1103/PhysRevD.91.094035. arXiv:1504.02540
194. P. Pietrulewicz, F.J. Tackmann, W.J. Waalewijn, Factorization and resummation for generic hierarchies between jets. JHEP **08**, 002 (2016). https://doi.org/10.1007/JHEP08(2016)002. arXiv:1601.05088
195. D.W. Kolodrubetz, P. Pietrulewicz, I.W. Stewart, F.J. Tackmann, W.J. Waalewijn, Factorization for jet radius logarithms in jet mass spectra at the LHC. JHEP **12**, 054 (2016). https://doi.org/10.1007/JHEP12(2016)054. arXiv:1605.08038
196. T.T. Jouttenus, I.W. Stewart, F.J. Tackmann, W.J. Waalewijn, Jet mass spectra in Higgs boson plus one jet at next-to-next-to-leading logarithmic order. Phys. Rev. D **88**, 054031 (2013). https://doi.org/10.1103/PhysRevD.88.054031. arXiv:1302.0846
197. Z.-B. Kang, X. Liu, S. Mantry, 1-jettiness DIS event shape: NNLL+NLO results. Phys. Rev. D **90**, 014041 (2014). https://doi.org/10.1103/PhysRevD.90.014041. arXiv:1312.0301

198. S. Alioli, C.W. Bauer, C. Berggren, F.J. Tackmann, J.R. Walsh, Drell-Yan production at NNLL'+NNLO matched to parton showers. Phys. Rev. D **92**, 094020 (2015). https://doi.org/10.1103/PhysRevD.92.094020. arXiv:1508.01475
199. S.-F. Ge, The Georgi algorithms of jet clustering. JHEP **05**, 066 (2015). https://doi.org/10.1007/JHEP05(2015)066. arXiv:1408.3823
200. T. Kaufmann, A. Mukherjee, W. Vogelsang, Next-to-leading order calculation for jets defined by a maximized jet function. Phys. Rev. D **91**, 034001 (2015). https://doi.org/10.1103/PhysRevD.91.034001. arXiv:1412.0298
201. J. Thaler, Jet maximization, axis minimization, and stable cone finding. Phys. Rev. D **92**, 074001 (2015). https://doi.org/10.1103/PhysRevD.92.074001. arXiv:1506.07876
202. G.C. Fox, S. Wolfram, Observables for the analysis of event shapes in e^+e^- annihilation and other processes. Phys. Rev. Lett. **41**, 1581 (1978). https://doi.org/10.1103/PhysRevLett.41.1581
203. G.C. Fox, S. Wolfram, Event shapes in e^+e^- annihilation. Nucl. Phys. B **149**, 413 (1979). https://doi.org/10.1016/0550-3213(79)90003-8
204. S.D. Ellis, C.K. Vermilion, J.R. Walsh, Recombination algorithms and jet substructure: Pruning as a tool for heavy particle searches. Phys. Rev. D **81**, 094023 (2010). https://doi.org/10.1103/PhysRevD.81.094023. arXiv:0912.0033
205. M. Dasgupta, G.P. Salam, Resummation of nonglobal QCD observables. Phys. Lett. B **512**, 323 (2001). https://doi.org/10.1016/S0370-2693(01)00725-0. arXiv:hep-ph/0104277
206. M. Dasgupta, G.P. Salam, Accounting for coherence in interjet E_t flow: a case study. JHEP **03**, 017 (2002). https://doi.org/10.1088/1126-6708/2002/03/017. arXiv:hep-ph/0203009
207. A. Banfi, M. Dasgupta, K. Khelifa-Kerfa, S. Marzani, Non-global logarithms and jet algorithms in high-p_T jet shapes. JHEP **08**, 064 (2010). https://doi.org/10.1007/JHEP08(2010)064. arXiv:1004.3483
208. M. Dasgupta, G.P. Salam, Resummed event shape variables in DIS. JHEP **08**, 032 (2002). https://doi.org/10.1088/1126-6708/2002/08/032. arXiv:hep-ph/0208073
209. J. Thaler, K. Van Tilburg, Identifying boosted objects with N-subjettiness. JHEP **03**, 015 (2011). https://doi.org/10.1007/JHEP03(2011)015. arXiv:1011.2268
210. J. Thaler, K. Van Tilburg, Maximizing boosted top identification by minimizing N-subjettiness. JHEP **02**, 093 (2012). https://doi.org/10.1007/JHEP02(2012)093. arXiv:1108.2701
211. I. Feige, M.D. Schwartz, I.W. Stewart, J. Thaler, Precision jet substructure from boosted event shapes. Phys. Rev. Lett. **109**, 092001 (2012). https://doi.org/10.1103/PhysRevLett.109.092001. arXiv:1204.3898
212. A.J. Larkoski, J. Thaler, Unsafe but calculable: ratios of angularities in perturbative QCD. JHEP **09**, 137 (2013). https://doi.org/10.1007/JHEP09(2013)137. arXiv:1307.1699
213. A.J. Larkoski, S. Marzani, J. Thaler, Sudakov safety in perturbative QCD. Phys. Rev. D **91**, 111501 (2015). https://doi.org/10.1103/PhysRevD.91.111501. arXiv:1502.01719
214. A.J. Larkoski, I. Moult, The singular behavior of jet substructure observables. Phys. Rev. D **93**, 014017 (2016). https://doi.org/10.1103/PhysRevD.93.014017. arXiv:1510.08459
215. A.J. Larkoski, G.P. Salam, J. Thaler, Energy correlation functions for jet substructure. JHEP **06**, 108 (2013). https://doi.org/10.1007/JHEP06(2013)108. arXiv:1305.0007
216. G. Parisi, Super inclusive cross-sections. Phys. Lett. B **74**, 65 (1978). https://doi.org/10.1016/0370-2693(78)90061-8
217. J.F. Donoghue, F.E. Low, S.-Y. Pi, Tensor analysis of hadronic jets in quantum chromodynamics. Phys. Rev. D **20**, 2759 (1979). https://doi.org/10.1103/PhysRevD.20.2759
218. A.J. Larkoski, I. Moult, D. Neill, Power counting to better jet observables. JHEP **12**, 009 (2014). https://doi.org/10.1007/JHEP12(2014)009. arXiv:1409.6298
219. A.J. Larkoski, I. Moult, D. Neill, Analytic boosted boson discrimination. JHEP **05**, 117 (2016). https://doi.org/10.1007/JHEP05(2016)117. arXiv:1507.03018
220. A.J. Larkoski, I. Moult, D. Neill, Building a better boosted top tagger. Phys. Rev. D **91**, 034035 (2015). https://doi.org/10.1103/PhysRevD.91.034035. arXiv:1411.0665

221. I. Moult, L. Necib, J. Thaler, New angles on energy correlation functions. JHEP **12**, 153 (2016). https://doi.org/10.1007/JHEP12(2016)153. arXiv:1609.07483
222. P.T. Komiske, E.M. Metodiev, J. Thaler, Energy flow polynomials: a complete linear basis for jet substructure. JHEP **04**, 013 (2018). https://doi.org/10.1007/JHEP04(2018)013. arXiv:1712.07124
223. A.J. Larkoski, J. Thaler, W.J. Waalewijn, Gaining (mutual) information about quark/gluon discrimination. JHEP **11**, 129 (2014). https://doi.org/10.1007/JHEP11(2014)129. arXiv:1408.3122
224. P.T. Komiske, E.M. Metodiev, J. Thaler, Cutting multiparticle correlators down to size. Phys. Rev. D **101**, 036019 (2020). https://doi.org/10.1103/PhysRevD.101.036019. arXiv:1911.04491
225. P.T. Komiske, E.M. Metodiev, J. Thaler, Energy flow networks: deep sets for particle jets. JHEP **01**, 121 (2019). https://doi.org/10.1007/JHEP01(2019)121. arXiv:1810.05165
226. J. Gallicchio, M.D. Schwartz, Seeing in color: jet superstructure. Phys. Rev. Lett. **105**, 022001 (2010). https://doi.org/10.1103/PhysRevLett.105.022001. arXiv:1001.5027
227. Y. Bao, A.J. Larkoski, Calculating pull for non-singlet jets. JHEP **12**, 035 (2019). https://doi.org/10.1007/JHEP12(2019)035. arXiv:1910.02085
228. A. Larkoski, S. Marzani, C. Wu, Safe use of jet pull. JHEP **01**, 104 (2020). https://doi.org/10.1007/JHEP01(2020)104. arXiv:1911.05090
229. J. Conway, R. Bhaskar, R. Erbacher, J. Pilot, Identification of high-momentum top quarks, Higgs bosons, and W and Z bosons using boosted event shapes. Phys. Rev. D **94**, 094027 (2016). https://doi.org/10.1103/PhysRevD.94.094027. arXiv:1606.06859
230. D. Krohn, J. Thaler, L.-T. Wang, Jet trimming. JHEP **02**, 084 (2010). https://doi.org/10.1007/JHEP02(2010)084. arXiv:0912.1342
231. A.J. Larkoski, S. Marzani, G. Soyez, J. Thaler, Soft drop. JHEP **05**, 146 (2014). https://doi.org/10.1007/JHEP05(2014)146. arXiv:1402.2657
232. T. Plehn, G.P. Salam, M. Spannowsky, Fat jets for a light Higgs boson. Phys. Rev. Lett. **104**, 111801 (2010). https://doi.org/10.1103/PhysRevLett.104.111801. arXiv:0910.5472
233. T. Plehn, M. Spannowsky, M. Takeuchi, D. Zerwas, Stop reconstruction with tagged tops. JHEP **10**, 078 (2010). https://doi.org/10.1007/JHEP10(2010)078. arXiv:1006.2833
234. G. Kasieczka, T. Plehn, T. Schell, T. Strebler, G.P. Salam, Resonance searches with an updated top tagger. JHEP **06**, 203 (2015). https://doi.org/10.1007/JHEP06(2015)203. arXiv:1503.05921
235. M. Rubin, Non-global logarithms in filtered jet algorithms. JHEP **05**, 005 (2010). https://doi.org/10.1007/JHEP05(2010)005. arXiv:1002.4557
236. S.D. Ellis, C.K. Vermilion, J.R. Walsh, Techniques for improved heavy particle searches with jet substructure. Phys. Rev. D **80**, 051501 (2009). https://doi.org/10.1103/PhysRevD.80.051501. arXiv:0903.5081
237. M. Dasgupta, A. Fregoso, S. Marzani, G.P. Salam, Towards an understanding of jet substructure. JHEP **09**, 029 (2013). https://doi.org/10.1007/JHEP09(2013)029. arXiv:1307.0007
238. S. Marzani, L. Schunk, G. Soyez, A study of jet mass distributions with grooming. JHEP **07**, 132 (2017). https://doi.org/10.1007/JHEP07(2017)132. arXiv:1704.02210
239. F.A. Dreyer, L. Necib, G. Soyez, J. Thaler, Recursive soft drop. JHEP **06**, 093 (2018). https://doi.org/10.1007/JHEP06(2018)093. arXiv:1804.03657
240. CMS Collaboration, A Cambridge-Aachen (C-A) based jet algorithm for boosted top-jet tagging. CMS Physics Analysis Summary. CMS-PAS-JME-09-001 (2009), http://cdsweb.cern.ch/record/1194489
241. CMS Collaboration, Study of jet substructure in pp collisions at 7 TeV in CMS. CMS Physics Analysis Summary. CMS-PAS-JME-10-013 (2011), https://cds.cern.ch/record/1333700
242. M. Dasgupta, M. Guzzi, J. Rawling, G. Soyez, Top tagging: an analytical perspective. JHEP **09**, 170 (2018). https://doi.org/10.1007/JHEP09(2018)170. arXiv:1807.04767
243. CMS Collaboration, Boosted top jet tagging at CMS. CMS Physics Analysis Summary. CMS-PAS-JME-13-007 (2014), http://cds.cern.ch/record/1647419

244. M. Stoll, Vetoed jet clustering: the mass-jump algorithm. JHEP **04**, 111 (2015). https://doi.org/10.1007/JHEP04(2015)111. arXiv:1410.4637
245. T. Lapsien, R. Kogler and J. Haller, A new tagger for hadronically decaying heavy particles at the LHC. Eur. Phys. J. C **76**, 600 (2016). https://doi.org/10.1140/epjc/s10052-016-4443-8. arXiv:1606.04961
246. K. Hamaguchi, S.P. Liew, M. Stoll, How to decontaminate overlapping fat jets. Phys. Rev. D **92**, 015012 (2015). https://doi.org/10.1103/PhysRevD.92.015012. arXiv:1505.02930
247. T. Lapsien, Studies of top tagging identification methods and development of a new heavy object tagger. Ph.D. thesis, Universität Hamburg, Hamburg, Germany, 2016, https://bib-pubdb1.desy.de/record/299813
248. T. Sjostrand, S. Mrenna, P.Z. Skands, PYTHIA 6.4 physics and manual. JHEP **05**, 026, https://doi.org/10.1088/1126-6708/2006/05/026. arXiv:hep-ph/0603175
249. T. Sjöstrand, S. Mrenna, P.Z. Skands, A brief introduction to PYTHIA 8.1. Comput. Phys. Commun. **178**, 852 (2008). https://doi.org/10.1016/j.cpc.2008.01.036. arXiv:0710.3820
250. T. Sjöstrand, S. Ask, J.R. Christiansen, R. Corke, N. Desai, P. Ilten et al., An introduction to PYTHIA 8.2. Comput. Phys. Commun. **191**, 159 (2015). https://doi.org/10.1016/j.cpc.2015.01.024. arXiv:1410.3012
251. J. Dolen, P. Harris, S. Marzani, S. Rappoccio, N. Tran, Thinking outside the ROCs: Designing Decorrelated Taggers (DDT) for jet substructure. JHEP **05**, 156 (2016). https://doi.org/10.1007/JHEP05(2016)156. arXiv:1603.00027
252. I. Moult, B. Nachman, D. Neill, Convolved substructure: analytically decorrelating jet substructure observables. JHEP **05**, 002 (2018). https://doi.org/10.1007/JHEP05(2018)002. arXiv:1710.06859
253. D. Napoletano, G. Soyez, Computing N-subjettiness for boosted jets. JHEP **12**, 031 (2018). https://doi.org/10.1007/JHEP12(2018)031. arXiv:1809.04602
254. CMS Collaboration, A multi-dimensional search for new heavy resonances decaying to boosted WW, WZ, or ZZ boson pairs in the dijet final state at 13 TeV. Eur. Phys. J. C **80**, 237 (2020). https://doi.org/10.1140/epjc/s10052-020-7773-5. arXiv:1906.05977
255. S.D. Ellis, A. Hornig, T.S. Roy, D. Krohn, M.D. Schwartz, Nondeterministic approach to tree-baased jet substructure. Phys. Rev. Lett. **108**, 182003 (2012). https://doi.org/10.1103/PhysRevLett.108.182003. arXiv:1201.1914
256. S.D. Ellis, A. Hornig, D. Krohn, T.S. Roy, On statistical aspects of Qjets. JHEP **01**, 022 (2015). https://doi.org/10.1007/JHEP01(2015)022. arXiv:1409.6785
257. D.E. Soper, M. Spannowsky, Finding physics signals with shower deconstruction. Phys. Rev. D **84**, 074002 (2011). https://doi.org/10.1103/PhysRevD.84.074002. arXiv:1102.3480
258. D.E. Soper, M. Spannowsky, Finding top quarks with shower deconstruction. Phys. Rev. D **87**, 054012 (2013). https://doi.org/10.1103/PhysRevD.87.054012. arXiv:1211.3140
259. D. Ferreira de Lima, P. Petrov, D. Soper, M. Spannowsky, Quark-gluon tagging with shower deconstruction: unearthing dark matter and Higgs couplings. Phys. Rev. D **95**, 034001 (2017). https://doi.org/10.1103/PhysRevD.95.034001. arXiv:1607.06031
260. P.T. Komiske, E.M. Metodiev, J. Thaler, Metric space of collider events. Phys. Rev. Lett. **123**, 041801 (2019). https://doi.org/10.1103/PhysRevLett.123.041801. arXiv:1902.02346
261. T. Cover, P. Hart, Nearest neighbour pattern classification. IEEE Trans. Inform. Theory **13**, 21 (1967). https://doi.org/10.1109/TIT.1967.1053964
262. P.T. Komiske, R. Mastandrea, E.M. Metodiev, P. Naik, J. Thaler, Exploring the space of jets with CMS open data. Phys. Rev. D **101**, 034009 (2020). https://doi.org/10.1103/PhysRevD.101.034009. arXiv:1908.08542
263. G. Luisoni, S. Marzani, QCD resummation for hadronic final states. J. Phys. G **42**, 103101 (2015). https://doi.org/10.1088/0954-3899/42/10/103101. arXiv:1505.04084
264. C.W. Bauer, S. Fleming, M.E. Luke, Summing Sudakov logarithms in $B \to X_s + \gamma$ in effective field theory. Phys. Rev. D **63**, 014006 (2000). https://doi.org/10.1103/PhysRevD.63.014006. arXiv:hep-ph/0005275
265. C.W. Bauer, S. Fleming, D. Pirjol, I.W. Stewart, An effective field theory for collinear and soft gluons: heavy to light decays. Phys. Rev. D **63**, 114020 (2001). https://doi.org/10.1103/PhysRevD.63.114020. arXiv:hep-ph/0011336

266. C.W. Bauer, I.W. Stewart, Invariant operators in collinear effective theory. Phys. Lett. B **516**, 134, https://doi.org/10.1016/S0370-2693(01)00902-9. arXiv:hep-ph/0107001
267. C.W. Bauer, D. Pirjol, I.W. Stewart, Soft collinear factorization in effective field theory. Phys. Rev. D **65**, 054022 (2002). https://doi.org/10.1103/PhysRevD.65.054022. arXiv:hep-ph/0109045
268. M. Beneke, A.P. Chapovsky, M. Diehl, T. Feldmann, Soft collinear effective theory and heavy to light currents beyond leading power. Nucl. Phys. B **643**, 431 (2002). https://doi.org/10.1016/S0550-3213(02)00687-9. arXiv:hep-ph/0206152
269. M. Beneke, T. Feldmann, Multipole expanded soft collinear effective theory with non-Abelian gauge symmetry. Phys. Lett. B **553**, 267 (2003). https://doi.org/10.1016/S0370-2693(02)03204-5. arXiv:hep-ph/0211358
270. R.J. Hill, M. Neubert, Spectator interactions in soft collinear effective theory. Nucl. Phys. B **657**, 229 (2003). https://doi.org/10.1016/S0550-3213(03)00116-0. arXiv:hep-ph/0211018
271. T. Becher, A. Broggio, A. Ferroglia, Introduction to soft-collinear effective theory. Lect. Notes Phys. **896**, 337 (2015). https://doi.org/10.1007/978-3-319-14848-9. arXiv:1410.1892
272. M. Bonvini, S. Forte, M. Ghezzi, G. Ridolfi, Threshold resummation in SCET vs. perturbative QCD: an analytic comparison. Nucl. Phys. B **861**, 337 (2012). https://doi.org/10.1016/j.nuclphysb.2012.04.010. arXiv:1201.6364
273. M. Bonvini, S. Forte, M. Ghezzi, G. Ridolfi, The scale of soft resummation in SCET vs perturbative QCD. Nucl. Phys. Proc. Suppl. **241**, 121 (2013). https://doi.org/10.1016/j.nuclphysbps.2013.06.020. arXiv:1301.4502
274. M. Bonvini, S. Forte, G. Ridolfi, L. Rottoli, Resummation prescriptions and ambiguities in SCET vs. direct QCD: Higgs production as a case study. JHEP **01**, 046 (2015). https://doi.org/10.1007/JHEP01(2015)046. arXiv:1409.0864
275. L.G. Almeida, S.D. Ellis, C. Lee, G. Sterman, I. Sung, J.R. Walsh, Comparing and counting logs in direct and effective methods of QCD resummation. JHEP **04**, 174 (2014). https://doi.org/10.1007/JHEP04(2014)174. arXiv:1401.4460
276. F. Bloch, A. Nordsieck, Note on the radiation field of the electron. Phys. Rev. **52**, 54 (1937)
277. T. Kinoshita, Mass singularities of Feynman amplitudes. J. Math. Phys. **3**, 650 (1962)
278. T. Lee, M. Nauenberg, Degenerate systems and mass singularities. Phys. Rev. B **133**, 1549 (1964)
279. A. Banfi, G. Marchesini, G. Smye, Away from jet energy flow. JHEP **08**, 006 (2002). https://doi.org/10.1088/1126-6708/2002/08/006. arXiv:hep-ph/0206076
280. M.D. Schwartz, H.X. Zhu, Nonglobal logarithms at three loops, four loops, five loops, and beyond. Phys. Rev. D **90**, 065004 (2014). https://doi.org/10.1103/PhysRevD.90.065004. arXiv:1403.4949
281. K. Khelifa-Kerfa, Y. Delenda, Non-global logarithms at finite N_c beyond leading order. JHEP **03**, 094 (2015). https://doi.org/10.1007/JHEP03(2015)094. arXiv:1501.00475
282. M. Dasgupta, K. Khelifa-Kerfa, S. Marzani, M. Spannowsky, On jet mass distributions in Z +jet and dijet processes at the LHC. JHEP **10**, 126 (2012). https://doi.org/10.1007/JHEP10(2012)126. arXiv:1207.1640
283. I.W. Stewart, F.J. Tackmann, W.J. Waalewijn, Factorization at the LHC: from PDFs to initial state jets. Phys. Rev. D **81**, 094035 (2010). https://doi.org/10.1103/PhysRevD.81.094035. arXiv:0910.0467
284. C.W. Bauer, A. Hornig, F.J. Tackmann, Factorization for generic jet production. Phys. Rev. D **79**, 114013 (2009). https://doi.org/10.1103/PhysRevD.79.114013. arXiv:0808.2191
285. A.J. Larkoski, I. Moult, D. Neill, Non-global logarithms, factorization, and the soft substructure of jets. JHEP **09**, 143(2015). https://doi.org/10.1007/JHEP09(2015)143. arXiv:1501.04596
286. Y.-T. Chien, I. Vitev, Jet shape resummation using soft-collinear effective theory. JHEP **12**, 061 (2014). https://doi.org/10.1007/JHEP12(2014)061. arXiv:1405.4293
287. A. Hornig, Y. Makris, T. Mehen, Jet shapes in dijet events at the LHC in SCET. JHEP **04**, 097 (2016). https://doi.org/10.1007/JHEP04(2016)097. arXiv:1601.01319

288. C. Frye, A.J. Larkoski, M.D. Schwartz, K. Yan, Factorization for groomed jet substructure beyond the next-to-leading logarithm. JHEP **07**, 064 (2016). https://doi.org/10.1007/JHEP07(2016)064. arXiv:1603.09338
289. M. Dasgupta, Y. Delenda, On the universality of hadronisation corrections to QCD jets. JHEP **07**, 004 (2009). https://doi.org/10.1088/1126-6708/2009/07/004. arXiv:0903.2187
290. Y.-T. Chien, A. Hornig, C. Lee, Soft-collinear mode for jet cross sections in soft collinear effective theory. Phys. Rev. D **93**, 014033 (2016). https://doi.org/10.1103/PhysRevD.93.014033. arXiv:1509.04287
291. P.Z. Skands, Tuning Monte Carlo generators: the Perugia Tunes. Phys. Rev. D **82**, 074018 (2010). https://doi.org/10.1103/PhysRevD.82.074018. arXiv:1005.3457
292. ATLAS Collaboration, Summary of ATLAS PYTHIA 8 tunes. ATLAS Physics Note. ATL-PHYS-PUB-2012-003 (2012), http://cds.cern.ch/record/1474107
293. S. Gieseke, C. Rohr, A. Siodmok, Colour reconnections in Herwig++. Eur. Phys. J. C **72**, 2225 (2012). https://doi.org/10.1140/epjc/s10052-012-2225-5. arXiv:1206.0041
294. P. Skands, S. Carrazza, J. Rojo, Tuning PYTHIA 8.1: the Monash 2013 Tune. Eur. Phys. J. C **74**, 3024 (2014). https://doi.org/10.1140/epjc/s10052-014-3024-y. arXiv:1404.5630
295. CMS Collaboration, V. Khachatryan et al., Event generator tunes obtained from underlying event and multiparton scattering measurements. Eur. Phys. J. C **76**, 155 (2016). https://doi.org/10.1140/epjc/s10052-016-3988-x. arXiv:1512.00815
296. T. Sjöstrand, P. Eden, C. Friberg, L. Lonnblad, G. Miu, S. Mrenna et al., High-energy physics event generation with PYTHIA 6.1. Comput. Phys. Commun. **135**, 238 (2001). https://doi.org/10.1016/S0010-4655(00)00236-8. arXiv:hep-ph/0010017
297. G. Corcella, I.G. Knowles, G. Marchesini, S. Moretti, K. Odagiri, P. Richardson et al., HERWIG 6: an event generator for hadron emission reactions with interfering gluons (including supersymmetric processes). JHEP **01**, 010 (2001). https://doi.org/10.1088/1126-6708/2001/01/010. arXiv:hep-ph/0011363
298. M. Bahr et al., Herwig++ physics and manual. Eur. Phys. J. C **58**, 639–707 (2008). https://doi.org/10.1140/epjc/s10052-008-0798-9. arXiv:0803.0883
299. J. Bellm et al., Herwig++ 2.7 Release Note, arXiv:1310.6877
300. J. Bellm et al., Herwig 7.0/Herwig++ 3.0 release note. Eur. Phys. J. C **76**, 196 (2016). https://doi.org/10.1140/epjc/s10052-016-4018-8. arXiv:1512.01178
301. T. Gleisberg, S. Höche, F. Krauss, A. Schalicke, S. Schumann, J.-C. Winter, SHERPA 1. alpha: a proof of concept version. JHEP **02**, 056 (2004). https://doi.org/10.1088/1126-6708/2004/02/056. arXiv:hep-ph/0311263
302. T. Gleisberg, S. Höche, F. Krauss, M. Schonherr, S. Schumann, F. Siegert et al., Event generation with SHERPA 1.1. JHEP **02**, 007 (2009). https://doi.org/10.1088/1126-6708/2009/02/007. arXiv:0811.4622
303. F. Krauss, R. Kuhn, G. Soff, AMEGIC++ 1.0: a matrix element generator in C++. JHEP **02**, 044 (2002). https://doi.org/10.1088/1126-6708/2002/02/044. arXiv:hep-ph/0109036
304. T. Gleisberg, S. Höche, Comix, a new matrix element generator. JHEP **12**, 039 (2008). https://doi.org/10.1088/1126-6708/2008/12/039. arXiv:0808.3674
305. R. Brun, F. Bruyant et al., GEANT: simulation program for particle physics experiments. User guide and reference manual. CERN-DD/EE 84–1 (1987)
306. GEANT4 Collaboration, S. Agostinelli et al., GEANT4-a simulation toolkit. Nucl. Instrum. Meth. A **506**, 250 (2003). https://doi.org/10.1016/S0168-9002(03)01368-8
307. J. Conway, PGS: Pretty Good Simulator. http://www.physics.ucdavis.edu/conway/research/software/pgs/pgs.html
308. S. Ovyn, X. Rouby, V. Lemaitre, DELPHES, a framework for fast simulation of a generic collider experiment, arXiv:0903.2225
309. Y.L. Dokshitzer, Calculation of the structure functions for deep inelastic scattering and e^+e^- annihilation by perturbation theory in quantum chromodynamics. Sov. Phys. JETP **46**, 641 (1977)
310. V.N. Gribov, L.N. Lipatov, Deep inelastic *ep* scattering in perturbation theory. Sov. J. Nucl. Phys. **15**, 438 (1972)

311. L.N. Lipatov, The parton model and perturbation theory, Sov. J. Nucl. Phys. **20**, 94 (1975). Yad. Fiz. **20**, 181 (1974)
312. G. Altarelli, G. Parisi, Asymptotic freedom in parton language. Nucl. Phys. B **126**, 298 (1977)
313. A.D. Martin, W.J. Stirling, R.S. Thorne, G. Watt, Parton distributions for the LHC. Eur. Phys. J. C **63**, 189–285 (2009). https://doi.org/10.1140/epjc/s10052-009-1072-5. arXiv:0901.0002
314. A.D. Martin, R.G. Roberts, W.J. Stirling, R.S. Thorne, MRST2001: partons and α_s from precise deep inelastic scattering and Tevatron jet data. Eur. Phys. J. C **23**, 73 (2002). https://doi.org/10.1007/s100520100842. arXiv:hep-ph/0110215
315. S. Dulat, T.-J. Hou, J. Gao, M. Guzzi, J. Huston, P. Nadolsky et al., New parton distribution functions from a global analysis of quantum chromodynamics. Phys. Rev. D **93**, 033006 (2016). https://doi.org/10.1103/PhysRevD.93.033006. arXiv:1506.07443
316. NNPDF Collaboration, R.D. Ball et al., Parton distributions for the LHC Run II. JHEP **04**, 040 (2015). https://doi.org/10.1007/JHEP04(2015)040. arXiv:1410.8849
317. L.A. Harland-Lang, A.D. Martin, P. Motylinski, R.S. Thorne, Parton distributions in the LHC era: MMHT 2014 PDFs. Eur. Phys. J. C **75**, 204 (2015). https://doi.org/10.1140/epjc/s10052-015-3397-6. arXiv:1412.3989
318. ZEUS, H1 Collaborations, H. Abramowicz et al., Combination of measurements of inclusive deep inelastic $e^{\pm}p$ scattering cross sections and QCD analysis of HERA data. Eur. Phys. J. C **75**, 580 (2015). https://doi.org/10.1140/epjc/s10052-015-3710-4. arXiv:1506.06042
319. S. Alekhin, J. Blümlein, S. Moch, The ABM parton distributions tuned to LHC data. Phys. Rev. D **89**, 054028 (2014). https://doi.org/10.1103/PhysRevD.89.054028. arXiv:1310.3059
320. S. Alekhin et al., The PDF4LHC working group interim report, arXiv:1101.0536
321. J. Butterworth et al., PDF4LHC recommendations for LHC Run II. J. Phys. G **43**, 023001 (2016). https://doi.org/10.1088/0954-3899/43/2/023001. arXiv:1510.03865
322. G. Watt, Parton distribution function dependence of benchmark Standard Model total cross sections at the 7 TeV LHC. JHEP **09**, 069 (2011). https://doi.org/10.1007/JHEP09(2011)069. arXiv:1106.5788
323. S. Alekhin, J. Blümlein, S. Moch, Parton distribution functions and benchmark cross sections at NNLO. Phys. Rev. D **86**, 054009 (2012). https://doi.org/10.1103/PhysRevD.86.054009. arXiv:1202.2281
324. R.D. Ball et al., Parton distribution benchmarking with LHC data. JHEP **04**, 125 (2013). https://doi.org/10.1007/JHEP04(2013)125. arXiv:1211.5142
325. R.S. Thorne, The effect on PDFs and $\alpha_S(M_Z^2)$ due to changes in flavour scheme and higher twist contributions. Eur. Phys. J. C **74**, 2958 (2014). https://doi.org/10.1140/epjc/s10052-014-2958-4. arXiv:1402.3536
326. J. Gao, P. Nadolsky, A meta-analysis of parton distribution functions. JHEP **07**, 035 (2014). https://doi.org/10.1007/JHEP07(2014)035. arXiv:1401.0013
327. S. Carrazza, S. Forte, Z. Kassabov, J.I. Latorre, J. Rojo, An unbiased Hessian representation for Monte Carlo PDFs. Eur. Phys. J. C **75**, 369 (2015). https://doi.org/10.1140/epjc/s10052-015-3590-7. arXiv:1505.06736
328. S. Carrazza, J.I. Latorre, J. Rojo, G. Watt, A compression algorithm for the combination of PDF sets. Eur. Phys. J. C **75**, 474 (2015). https://doi.org/10.1140/epjc/s10052-015-3703-3. arXiv:1504.06469
329. S. Libby, G. Sterman, Jet and lepton-pair production in high-energy lepton-hadron and hadron-hadron scattering. Phys. Rev. D **18**, 3252 (1978)
330. R.K. Ellis, H. Georgi et al., Perturbation theory and the parton model in QCD. Nucl. Phys. B **152**, 285 (1979)
331. J.C. Collins, D.E. Soper, G. Sterman, Factorization for short distance hadron-hadron scattering. Nucl. Phys. B **261**, 104 (1985). https://doi.org/10.1016/0550-3213(85)90565-6
332. S. Höche, Introduction to parton-shower event generators, in *Journeys Through the Precision Frontier: Amplitudes for Colliders. Proceedings, Theoretical Advanced Study Institute in Elementary Particle Physics (TASI)*, Boulder, USA, 2–27 June 2014 (2014), arXiv:1411.4085

333. S. Höche, S. Prestel, The midpoint between dipole and parton showers. Eur. Phys. J. C **75**, 461 (2015). https://doi.org/10.1140/epjc/s10052-015-3684-2. arXiv:1506.05057
334. T. Sjöstrand, A model for initial state parton showers. Phys. Lett. B **157**, 321 (1985). https://doi.org/10.1016/0370-2693(85)90674-4
335. M. Bengtsson, T. Sjostrand, A comparative study of coherent and noncoherent parton shower evolution. Nucl. Phys. B **289**, 810 (1987). https://doi.org/10.1016/0550-3213(87)90407-X
336. T. Sjostrand, P.Z. Skands, Transverse-momentum-ordered showers and interleaved multiple interactions. Eur. Phys. J. C **39**, 129 (2005). https://doi.org/10.1140/epjc/s2004-02084-y. arXiv:hep-ph/0408302
337. G. Marchesini, B.R. Webber, Monte Carlo Simulation of general hard processes with coherent QCD radiation. Nucl. Phys. B **310**, 461 (1988). https://doi.org/10.1016/0550-3213(88)90089-2
338. S. Gieseke, P. Stephens, B. Webber, New formalism for QCD parton showers. JHEP **12**, 045 (2003). https://doi.org/10.1088/1126-6708/2003/12/045. arXiv:hep-ph/0310083
339. S. Platzer, S. Gieseke, Coherent parton showers with local recoils. JHEP **01**, 024 (2011). https://doi.org/10.1007/JHEP01(2011)024. arXiv:0909.5593
340. R. Kuhn, F. Krauss, B. Ivanyi, G. Soff, APACIC++: a PArton Cascade in C++, version 1.0. Comput. Phys. Commun. **134**, 223 (2001). https://doi.org/10.1016/S0010-4655(00)00201-0. arXiv:hep-ph/0004270
341. S. Schumann, F. Krauss, A Parton shower algorithm based on Catani-Seymour dipole factorization. JHEP **03**, 038 (2008). https://doi.org/10.1088/1126-6708/2008/03/038. arXiv:0709.1027
342. M. Bengtsson, T. Sjostrand, M. van Zijl, Initial state radiation effects on W and jet production. Z. Phys. C **32**, 67 (1986). https://doi.org/10.1007/BF01441353
343. M. Dasgupta, F.A. Dreyer, K. Hamilton, P.F. Monni, G.P. Salam, G. Soyez, Parton showers beyond leading logarithmic accuracy. Phys. Rev. Lett. **125**, 052002 (2020). https://doi.org/10.1103/PhysRevLett.125.052002. arXiv:2002.11114
344. Z. Nagy, D.E. Soper, A parton shower based on factorization of the quantum density matrix. JHEP **06**, 097 (2014). https://doi.org/10.1007/JHEP06(2014)097. arXiv:1401.6364
345. Z. Nagy, D.E. Soper, Effects of subleading color in a parton shower. JHEP **07**, 119 (2015). https://doi.org/10.1007/JHEP07(2015)119. arXiv:1501.00778
346. H.T. Li, P. Skands, A framework for second-order parton showers. Phys. Lett. B **771**, 59 (2017). https://doi.org/10.1016/j.physletb.2017.05.011. arXiv:1611.00013
347. S. Plätzer, M. Sjodahl, J. Thorén, Color matrix element corrections for parton showers. JHEP **11**, 009 (2018). https://doi.org/10.1007/JHEP11(2018)009. arXiv:1808.00332
348. J.R. Forshaw, J. Holguin, S. Plätzer, Parton branching at amplitude level. JHEP **08**, 145 (2019). https://doi.org/10.1007/JHEP08(2019)145. arXiv:1905.08686
349. Z. Nagy, D.E. Soper, Summations of large logarithms by parton showers, arXiv:2011.04773
350. C.F. Berger, Z. Bern, L.J. Dixon, F. Febres Cordero, D. Forde, H. Ita et al., An automated implementation of on-shell methods for one-loop amplitudes. Phys. Rev. D **78**, 036003 (2008). https://doi.org/10.1103/PhysRevD.78.036003. arXiv:0803.4180
351. G. Cullen et al., GOSAM-2.0: a tool for automated one-loop calculations within the standard model and beyond. Eur. Phys. J. C **74**, 3001 (2014). https://doi.org/10.1140/epjc/s10052-014-3001-5. arXiv:1404.7096
352. F. Cascioli, P. Maierhofer, S. Pozzorini, Scattering amplitudes with Open Loops. Phys. Rev. Lett. **108**, 11601 (2012). https://doi.org/10.1103/PhysRevLett.108.111601. arXiv:1111.5206
353. S. Badger, B. Biedermann, P. Uwer, V. Yundin, Numerical evaluation of virtual corrections to multi-jet production in massless QCD. Comput. Phys. Commun. **184**, 1981 (2013). https://doi.org/10.1016/j.cpc.2013.03.018. arXiv:1209.0100
354. V. Hirschi, R. Frederix, S. Frixione, M.V. Garzelli, F. Maltoni, R. Pittau, Automation of one-loop QCD corrections. JHEP **05**, 044 (2011). https://doi.org/10.1007/JHEP05(2011)044. arXiv:1103.0621

355. G. Ossola, C.G. Papadopoulos, R. Pittau, CutTools: a program implementing the OPP reduction method to compute one-loop amplitudes. JHEP **03**, 042 (2008). https://doi.org/10.1088/1126-6708/2008/03/042. arXiv:0711.3596
356. P. Bolzoni, F. Maltoni, S.-O. Moch, M. Zaro, Higgs production via vector-boson fusion at NNLO in QCD. Phys. Rev. Lett. **105**, 011801 (2010). https://doi.org/10.1103/PhysRevLett.105.011801. arXiv:1003.4451
357. M. Czakon, P. Fiedler, A. Mitov, Total top-quark pair-production cross section at hadron colliders through $O(\alpha_s^4)$. Phys. Rev. Lett. **110**, 252004 (2013). https://doi.org/10.1103/PhysRevLett.110.252004. arXiv:1303.6254
358. F. Cascioli, T. Gehrmann, M. Grazzini, S. Kallweit, P. Maierhöfer, A. von Manteuffel et al., ZZ production at hadron colliders in NNLO QCD. Phys. Lett. B **735**, 311 (2014). https://doi.org/10.1016/j.physletb.2014.06.056. arXiv:1405.2219
359. T. Gehrmann, M. Grazzini, S. Kallweit, P. Maierhöfer, A. von Manteuffel, S. Pozzorini et al., W^+W^- production at hadron colliders in next-to-next-to-leading order QCD. Phys. Rev. Lett. **113**, 212001 (2014). https://doi.org/10.1103/PhysRevLett.113.212001. arXiv:1408.5243
360. M. Grazzini, S. Kallweit, D. Rathlev, $W\gamma$ and $Z\gamma$ production at the LHC in NNLO QCD. JHEP **07**, 085 (2015). https://doi.org/10.1007/JHEP07(2015)085. arXiv:1504.01330
361. R. Boughezal, F. Caola, K. Melnikov, F. Petriello, M. Schulze, Higgs boson production in association with a jet at next-to-next-to-leading order. Phys. Rev. Lett. **115**, 082003 (2015). https://doi.org/10.1103/PhysRevLett.115.082003. arXiv:1504.07922
362. R. Boughezal, C. Focke, X. Liu, F. Petriello, W boson production in association with a jet at next-to-next-to-leading order in perturbative QCD. Phys. Rev. Lett. **115**, (2015). https://doi.org/10.1103/PhysRevLett.115.062002. arXiv:1504.02131
363. R. Boughezal, J.M. Campbell, R.K. Ellis, C. Focke, W.T. Giele, X. Liu et al., Z boson production in association with a jet at next-to-next-to-leading order in perturbative QCD. Phys. Rev. Lett. **116**, 152001 (2016). https://doi.org/10.1103/PhysRevLett.116.152001. arXiv:1512.01291
364. M. Grazzini, S. Kallweit, D. Rathlev, M. Wiesemann, $W^{\pm}Z$ production at hadron colliders in NNLO QCD. Phys. Lett. B **761**, 179 (2016). https://doi.org/10.1016/j.physletb.2016.08.017. arXiv:1604.08576
365. S. Catani, F. Krauss, R. Kuhn, B.R. Webber, QCD matrix elements + parton showers. JHEP **11**, 063 (2001). https://doi.org/10.1088/1126-6708/2001/11/063. arXiv:hep-ph/0109231
366. L. Lönnblad, Correcting the color dipole cascade model with fixed order matrix elements. JHEP **05**, 046 (2002). https://doi.org/10.1088/1126-6708/2002/05/046. arXiv:hep-ph/0112284
367. M.L. Mangano, M. Moretti, F. Piccinini, M. Treccani, Matching matrix elements and shower evolution for top-quark production in hadronic collisions. JHEP **01**, 013 (2007). https://doi.org/10.1088/1126-6708/2007/01/013. arXiv:hep-ph/0611129
368. L. Lönnblad, S. Prestel, Unitarising matrix element + parton shower merging. JHEP **02**, 094 (2013). https://doi.org/10.1007/JHEP02(2013)094. arXiv:1211.4827
369. S. Frixione, B.R. Webber, Matching NLO QCD computations and parton shower simulations. JHEP **06**, 029 (2002). https://doi.org/10.1088/1126-6708/2002/06/029. arXiv:hep-ph/0204244
370. P. Nason, A new method for combining NLO QCD with shower Monte Carlo algorithms. JHEP **11**, 040 (2004). https://doi.org/10.1088/1126-6708/2004/11/040. arXiv:hep-ph/0409146
371. S. Frixione, P. Nason, C. Oleari, Matching NLO QCD computations with parton shower simulations: the POWHEG method. JHEP **11**, 070 (2007). https://doi.org/10.1088/1126-6708/2007/11/070. arXiv:0709.2092
372. J. Alwall, R. Frederix, S. Frixione, V. Hirschi, F. Maltoni, O. Mattelaer et al., The automated computation of tree-level and next-to-leading order differential cross sections, and their matching to parton shower simulations. JHEP **07**, 079 (2014). https://doi.org/10.1007/JHEP07(2014)079. arXiv:1405.0301

373. S. Alioli, P. Nason, C. Oleari, E. Re, A general framework for implementing NLO calculations in shower Monte Carlo programs: the POWHEG BOX. JHEP **06**, 043 (2010). https://doi.org/10.1007/JHEP06(2010)043. arXiv:1002.2581
374. K. Hamilton, P. Nason, G. Zanderighi, MINLO: multi-scale improved NLO. JHEP **10**, 155 (2012). https://doi.org/10.1007/JHEP10(2012)155. arXiv:1206.3572
375. S. Höche, F. Krauss, M. Schönherr, F. Siegert, QCD matrix elements + parton showers: the NLO case. JHEP **04**, 027 (2013). https://doi.org/10.1007/JHEP04(2013)027. arXiv:1207.5030
376. L. Lönnblad, S. Prestel, Merging multi-leg NLO matrix elements with parton showers. JHEP **03**, 166 (2013). https://doi.org/10.1007/JHEP03(2013)166. arXiv:1211.7278
377. K. Hamilton, P. Nason, C. Oleari, G. Zanderighi, Merging H/W/Z + 0 and 1 jet at NLO with no merging scale: a path to parton shower + NNLO matching. JHEP **05**, 082 (2013). https://doi.org/10.1007/JHEP05(2013)082. arXiv:1212.4504
378. T. Sjöstrand, P.Z. Skands, Multiple interactions and the structure of beam remnants. JHEP **03**, 053 (2004). https://doi.org/10.1088/1126-6708/2004/03/053. arXiv:hep-ph/0402078
379. T. Sjöstrand, M. van Zijl, A multiple interaction model for the event structure in hadron collisions **36**, 2019 (1987). https://doi.org/10.1103/PhysRevD.36.2019
380. R. Corke, T. Sjostrand, Interleaved parton showers and tuning prospects. JHEP **03**, 032 (2011). https://doi.org/10.1007/JHEP03(2011)032. arXiv:1011.1759
381. B. Humpert, Are there multi-quark interactions? Phys. Lett. B **131**, 461 (1983). https://doi.org/10.1016/0370-2693(83)90540-3
382. N. Paver, D. Treleani, Multiple parton interactions and multi-jet events at collider and Tevatron energies. Phys. Lett. B **146**, 252 (1984). https://doi.org/10.1016/0370-2693(84)91029-3
383. T. Sjöstrand, The development of MPI modeling in Pythia. Adv. Ser. Direct. High Energy Phys. **29**, 191 (2018). https://doi.org/10.1142/9789813227767_0010. arXiv:1706.02166
384. B. Andersson, G. Gustafson, B. Soderberg, A general model for jet fragmentation. Z. Phys. C **20**, 317 (1983). https://doi.org/10.1007/BF01407824
385. B. Andersson, G. Gustafson, B. Soderberg, A probability measure on parton and string states. Nucl. Phys. B **264**, 29 (1986). https://doi.org/10.1016/0550-3213(86)90471-2
386. B. Andersson, The Lund Model (Cambridge University Press, Cambridge, 1998). https://doi.org/10.1017/CBO9780511524363
387. G.S. Bali, Casimir scaling of SU(3) static potentials. Phys. Rev. D **62**, 114503 (2000). https://doi.org/10.1103/PhysRevD.62.114503. arXiv:hep-lat/0006022
388. T. Sjöstrand, Jet fragmentation of nearby partons. Nucl. Phys. B **248**, 469 (1984). https://doi.org/10.1016/0550-3213(84)90607-2
389. T. Sjöstrand, Status and developments of event generators, in *4th Large Hadron Collider Physics Conference (LHCP 2016)*, Lund, Sweden, 13–18 June 2016 (2016), arXiv:1608.06425
390. T. Sjöstrand, P.Z. Skands, Baryon number violation and string topologies. Nucl. Phys. B **659**, 243 (2003). https://doi.org/10.1016/S0550-3213(03)00193-7. arXiv:hep-ph/0212264
391. B. Andersson, G. Gustafson, T. Sjöstrand, A model for baryon production in quark and gluon jets. Nucl. Phys. B **197**, 45 (1982). https://doi.org/10.1016/0550-3213(82)90153-5
392. B. Andersson, G. Gustafson, T. Sjöstrand, Baryon production in jet fragmentation and t decay. Phys. Scripta **32**, 574 (1985). https://doi.org/10.1088/0031-8949/32/6/003
393. P. Edén, G. Gustafson, Baryon production in the string fragmentation picture. Z. Phys. C **75**, 41 (1997). https://doi.org/10.1007/s002880050445
394. D. Amati, G. Veneziano, Preconfinement as a property of perturbative QCD. Phys. Lett. B **83**, 87 (1979). https://doi.org/10.1016/0370-2693(79)90896-7
395. A. Bassetto, M. Ciafaloni, G. Marchesini, Color singlet distributions and mass damping in perturbative QCD. Phys. Lett. B **83**, 207 (1979). https://doi.org/10.1016/0370-2693(79)90687-7
396. B.R. Webber, A QCD Model for jet fragmentation including soft gluon interference. Nucl. Phys. B **238**, 492 (1984). https://doi.org/10.1016/0550-3213(84)90333-X

397. J.-C. Winter, F. Krauss, G. Soff, A modified cluster hadronization model. Eur. Phys. J. C **36**, 381 (2004). https://doi.org/10.1140/epjc/s2004-01960-8. arXiv:hep-ph/0311085
398. Particle Data Group, C. Patrignani et al., Review of particle physics. Chin. Phys. **C40**, 100001 (2016). https://doi.org/10.1088/1674-1137/40/10/100001
399. A. Buckley, H. Hoeth, H. Lacker, H. Schulz, J.E. von Seggern, Systematic event generator tuning for the LHC. Eur. Phys. J. C **65**, 331–357 (2010). https://doi.org/10.1140/epjc/s10052-009-1196-7. arXiv:0907.2973
400. J.R. Andersen et al., Les Houches 2015: Physics at TeV colliders standard model working group report, in *9th Les Houches Workshop on Physics at TeV Colliders (PhysTeV 2015). Proceedings*, Les Houches, France, 1–19 June 2015 (2016), arXiv:1605.04692
401. ATLAS Collaboration, The ATLAS experiment at the CERN Large Hadron Collider. JINST **3**, S08003 (2008). https://doi.org/10.1088/1748-0221/3/08/S08003
402. C.M.S. Collaboration, The CMS experiment at the CERN LHC. JINST **3**, S08004 (2008). https://doi.org/10.1088/1748-0221/3/08/S08004
403. ATLAS Collaboration, The ATLAS inner detector commissioning and calibration. Eur. Phys. J. C **70**, 787 (2010). https://doi.org/10.1140/epjc/s10052-010-1366-7. arXiv:1004.5293
404. C.M.S. Collaboration, Description and performance of track and primary-vertex reconstruction with the CMS tracker. JINST **9**, P10009 (2014). https://doi.org/10.1088/1748-0221/9/10/P10009. arXiv:1405.6569
405. ATLAS IBL Collaborations, B. Abbott et al., Production and integration of the ATLAS insertable B-layer. JINST **13**, T05008 (2018). https://doi.org/10.1088/1748-0221/13/05/T05008. arXiv:1803.00844
406. CMS Collaboration, The CMS hadron calorimeter project: Technical design report, Technical report. CERN-LHCC-97-031 (1997), https://cds.cern.ch/record/357153
407. CMS Collaboration, Particle-flow reconstruction and global event description with the CMS detector. JINST **12**, P10003 (2017). https://doi.org/10.1088/1748-0221/12/10/P10003. arXiv:1706.04965
408. ATLAS Collaboration, Jet reconstruction and performance using particle flow with the ATLAS Detector. Eur. Phys. J. C **77**, 466 (2017). https://doi.org/10.1140/epjc/s10052-017-5031-2. arXiv:1703.10485
409. ATLAS Collaboration, Topological cell clustering in the ATLAS calorimeters and its performance in LHC Run 1. Eur. Phys. J. C **77**, 490 (2017). https://doi.org/10.1140/epjc/s10052-017-5004-5. arXiv:1603.02934
410. ATLAS Collaboration, Improving jet substructure performance in ATLAS using Track-CaloClusters, ATLAS Physics Note. ATL-PHYS-PUB-2017-015 (2017), https://cds.cern.ch/record/2275636
411. CMS Collaboration, V tagging observables and correlations, CMS Physics Analysis Summary. CMS-PAS-JME-14-002 (2014), http://cds.cern.ch/record/1754913
412. ATLAS Collaboration, M. Aaboud et al., A measurement of the calorimeter response to single hadrons and determination of the jet energy scale uncertainty using LHC Run-1 pp-collision data with the ATLAS detector. Eur. Phys. J. C **77**, 26 (2017). https://doi.org/10.1140/epjc/s10052-016-4580-0. arXiv:1607.08842
413. ATLAS Collaboration, Charged-particle distributions at low transverse momentum in $\sqrt{s} =$ 13 TeV pp interactions measured with the ATLAS detector at the LHC. Eur. Phys. J. C **76**, 502 (2016). https://doi.org/10.1140/epjc/s10052-016-4335-y. arXiv:1606.01133
414. ATLAS Collaboration, M. Aaboud et al., Performance of the ATLAS transition radiation tracker in Run 1 of the LHC: tracker properties. JINST **12**, P05002 (2017). https://doi.org/10.1088/1748-0221/12/05/P05002. arXiv:1702.06473
415. ATLAS Collaboration, A neural network clustering algorithm for the ATLAS silicon pixel detector. JINST **9**, P09009 (2014). https://doi.org/10.1088/1748-0221/9/09/P09009. arXiv:1406.7690
416. ATLAS Collaboration, Performance of the ATLAS track reconstruction algorithms in dense environments in LHC Run 2. Eur. Phys. J. C **77**, 673 (2017). https://doi.org/10.1140/epjc/s10052-017-5225-7., arXiv:1704.07983

417. ATLAS Collaboration, Measurement of track reconstruction inefficiencies in the core of jets via pixel dE/dx with the ATLAS experiment using $\sqrt{s} = 13$ TeV pp collision data, ATLAS Physics Note. ATL-PHYS-PUB-2016-007 (2016), https://cds.cern.ch/record/2140460
418. ATLAS Collaboration, Modelling of track reconstruction inside jets with the 2016 ATLAS $\sqrt{s} = 13$ TeV pp dataset, ATLAS Physics Note. ATL-PHYS-PUB-2017-016 (2017), https://atlas.web.cern.ch/Atlas/GROUPS/PHYSICS/PUBNOTES/ATL-PHYS-PUB-2017-016/
419. ATLAS Collaboration, Measurement of performance of the pixel neural network clustering algorithm of the ATLAS experiment at $\sqrt{s} = 13$ TeV, ATLAS Physics Note. ATL-PHYS-PUB-2015-044 (2015), https://cds.cern.ch/record/2054921
420. ATLAS Collaboration, Jet energy measurement and its systematic uncertainty in proton-proton collisions at $\sqrt{s} = 7$ TeV with the ATLAS detector. Eur. Phys. J. C **75**, 17 (2015). https://doi.org/10.1140/epjc/s10052-014-3190-y. arXiv:1406.0076
421. ATLAS Collaboration, Jet energy scale measurements and their systematic uncertainties in proton-proton collisions at $\sqrt{s} = 13$ TeV with the ATLAS detector. Phys. Rev. D **96**, 072002 (2017). https://doi.org/10.1103/PhysRevD.96.072002. arXiv:1703.09665
422. C.M.S. Collaboration, Determination of jet energy calibration and transverse momentum resolution in CMS. JINST **6**, P11002 (2011). https://doi.org/10.1088/1748-0221/6/11/P11002. arXiv:1107.4277
423. A. Cukierman, B. Nachman, Mathematical properties of numerical inversion for jet calibrations. Nucl. Instrum. Meth. **A858**, 1 (2017). https://doi.org/10.1016/j.nima.2017.03.038. arXiv:1609.05195
424. ATLAS Collaboration, Jet global sequential corrections with the ATLAS detector in proton-proton collisions at $\sqrt{s} = 8$ TeV, ATLAS Conference Note. ATLAS-CONF-2015-002 (2015), https://cds.cern.ch/record/2001682
425. ATLAS Collaboration, Determination of jet calibration and energy resolution in proton-proton collisions at $\sqrt{s} = 8$ TeV using the ATLAS detector, arXiv:1910.04482
426. CMS Collaboration, Jet energy scale uncertainty correlations between ATLAS and CMS. CMS Physics Analysis Summary. CMS-PAS-JME-14-003 (2014), https://cds.cern.ch/record/1967369
427. ATLAS Collaboration, Jet mass reconstruction with the ATLAS detector in early Run 2 data. ATLAS Conference Note. ATLAS-CONF-2016-035 (2016), http://cds.cern.ch/record/2200211
428. ATLAS Collaboration, Measurement of large radius jet mass reconstruction performance at $\sqrt{s} = 8$ TeV using the ATLAS detector. ATLAS Conference Note. ATLAS-CONF-2016-008 (2016), https://cds.cern.ch/record/2139642
429. ATLAS Collaboration, In situ calibration of large-radius jet energy and mass in 13 TeV proton-proton collisions with the ATLAS detector. Eur. Phys. J. C **79**, 135 (2019). https://doi.org/10.1140/epjc/s10052-019-6632-8. arXiv:1807.09477
430. ATLAS Collaboration, Track assisted techniques for jet substructure. ATLAS Physics Note. ATL-PHYS-PUB-2018-012 (2018), https://cds.cern.ch/record/2630864
431. B. Nachman, P. Nef, A. Schwartzman, M. Swiatlowski, C. Wanotayaroj, Jets from jets: re-clustering as a tool for large radius jet reconstruction and grooming at the LHC. JHEP **02**, 075 (2015). https://doi.org/10.1007/JHEP02(2015)075. arXiv:1407.2922
432. ATLAS Collaboration, Jet reclustering and close-by effects in ATLAS Run 2. ATLAS Conference Note. ATLAS-CONF-2017-062 (2017), http://cds.cern.ch/record/2275649
433. ATLAS Collaboration, Performance of jet substructure techniques for large-R jets in proton-proton collisions at $\sqrt{s} = 7$ TeV using the ATLAS detector. JHEP **09**, 076 (2013). https://doi.org/10.1007/JHEP09(2013)076. arXiv:1306.4945
434. ATLAS Collaboration, Optimisation of large-radius jet reconstruction for the ATLAS detector in 13 TeV proton-proton collisions, arXiv:2009.04986
435. C.M.S. Collaboration, Identification techniques for highly boosted W bosons that decay into hadrons. JHEP **12**, 017 (2014). https://doi.org/10.1007/JHEP12(2014)017. arXiv:1410.4227
436. CMS Collaboration, Jet algorithms performance in 13 TeV data. CMS Physics Analysis Summary. CMS-PAS-JME-16-003 (2017), https://cds.cern.ch/record/2256875

437. CMS Collaboration, Search for massive resonances decaying into WW, WZ, ZZ, qW , and qZ with dijet final states at $\sqrt{s} = 13$TeV. Phys. Rev. D **97**, 072006 (2018). https://doi.org/10.1103/PhysRevD.97.072006. arXiv:1708.05379
438. CMS Collaboration, Measurement of the jet mass in highly boosted $t\bar{t}$ events from pp collisions at $\sqrt{s} = 8$ TeV. Eur. Phys. J. C **77**, 467 (2017). https://doi.org/10.1140/epjc/s10052-017-5030-3. arXiv:1703.06330
439. CMS Collaboration, Measurements of the differential jet cross section as a function of the jet mass in dijet events from proton-proton collisions at $\sqrt{s} = 13$ TeV. JHEP **11**, 113 (2018). https://doi.org/10.1007/JHEP11(2018)113. arXiv:1807.05974
440. CMS Collaboration, Pileup mitigation at CMS in 13 TeV data. JINST **15**, P09018 (2020). https://doi.org/10.1088/1748-0221/15/09/p09018. arXiv:2003.00503
441. CMS Collaboration, Pileup removal algorithms. CMS Physics Analysis Summary. CMS-PAS-JME-14-001 (2014), http://cds.cern.ch/record/1751454
442. ATLAS Collaboration, Performance of pile-up mitigation techniques for jets in pp collisions at $\sqrt{s} = 8$ TeV using the ATLAS detector. Eur. Phys. J. C **76**, 581 (2016). https://doi.org/10.1140/epjc/s10052-016-4395-z. arXiv:1510.03823
443. G. Soyez, G.P. Salam, J. Kim, S. Dutta, M. Cacciari, Pileup subtraction for jet shapes. Phys. Rev. Lett. **110**, 162001 (2013). https://doi.org/10.1103/PhysRevLett.110.162001. arXiv:1211.2811
444. CMS Collaboration, Pileup jet identification. CMS Physics Analysis Summary. CMS-PAS-JME-13-005 (2013), https://cds.cern.ch/record/1581583
445. ATLAS Collaboration, Identification and rejection of pile-up jets at high pseudorapidity with the ATLAS detector. Eur. Phys. J. C **77**, 580 (2017). https://doi.org/10.1140/epjc/s10052-017-5081-5. arXiv:1705.02211. Erratum: Eur. Phys. J. C **77**, 712 (2017)
446. D. Krohn, M.D. Schwartz, M. Low, L.-T. Wang, Jet cleansing: Pileup removal at high luminosity. Phys. Rev. D **90**, 065020 (2014). https://doi.org/10.1103/PhysRevD.90.065020. arXiv:1309.4777
447. D. Bertolini, T. Chan, J. Thaler, Jet observables without jet algorithms. JHEP **04**, 013 (2014). https://doi.org/10.1007/JHEP04(2014)013. arXiv:1310.7584
448. M. Cacciari, G.P. Salam, G. Soyez, SoftKiller, a particle-level pileup removal method. Eur. Phys. J. C **75**, 59 (2015). https://doi.org/10.1140/epjc/s10052-015-3267-2. arXiv:1407.0408
449. P. Berta, M. Spousta, D.W. Miller, R. Leitner, Particle-level pileup subtraction for jets and jet shapes. JHEP **06**, 092 (2014). https://doi.org/10.1007/JHEP06(2014)092. arXiv:1403.3108
450. D. Bertolini, P. Harris, M. Low, N. Tran, Pileup per particle identification. JHEP **10**, 059 (2014). https://doi.org/10.1007/JHEP10(2014)059. arXiv:1407.6013
451. P.T. Komiske, E.M. Metodiev, B. Nachman, M.D. Schwartz, Pileup mitigation with machine learning (PUMML). JHEP **12**, 051 (2017). https://doi.org/10.1007/JHEP12(2017)051. arXiv:1707.08600
452. ATLAS Collaboration, Impact of alternative inputs and grooming methods on large-R jet reconstruction in ATLAS. ATLAS Physics Note. ATL-PHYS-PUB-2017-020 (2017), https://cds.cern.ch/record/2297485
453. ATLAS Collaboration, Impact of alternative inputs and jet grooming on large-R jet performance. ATLAS Physics Note. ATL-PHYS-PUB-2019-027 (2019), https://cds.cern.ch/record/2683619
454. D. Contardo, M. Klute, J. Mans, L. Silvestris, J. Butler, Technical proposal for the Phase-II upgrade of the CMS detector. CMS Technical Design Report. CMS-TDR-15-02 (2015), https://cds.cern.ch/record/2020886
455. ATLAS Collaboration, Constituent-level pileup mitigation performance using 2015 data. ATLAS Conference Note. ATLAS-CONF-2017-065 (2017), https://cds.cern.ch/record/2281055
456. ATLAS Collaboration, Search for pair production of gluinos decaying via stop and sbottom in events with b-jets and large missing transverse momentum in pp collisions at $\sqrt{s} = 13$ TeV with the ATLAS detector. Phys. Rev. D **94**, 032003 (2016). https://doi.org/10.1103/PhysRevD.94.032003. arXiv:1605.09318

457. ATLAS Collaboration, Search for top squarks in final states with one isolated lepton, jets, and missing transverse momentum in $\sqrt{s} = 13$ TeV pp collisions with the ATLAS detector. Phys. Rev. D **94**, 052009 (2016). https://doi.org/10.1103/PhysRevD.94.052009 arXiv:1606.03903
458. ATLAS Collaboration, Search for new phenomena with large jet multiplicities and missing transverse momentum using large-radius jets and flavour-tagging at ATLAS in 13 TeV pp collisions. JHEP **12**, 034 (2017). https://doi.org/10.1007/JHEP12(2017)034. arXiv:1708.02794
459. CMS Collaboration, Top tagging with new approaches. CMS Physics Analysis Summary. CMS-PAS-JME-15-002 (2016), http://cds.cern.ch/record/2126325
460. G. Altarelli, G. Parisi, Asymptotic freedom in parton language. Nucl. Phys. B **126**, 298 (1977). https://doi.org/10.1016/0550-3213(77)90384-4
461. A. Banfi, G.P. Salam, G. Zanderighi, Infrared safe definition of jet flavor. Eur. Phys. J. C **47**, 113 (2006). https://doi.org/10.1140/epjc/s2006-02552-4. arXiv:hep-ph/0601139
462. A. Buckley, C. Pollard, QCD-aware partonic jet clustering for truth-jet flavour labelling. Eur. Phys. J. C **76**, 71 (2016). https://doi.org/10.1140/epjc/s10052-016-3925-z.arXiv:1507.00508
463. E. M. Metodiev, J. Thaler, Jet topics: disentangling quarks and gluons at colliders. Phys. Rev. Lett. **120**, 241602 (2018). https://doi.org/10.1103/PhysRevLett.120.241602. arXiv:1802.00008
464. P.T. Komiske, E.M. Metodiev, J. Thaler, An operational definition of quark and gluon jets. JHEP **11**, 059 (2018). https://doi.org/10.1007/JHEP11(2018)059. arXiv:1809.01140
465. S. Bright-Thonney, B. Nachman, Investigating the topology dependence of quark and gluon jets. JHEP **03**, 098 (2019). https://doi.org/10.1007/JHEP03(2019)098. arXiv:1810.05653
466. ATLAS Collaboration, Light-quark and gluon jet discrimination in pp collisions at $\sqrt{s} =$ 7 TeV with the ATLAS detector. Eur. Phys. J. C **74**, 3023 (2014). https://doi.org/10.1140/epjc/s10052-014-3023-z. arXiv:1405.6583
467. P. Gras, S. Höche, D. Kar, A. Larkoski, L. Lönnblad, S. Plätzer et al., Systematics of quark/gluon tagging. JHEP **07**, 091 (2017). https://doi.org/10.1007/JHEP07(2017)091. arXiv:1704.03878
468. B. Andersson, G. Gustafson, L. Lönnblad, U. Pettersson, Coherence effects in deep inelastic scattering. Z. Phys. C **43**, 625 (1989). https://doi.org/10.1007/BF01550942
469. F.A. Dreyer, G.P. Salam, G. Soyez, The Lund jet plane. JHEP **12**, 064 (2018). https://doi.org/10.1007/JHEP12(2018)064. arXiv:1807.04758
470. ATLAS Collaboration, Measurement of the Lund Jet Plane using charged particles in 13 TeV proton-proton collisions with the ATLAS detector . Phys. Rev. Lett. **124**, 222002 (2020). https://doi.org/10.1103/PhysRevLett.124.222002. arXiv:2004.03540
471. P.E. Rakow, B. Webber, Transverse momentum moments of hadron distributions in QCD jets. Nucl. Phys. B **191**, 63–74 (1981). https://doi.org/10.1016/0550-3213(81)90286-8
472. S. Catani, G. Turnock, B. Webber, Jet broadening measures in e^+e^- annihilation. Phys. Lett. B **295**, 269–276 (1992). https://doi.org/10.1016/0370-2693(92)91565-Q
473. J. Gallicchio, J. Huth, M. Kagan, M.D. Schwartz, K. Black, B. Tweedie, Multivariate discrimination and the Higgs+W/Z search. JHEP **04**, 069 (2011). https://doi.org/10.1007/JHEP04(2011)069. arXiv:1010.3698
474. C.M.S. Collaboration, Search for a Higgs boson in the decay channel H to $ZZ^{(*)}$ to $q\bar{q}\ell^-\ell^+$ in pp collisions at $\sqrt{s} = 7$ TeV. JHEP **04**, 036 (2012). https://doi.org/10.1007/JHEP04(2012)036. arXiv:1202.1416
475. C. Frye, A.J. Larkoski, J. Thaler, K. Zhou, Casimir meets poisson: Improved quark/gluon discrimination with counting observables. JHEP **09**, 083 (2017). https://doi.org/10.1007/JHEP09(2017)083. arXiv:1704.06266
476. J. Gallicchio, M.D. Schwartz, Quark and gluon tagging at the LHC. Phys. Rev. Lett. **107**, 172001 (2011). https://doi.org/10.1103/PhysRevLett.107.172001. arXiv:1106.3076
477. P.T. Komiske, E.M. Metodiev, D. Schwartz, Deep learning in color: towards automated quark/gluon jet discrimination. JHEP **01**, 110 (2016). https://doi.org/10.1007/JHEP01(2017)110. arXiv:1612.01551

478. ATLAS Collaboration, Discrimination of light quark and gluon jets in pp collisions at $\sqrt{s} = 8$ TeV with the ATLAS detector. ATLAS Conference Note. ATLAS-CONF-2016-034 (2016), http://cds.cern.ch/record/2200202
479. ATLAS Collaboration, Quark versus gluon jet tagging using charged particle multiplicity with the ATLAS detector. ATLAS Physics Note. ATL-PHYS-PUB-2017-009 (2017), https://cds.cern.ch/record/2263679
480. ATLAS Collaboration, Quark versus gluon jet tagging using jet images with the ATLAS detector. ATLAS Physics Note. ATL-PHYS-PUB-2017-017 (2017), http://cds.cern.ch/record/2275641
481. CMS Collaboration, Performance of quark/gluon discrimination in 8 TeV pp data. CMS Physics Analysis Summary. CMS-PAS-JME-13-002 (2013), https://cds.cern.ch/record/1599732
482. CMS Collaboration, Performance of quark/gluon discrimination in 13 TeV data. CMS Detector Performance Summary. CMS-DP-2016-070 (2016), https://cds.cern.ch/record/2234117
483. ATLAS Collaboration, Identification of boosted, hadronically decaying W bosons and comparisons with ATLAS data taken at $\sqrt{s} = 8$ TeV. Eur. Phys. J. C **76**, 154 (2016). https://doi.org/10.1140/epjc/s10052-016-3978-z. arXiv:1510.05821
484. ATLAS Collaboration, Identification of boosted, hadronically-decaying W and Z bosons in $\sqrt{s} = 13$ TeV Monte Carlo simulations for ATLAS. ATLAS Physics Note. ATL-PHYS-PUB-2015-033 (2015), http://cds.cern.ch/record/2041461
485. L.G. Almeida, S.J. Lee, G. Perez, G.F. Sterman, I. Sung, J. Virzi, Substructure of high-p_{T} jets at the LHC. Phys. Rev. D **79**, 074017 (2009). https://doi.org/10.1103/PhysRevD.79.074017. arXiv:0807.0234
486. J. Thaler, L.-T. Wang, Strategies to identify boosted tops. JHEP **07**, 092 (2008). https://doi.org/10.1088/1126-6708/2008/07/092. arXiv:0806.0023
487. ATLAS Collaboration, Identification of hadronically-decaying W bosons and top quarks using high-level features as input to boosted decision trees and deep neural networks in ATLAS at $\sqrt{s} = 13$ TeV. ATLAS Physics Note. ATL-PHYS-PUB-2017-004 (2017), https://cds.cern.ch/record/2259646
488. A.J. Larkoski, D. Neill, J. Thaler, Jet shapes with the broadening axis. JHEP **04**, 017 (2014). https://doi.org/10.1007/JHEP04(2014)017. arXiv:1401.2158
489. CMS Collaboration, Search for low mass vector resonances decaying into quark-antiquark pairs in proton-proton collisions at $\sqrt{s} = 13$ TeV. JHEP **01**, 097 (2018). https://doi.org/10.1007/JHEP01(2018)097. arXiv:1710.00159
490. C.M.S. Collaboration, Search for massive resonances decaying into pairs of boosted bosons in semi-leptonic final states at $\sqrt{s} = 8$ TeV. JHEP **08**, 174 (2014). https://doi.org/10.1007/JHEP08(2014)174. arXiv:1405.3447
491. CMS Collaboration, Search for massive WH resonances decaying into the $\ell\nu\mathrm{b}\bar{\mathrm{b}}$ final state at $\sqrt{s} = 8$ TeV. Eur. Phys. J. C **76**, 237 (2016). https://doi.org/10.1140/epjc/s10052-016-4067-z. arXiv:1601.06431
492. CMS Collaboration, Search for massive resonances decaying into WW, WZ or ZZ bosons in proton-proton collisions at $\sqrt{s} = 13$ TeV. JHEP **03**, 162 (2017). https://doi.org/10.1007/JHEP03(2017)162. arXiv:1612.09159
493. ATLAS Collaboration, Search for heavy resonances decaying to a W or Z boson and a Higgs boson in the $q\bar{q}^{(\prime)}b\bar{b}$ final state in pp collisions at $\sqrt{s} = 13$ TeV with the ATLAS detector. Phys. Lett. B **774**, 494 (2017). https://doi.org/10.1016/j.physletb.2017.09.066. arXiv:1707.06958
494. ATLAS Collaboration, Search for diboson resonances with boson-tagged jets in pp collisions at $\sqrt{s} = 13$ TeV with the ATLAS detector. Phys. Lett. B **777**, 91 (2018). https://doi.org/10.1016/j.physletb.2017.12.011. arXiv:1708.04445
495. ATLAS Collaboration, Search for WW/WZ resonance production in $\ell\nu qq$ final states in pp collisions at $\sqrt{s} = 13$ TeV with the ATLAS detector. JHEP **03**, 042 (2018). https://doi.org/10.1007/JHEP03(2018)042. arXiv:1710.07235

496. C. Shimmin, P. Sadowski, P. Baldi, E. Weik, D. Whiteson, E. Goul et al., Decorrelated jet substructure tagging using adversarial neural networks. Phys. Rev. D **96**, 074034 (2017). https://doi.org/10.1103/PhysRevD.96.074034. arXiv:1703.03507
497. J.A. Aguilar-Saavedra, J.H. Collins, R.K. Mishra, A generic anti-QCD jet tagger. JHEP **11**, 163 (2017). https://doi.org/10.1007/JHEP11(2017)163. arXiv:1709.01087
498. ATLAS Collaboration, Boosted object tagging with variable-R jets in the ATLAS detector. ATLAS Physics Note. ATL-PHYS-PUB-2016-013 (2016), https://cds.cern.ch/record/2199360
499. ATLAS Collaboration, A new method to distinguish hadronically decaying boosted Z bosons from W bosons using the ATLAS detector. Eur. Phys. J. C **76**, 238 (2016). https://doi.org/10.1140/epjc/s10052-016-4065-1. arXiv:1509.04939
500. ATLAS Collaboration, Evidence for the spin-0 nature of the Higgs boson using ATLAS data. Phys. Lett. B **726**, 120 (2013). https://doi.org/10.1016/j.physletb.2013.08.026. arXiv:1307.1432
501. ATLAS Collaboration, Observation and measurement of Higgs boson decays to WW^* with the ATLAS detector. Phys. Rev. D **92**, 012006 (2015). https://doi.org/10.1103/PhysRevD.92.012006. arXiv:1412.2641
502. C.M.S. Collaboration, Measurement of Higgs boson production and properties in the WW decay channel with leptonic final states. JHEP **1401**, 096 (2014). https://doi.org/10.1007/JHEP01(2014)096. arXiv:1312.1129
503. C.M.S. Collaboration, Constraints on the spin-parity and anomalous HVV couplings of the Higgs boson in proton collisions at 7 and 8 TeV. Phys. Rev. D **92**, 012004 (2015). https://doi.org/10.1103/PhysRevD.92.012004. arXiv:1411.3441
504. CMS Collaboration, Identification of heavy-flavour jets with the CMS detector in pp collisions at 13 TeV. JINST **13**, P05011 (2018). https://doi.org/10.1088/1748-0221/13/05/P05011. arXiv:1712.07158
505. ATLAS Collaboration, Expected performance of the ATLAS b-tagging algorithms in Run 2. ATLAS Physics Note. ATL-PHYS-PUB-2015-022 (July 2015), https://cds.cern.ch/record/2037697
506. ATLAS Collaboration, Optimisation of the ATLAS b-tagging performance for the 2016 LHC Run. ATLAS Physics Note. ATL-PHYS-PUB-2016-012 (June 2016), https://cds.cern.ch/record/2160731
507. CMS Collaboration, Identification of b quark jets at the CMS experiment in the LHC Run 2. CMS Physics Analysis Summary. CMS-PAS-BTV-15-001 (2016), http://cds.cern.ch/record/2138504
508. ATLAS Collaboration, Measurements of b-jet tagging efficiency with the ATLAS detector using $t\bar{t}$ events at $\sqrt{s} = 13$ TeV. JHEP **08**, 089 (2018). https://doi.org/10.1007/JHEP08(2018)089. arXiv:1805.01845
509. C.M.S. Collaboration, Identification of b-quark jets with the CMS experiment. JINST **8**, P04013 (2013). https://doi.org/10.1088/1748-0221/8/04/P04013. arXiv:1211.4462
510. CMS Collaboration, Performance of b tagging at $\sqrt{s} = 8$ TeV in multijet, $t\bar{t}$ and boosted topology events. CMS Physics Analysis Summary. CMS-PAS-BTV-13-001 (2013), http://cds.cern.ch/record/1581306
511. ATLAS Collaboration, Flavor tagging with track jets in boosted topologies with the ATLAS detector. ATLAS Physics Note. ATL-PHYS-PUB-2014-013 (2014), https://cds.cern.ch/record/1750681
512. ATLAS Collaboration, Studies of b-tagging performance and jet substructure in a high p_T $g \to b\bar{b}$ rich sample of large-R jets from pp collisions at $\sqrt{s} = 8$ TeV with the ATLAS detector. ATLAS Conference Note. ATLAS-CONF-2016-002 (2016), http://cds.cern.ch/record/2135187
513. ATLAS Collaboration, Performance of b-jet identification in the ATLAS experiment. JINST **11**, P04008 (2016). https://doi.org/10.1088/1748-0221/11/04/P04008. arXiv:1512.01094
514. ATLAS Collaboration, Identification of boosted Higgs bosons decaying into b-quark pairs with the ATLAS detector at 13 TeV. Eur. Phys. J. C **79**, 836 (2019). https://doi.org/10.1140/epjc/s10052-019-7335-x. arXiv:1906.11005

515. ATLAS Collaboration, Variable radius, exclusive-k_T, and center-of-mass subjet reconstruction for Higgs($\rightarrow b\bar{b}$) tagging in ATLAS. ATLAS Physics Note. ATL-PHYS-PUB-2017-010 (2017), https://cds.cern.ch/record/2268678
516. CMS Collaboration, Identification of double-b quark jets in boosted event topologies. CMS Physics Analysis Summary. CMS-PAS-BTV-15-002 (2016), http://cds.cern.ch/record/2195743
517. C. Chen, Identification of a bottom quark-antiquark pair in a single jet with high transverse momentum and its application. Phys. Rev. D **92**, 093010 (2015). https://doi.org/10.1103/PhysRevD.92.093010. arXiv:1507.06913
518. CMS Collaboration, Search for a massive resonance decaying into a Higgs boson and a W or Z boson in hadronic final states in proton-proton collisions at $\sqrt{s} = 8$ TeV. JHEP **02**, 145 (2016). https://doi.org/10.1007/JHEP02(2016)145. arXiv:1506.01443
519. K. Agashe, J.H. Collins, P. Du, S. Hong, D. Kim, R.K. Mishra, Detecting a boosted diboson resonance. JHEP **11**, 027 (2018). https://doi.org/10.1007/JHEP11(2018)027. arXiv:1809.07334
520. ATLAS Collaboration, Identification of high transverse momentum top quarks in pp collisions at $\sqrt{s} = 8$ TeV with the ATLAS detector. JHEP **06**, 093 (2016). https://doi.org/10.1007/JHEP06(2016)093. arXiv:1603.03127
521. ATLAS Collaboration, Search for $W' \rightarrow tb$ decays in the hadronic final state using pp collisions at $\sqrt{s} = 13$ TeV with the ATLAS detector. Phys. Lett. B **781**, 327 (2018). https://doi.org/10.1016/j.physletb.2018.03.036. arXiv:1801.07893
522. ATLAS Collaboration, Performance of top quark and W boson tagging in Run 2 with ATLAS. ATLAS Conference Note. ATLAS-CONF-2017-064 (2017), https://atlas.web.cern.ch/Atlas/GROUPS/PHYSICS/CONFNOTES/ATLAS-CONF-2017-064/
523. G.P. Salam, L. Schunk, G. Soyez, Dichroic subjettiness ratios to distinguish colour flows in boosted boson tagging. JHEP **03**, 022 (2017). https://doi.org/10.1007/JHEP03(2017)022. arXiv:1612.03917
524. ATLAS Collaboration, Boosted hadronic top identification at ATLAS for early 13 TeV data. ATLAS Physics Note. ATL-PHYS-PUB-2015-053 (2015), http://cds.cern.ch/record/2116351
525. ATLAS Collaboration, Performance of top-quark and W-boson tagging with ATLAS in Run 2 of the LHC. Eur. Phys. J. C **79**, 375 (2019). https://doi.org/10.1140/epjc/s10052-019-6847-8. arXiv:1808.07858
526. CMS Collaboration, Identification of heavy, energetic, hadronically decaying particles using machine-learning techniques. JINST **15**, P06005 (2020). https://doi.org/10.1088/1748-0221/15/06/p06005. arXiv:2004.08262
527. ATLAS Collaboration, A search for $t\bar{t}$ resonances using lepton-plus-jets events in proton-proton collisions at $\sqrt{s} = 8$ TeV with the ATLAS detector. JHEP **08**, 148 (2015). https://doi.org/10.1007/JHEP08(2015)148. arXiv:1505.07018
528. ATLAS Collaboration, Search for top-squark pair production in final states with one lepton, jets, and missing transverse momentum using 36 fb^{-1} of $\sqrt{s} = 13$ TeV pp collision data with the ATLAS detector. JHEP **06**, 108 (2018). https://doi.org/10.1007/JHEP06(2018)108. arXiv:1711.11520
529. CMS Collaboration, Search for supersymmetry in proton-proton collisions at 13 TeV using identified top quarks. Phys. Rev. D **97**, 012007 (2018). https://doi.org/10.1103/PhysRevD.97.012007. arXiv:1710.11188
530. CMS Collaboration, W and top tagging scale factors. CMS Detector Performance Summary. CMS-DP-2017-026 (2017), https://cds.cern.ch/record/2275225
531. CMS Collaboration, W and top tagging scale factors for Run 2 data. CMS Detector Performance Summary. CMS-DP-2020-025 (2020), https://cds.cern.ch/record/2718978
532. L. Rokach, O.Z. *Maimon, Data Mining with Decision Trees: Theory and Applications*. Series in Machine Perception and Artificial Intelligence, vol. 69 (World Scientific Pub Co Inc., 2008)

533. Y. Freund, Boosting a weak learning algorithm by majority. Inf. Comput. **121**, 256 (1995). https://doi.org/10.1006/inco.1995.1136
534. S.S. Haykin, *Neural Networks and Learning Machines* (Pearson, 2008)
535. A. Hoecker et al., TMVA—Toolkit for multivariate data analysis, arXiv:physics/0703039
536. P. Baldi, P. Sadowski, D. Whiteson, Searching for exotic particles in high-energy physics with deep learning. Nat. Commun. **5**, 4308 (2014). https://doi.org/10.1038/ncomms5308. arXiv:1402.4735
537. L.G. Almeida, M. Backović, M. Cliche, S.J. Lee, M. Perelstein, Playing tag with ANN: boosted top identification with pattern recognition. JHEP **07**, 086 (2015). https://doi.org/10.1007/JHEP07(2015)086. arXiv:1501.05968
538. L. de Oliveira, M. Kagan, L. Mackey, B. Nachman, A. Schwartzman, Jet-images—deep learning edition. JHEP **07**, 069 (2016). https://doi.org/10.1007/JHEP07(2016)069. arXiv:1511.05190
539. P. Baldi, K. Bauer, C. Eng, P. Sadowski, D. Whiteson, Jet substructure classification in high-energy physics with deep neural networks. Phys. Rev. D **93**, 094034 (2016). https://doi.org/10.1103/PhysRevD.93.094034. arXiv:1603.09349
540. J. Cogan, M. Kagan, E. Strauss, A. Schwarztman, Jet-images: computer vision inspired techniques for jet tagging. JHEP **02**, 118 (2015). https://doi.org/10.1007/JHEP02(2015)118. arXiv:1407.5675
541. G. Louppe, K. Cho, C. Becot, K. Cranmer, QCD-aware recursive neural networks for jet physics. JHEP **01**, 057 (2019). https://doi.org/10.1007/JHEP01(2019)057. arXiv:1702.00748
542. CMS Collaboration, Performance of the DeepJet b tagging algorithm using 41.9 fb^{-1} of data from proton-proton collisions at 13 TeV with phase 1 CMS detector. CMS Detector Performance Summary (November 2018)
543. D. Guest, J. Collado, P. Baldi, S.-C. Hsu, G. Urban, D. Whiteson, Jet flavor classification in high-energy physics with deep neural networks. Phys. Rev. D **94**, 112002 (2016). https://doi.org/10.1103/PhysRevD.94.112002. arXiv:1607.08633
544. ATLAS Collaboration, Identification of jets containing b-hadrons with recurrent neural networks at the ATLAS experiment. ATLAS Physics Note. ATL-PHYS-PUB-2017-003 (2017), https://cds.cern.ch/record/2255226
545. J. Arjona Martínez, O. Cerri, M. Pierini, M. Spiropulu, J.-R. Vlimant, Pileup mitigation at the Large Hadron Collider with graph neural networks. Eur. Phys. J. Plus **134**, 333 (2019). https://doi.org/10.1140/epjp/i2019-12710-3. arXiv:1810.07988
546. H. Qu, L. Gouskos, ParticleNet: Jet tagging via particle clouds. Phys. Rev. D **101**, 056019 (2020). https://doi.org/10.1103/PhysRevD.101.056019. arXiv:1902.08570
547. A. Butter et al., The machine learning landscape of top taggers. SciPost Phys. **7**, 014 (2019). https://doi.org/10.21468/SciPostPhys.7.1.014. arXiv:1902.09914
548. J. Pearkes, W. Fedorko, A. Lister, C. Gay, Jet constituents for deep neural network based top quark tagging, arXiv:1704.02124
549. C. Chen, New approach to identifying boosted hadronically-decaying particle using jet substructure in its center-of-mass frame. Phys. Rev. D **85**, 034007 (2012). https://doi.org/10.1103/PhysRevD.85.034007. arXiv:1112.2567
550. C. Chen, Identifying the boosted hadronically decaying top quark using jet substructure in its center-of-mass frame. Phys. Rev. D **87**, 074007 (2013). https://doi.org/10.1103/PhysRevD.87.074007. arXiv:1303.3521
551. ATLAS Collaboration, Jet mass and substructure of inclusive jets in $\sqrt{s} = 7$ TeV pp collisions with the ATLAS experiment. JHEP **05**, 128 (2012). https://doi.org/10.1007/JHEP05(2012)128. arXiv:1203.4606
552. C.M.S. Collaboration, Studies of jet mass in dijet and W/Z+jet events. JHEP **05**, 090 (2013). https://doi.org/10.1007/JHEP05(2013)090. arXiv:1303.4811
553. ATLAS Collaboration, Measurement of the soft-drop jet mass in pp collisions at $\sqrt{s} = 13$ TeV with the ATLAS detector. Phys. Rev. Lett. **121**, 092001 (2018). https://doi.org/10.1103/PhysRevLett.121.092001. arXiv:1711.08341

554. Z.-B. Kang, K. Lee, X. Liu, F. Ringer, The groomed and ungroomed jet mass distribution for inclusive jet production at the LHC. JHEP **10**, 137 (2018). https://doi.org/10.1007/JHEP10(2018)137. arXiv:1803.03645
555. M. Freytsis, P. Harris, A. Hinzmann, I. Moult, N. Tran, C. Vernieri, Prospects for a measurement of the W boson mass in the all-jets final state at hadron colliders. JHEP **02**, 003 (2019). https://doi.org/10.1007/JHEP02(2019)003. arXiv:1807.07454
556. ATLAS Collaboration, Measurement of the jet mass in high transverse momentum $Z(\rightarrow b\bar{b})\gamma$ production at $\sqrt{s} = 13$ TeV using the ATLAS detector. Phys. Lett. B **812**, 135991 (2021). https://doi.org/10.1016/j.physletb.2020.135991. arXiv:1907.07093
557. CMS Collaboration, Measurement of the jet mass distribution and top quark mass in hadronic decays of boosted top quarks in pp collisions at $\sqrt{s} = 13$ TeV. Phys. Rev. Lett. **124**, 202001 (2020). https://doi.org/10.1103/PhysRevLett.124.202001. arXiv:1911.03800
558. CDF, D0 Collaborations, Tevatron Electroweak Working Group, Combination of CDF and D0 results on the mass of the top quark using up to 9.7 fb^{-1} at the Tevatron (2014)
559. CMS Collaboration, Measurement of the top quark mass using proton-proton data at $\sqrt{s} = 7$ and 8 TeV. Phys. Rev. D **93**, 072004 (2016). https://doi.org/10.1103/PhysRevD.93.072004. arXiv:1509.04044
560. ATLAS Collaboration, Measurement of the top quark mass in the $t\bar{t} \rightarrow$ lepton+jets and $t\bar{t} \rightarrow$ dilepton channels using $\sqrt{s} = 7$ TeV ATLAS data. Eur. Phys. J. C **75**, 330 (2015). https://doi.org/10.1140/epjc/s10052-015-3544-0. arXiv:1503.05427
561. ATLAS Collaboration, Measurement of the top quark mass in the $t\bar{t} \rightarrow$ dilepton channel from $\sqrt{s} = 8$ TeV ATLAS data. Phys. Lett. B **761**, 350 (2016). https://doi.org/10.1016/j.physletb.2016.08.042. arXiv:1606.02179
562. A.H. Hoang, S. Plätzer, D. Samitz, On the cutoff dependence of the quark mass parameter in angular ordered parton showers. JHEP **10**, 200 (2018). https://doi.org/10.1007/JHEP10(2018)200. arXiv:1807.06617
563. A.H. Hoang, What is the top quark mass? arXiv:2004.12915
564. M. Butenschoen, B. Dehnadi, A.H. Hoang, V. Mateu, M. Preisser, I.W. Stewart, Top quark mass calibration for Monte Carlo event generators. Phys. Rev. Lett. **117**, 232001 (2016). https://doi.org/10.1103/PhysRevLett.117.232001. arXiv:1608.01318
565. A.H. Hoang, S. Mantry, A. Pathak, I.W. Stewart, Extracting a short distance top mass with light grooming. Phys. Rev. D **100**, 074021 (2019). https://doi.org/10.1103/PhysRevD.100.074021. arXiv:1708.02586
566. S. Frixione, P. Nason, G. Ridolfi, A positive-weight next-to-leading-order Monte Carlo for heavy flavour hadroproduction. JHEP **09**, 126 (2007). https://doi.org/10.1088/1126-6708/2007/09/126. arXiv:0707.3088
567. S. Alioli, P. Nason, C. Oleari, E. Re, NLO single-top production matched with shower in POWHEG: s- and t-channel contributions. JHEP **09**, 111 (2009). https://doi.org/10.1088/1126-6708/2009/09/111. arXiv:0907.4076. Erratum: JHEP **02**, 011 (2010)
568. E. Re, Single-top Wt-channel production matched with parton showers using the POWHEG method. Eur. Phys. J. C **71**, 1547 (2011). https://doi.org/10.1140/epjc/s10052-011-1547-z. arXiv:1009.2450
569. J. Andersen et al., Les Houches 2017: physics at TeV Colliders Standard Model Working Group report, in *10th Les Houches Workshop on Physics at TeV Colliders (PhysTeV 2017). Proceedings*, Les Houches, France, 5–23 June 2017 (2018), arXiv:1803.07977
570. Z.-B. Kang, K. Lee, X. Liu, F. Ringer, Soft drop groomed jet angularities at the LHC. Phys. Lett. B **793**, 41 (2019). https://doi.org/10.1016/j.physletb.2019.04.018. arXiv:1811.06983
571. ATLAS Collaboration, Study of jet shapes in inclusive jet production in pp collisions at $\sqrt{s} = 7$ TeV using the ATLAS detector. Phys. Rev. D **83**, 052003 (2011). https://doi.org/10.1103/PhysRevD.83.052003. arXiv:1101.0070
572. ATLAS Collaboration, ATLAS measurements of the properties of jets for boosted particle searches. Phys. Rev. D **86**, 072006 (2012). https://doi.org/10.1103/PhysRevD.86.072006. arXiv:1206.5369

573. C.M.S. Collaboration, Shape, transverse size, and charged hadron multiplicity of jets in pp collisions at 7 TeV. JHEP **06**, 160 (2012). https://doi.org/10.1007/JHEP06(2012)160. arXiv:1204.3170
574. R. Field, Early LHC underlying event data—findings and surprises, in *21st Hadron Collider Physics Symposium (HCP2010). Proceedings*, Toronto, Canada, 23–27 August 2010 (2010), p. 10, arXiv:1010.3558
575. ATLAS Collaboration, Charged-particle multiplicities in pp interactions at $\sqrt{s} = 900$ GeV measured with the ATLAS detector at the LHC. Phys. Lett. B **688**, 21 (2010). https://doi.org/10.1016/j.physletb.2010.03.064. arXiv:1003.3124
576. ATLAS Collaboration, Charged particle multiplicities in pp interactions at $\sqrt{s} = 0.9$ and 7 TeV in a diffractive limited phase-space measured with the ATLAS detector at the LHC and new PYTHIA 6 tune. ATLAS Conference Note. ATLAS-CONF-2010-031 (July 2010), https://cds.cern.ch/record/1277665
577. CMS Collaboration, Measurement of the underlying event activity at the LHC with $\sqrt{s} = 7$ TeV and comparison with $\sqrt{s} = 0.9$ TeV. JHEP **09**, 109 (2011). https://doi.org/10.1007/JHEP09(2011)109. arXiv:1107.0330
578. ATLAS Collaboration, Properties of jets measured from tracks in proton-proton collisions at center-of-mass energy $\sqrt{s} = 7$ TeV with the ATLAS detector. Phys. Rev. D **84**, 054001 (2011). https://doi.org/10.1103/PhysRevD.84.054001. arXiv:1107.3311
579. ATLAS Collaboration, Measurement of the charged-particle multiplicity inside jets from $\sqrt{s} = 8$ TeV pp collisions with the ATLAS detector. Eur. Phys. J. C **76**, 322 (2016). https://doi.org/10.1140/epjc/s10052-016-4126-5. arXiv:1602.00988
580. ATLAS Collaboration, Properties of jet fragmentation using charged particles measured with the ATLAS detector in pp collisions at $\sqrt{s} = 13$ TeV. Phys. Rev. D **100**, 052011 (2019). https://doi.org/10.1103/PhysRevD.100.052011. arXiv:1906.09254
581. A. Capella, I. Dremin, J. Gary, V. Nechitailo, J. Tran Thanh Van, Evolution of average multiplicities of quark and gluon jets. Phys. Rev. D **61**, 074009 (2000). https://doi.org/10.1103/PhysRevD.61.074009. arXiv:hep-ph/9910226
582. I. Dremin, J. Gary, Energy dependence of mean multiplicities in gluon and quark jets at the next-to-next-to-next-to leading order. Phys. Lett. B **459**, 341 (1999). https://doi.org/10.1016/S0370-2693(99)00713-3. arXiv:hep-ph/9905477. Erratum: Phys. Lett. B **463**, 346 (1999)
583. ATLAS Collaboration, Measurement of the jet fragmentation function and transverse profile in proton-proton collisions at a center-of-mass energy of 7 TeV with the ATLAS detector. Eur. Phys. J. C **71**, 1795 (2011). https://doi.org/10.1140/epjc/s10052-011-1795-y. arXiv:1109.5816
584. ATLAS Collaboration, Properties of $g \to b\bar{b}$ at small opening angles in pp collisions with the ATLAS detector at $\sqrt{s} = 13$ TeV. Phys. Rev. D **99**, 052004 (2019). https://doi.org/10.1103/PhysRevD.99.052004. arXiv:1812.09283
585. ATLAS Collaboration, Measurement of jet charge in dijet events from $\sqrt{s} = 8$ TeV pp collisions with the ATLAS detector. Phys. Rev. D **93**, 052003 (2016). https://doi.org/10.1103/PhysRevD.93.052003. arXiv:1509.05190
586. CMS Collaboration, Measurements of jet charge with dijet events in pp collisions at $\sqrt{s} = 8$ TeV. JHEP **10**, 131 (2017). https://doi.org/10.1007/JHEP10(2017)131. arXiv:1706.05868
587. ATLAS Collaboration, Measurement of k_T splitting scales in $W \to \ell\nu$ events at $\sqrt{s} = 7$ TeV with the ATLAS detector. Eur. Phys. J. C **73**, 2432 (2013). https://doi.org/10.1140/epjc/s10052-013-2432-8. arXiv:1302.1415
588. CMS Collaboration, CERN Open Data Portal (2020), http://opendata.cern.ch
589. A. Larkoski, S. Marzani, J. Thaler, A. Tripathee and W. Xue, Exposing the QCD splitting function with CMS open data. Phys. Rev. Lett. **119**, 132003 (2017). https://doi.org/10.1103/PhysRevLett.119.132003. arXiv:1704.05066
590. A. Tripathee, W. Xue, A. Larkoski, S. Marzani, J. Thaler, Jet substructure studies with CMS open data. Phys. Rev. D **96**, 074003 (2017). https://doi.org/10.1103/PhysRevD.96.074003. arXiv:1704.05842

591. ATLAS Collaboration, Measurement of soft-drop jet observables in pp collisions with the ATLAS detector at $\sqrt{s} = 13$ TeV. Phys. Rev. D **101**, 052007 (2020). https://doi.org/10.1103/PhysRevD.101.052007. arXiv:1912.09837
592. Z.-B. Kang, K. Lee, X. Liu, D. Neill, F. Ringer, The soft drop groomed jet radius at NLL. JHEP **02**, 054 (2020). https://doi.org/10.1007/JHEP02(2020)054. arXiv:1908.01783
593. CMS Collaboration, Measurement of the splitting function in pp and Pb-Pb collisions at $\sqrt{s_{\rm NN}} = 5.02$ TeV. Phys. Rev. Lett. **120**, 142302 (2018). https://doi.org/10.1103/PhysRevLett.120.142302. arXiv:1708.09429
594. ATLAS Collaboration, Measurement of jet shapes in top-quark pair events at $\sqrt{s} = 7$ TeV using the ATLAS detector. Eur. Phys. J. C **73**, 2676 (2013). https://doi.org/10.1140/epjc/s10052-013-2676-3. arXiv:1307.5749
595. CMS Collaboration, Measurement of b jet shapes in proton-proton collisions at $\sqrt{s} = 5.02$ TeV, arXiv:2005.14219
596. CMS Collaboration, Measurement of jet substructure observables in $t\bar{t}$ events from proton-proton collisions at $\sqrt{s} = 13$ TeV. Phys. Rev. D **98**, 092014 (2018). https://doi.org/10.1103/PhysRevD.98.092014. arXiv:1808.07340
597. S. Chekanov, J. Proudfoot, Searches for TeV-scale particles at the LHC using jet shapes. Phys. Rev. D **81**, 114038 (2010). https://doi.org/10.1103/PhysRevD.81.114038. arXiv:1002.3982
598. ATLAS Collaboration, Measurement of colour flow with the jet pull angle in $t\bar{t}$ events using the ATLAS detector at $\sqrt{s} = 8$ TeV. Phys. Lett. B **750**, 475 (2015). https://doi.org/10.1016/j.physletb.2015.09.051. arXiv:1506.05629
599. A.J. Larkoski, S. Marzani, C. Wu, Theory predictions for the pull angle. Phys. Rev. D **99**, 091502 (2019). https://doi.org/10.1103/PhysRevD.99.091502. arXiv:1903.02275
600. ATLAS Collaboration, Measurement of colour flow using jet-pull observables in $t\bar{t}$ events with the ATLAS experiment at $\sqrt{s} = 13$ TeV. Eur. Phys. J. C **78**, 847 (2018). https://doi.org/10.1140/epjc/s10052-018-6290-2. arXiv:1805.02935
601. ATLAS Collaboration, Measurement of jet-substructure observables in top quark, W boson and light jet production in proton-proton collisions at $\sqrt{s} = 13$ TeV with the ATLAS detector. JHEP **08**, 033 (2019). https://doi.org/10.1007/JHEP08(2019)033 arXiv:1903.02942
602. T. Dreyer, First measurement of the jet mass in events with highly boosted top quarks and studies with top tagging at CMS. Ph.D. thesis, Universität Hamburg, Hamburg, Germany (2019), http://ediss.sub.uni-hamburg.de/volltexte/2019/10185
603. J. Skottke, Measurement of jet substructure in boosted top quark decays at CMS. Master's thesis, Universität Hamburg, Hamburg, Germany (2019), http://www.iexp.uni-hamburg.de/groups/pd/contents/sites/default/files/thesis/Masterarbeit_Jan_Skottke.pdf
604. ATLAS Collaboration, Measurement of the cross-section of high transverse momentum vector bosons reconstructed as single jets and studies of jet substructure in pp collisions at $\sqrt{s} = 7$ TeV with the ATLAS detector. New J. Phys. **16**, 113013 (2014). https://doi.org/10.1088/1367-2630/16/11/113013. arXiv:1407.0800
605. ATLAS Collaboration, Measurement of $WW/WZ \rightarrow \ell\nu qq'$ production with the hadronically decaying boson reconstructed as one or two jets in pp collisions at $\sqrt{s} = 8$ TeV with ATLAS, and constraints on anomalous gauge couplings. Eur. Phys. J. C **77**, 563 (2017). https://doi.org/10.1140/epjc/s10052-017-5084-2. arXiv:1706.01702
606. CMS Collaboration, Search for anomalous couplings in boosted $WW/WZ \rightarrow \ell\nu q\bar{q}$ production in proton-proton collisions at $\sqrt{s} = 8$ TeV. Phys. Lett. B **772**, 21 (2017). https://doi.org/10.1016/j.physletb.2017.06.009. arXiv:1703.06095
607. CMS Collaboration, Search for anomalous triple gauge couplings in WW and WZ production in lepton+jet events in proton-proton collisions at $\sqrt{s} = 13$ TeV. JHEP **12**, 062 (2019). https://doi.org/10.1007/JHEP12(2019)062. arXiv:1907.08354
608. K. Hagiwara, S. Ishihara, R. Szalapski, D. Zeppenfeld, Low energy effects of new interactions in the electroweak boson sector. Phys. Rev. D **48**, 2182 (1993). https://doi.org/10.1103/PhysRevD.48.2182
609. C. Degrande, N. Greiner, W. Kilian, O. Mattelaer, H. Mebane, T. Stelzer et al., Effective field theory: a modern approach to anomalous couplings. Annals Phys. **335**, 21 (2013). https://doi.org/10.1016/j.aop.2013.04.016. arXiv:1205.4231

610. ATLAS Collaboration, Measurement of the associated production of a Higgs boson decaying into b-quarks with a vector boson at high transverse momentum in pp collisions at $\sqrt{s} = 13$ TeV with the ATLAS detector, arXiv:2008.02508
611. C. Grojean, E. Salvioni, M. Schlaffer, A. Weiler, Very boosted Higgs in gluon fusion. JHEP **05**, 022 (2014). https://doi.org/10.1007/JHEP05(2014)022. arXiv:1312.3317
612. CMS Collaboration, Inclusive search for a highly boosted Higgs boson decaying to a bottom quark-antiquark pair. Phys. Rev. Lett. **120**, 071802 (2018). https://doi.org/10.1103/PhysRevLett.120.071802. arXiv:1709.05543
613. CMS Collaboration, Inclusive search for a highly boosted Higgs boson decaying to a bottom quark-antiquark pair at $\sqrt{s} = 13$ TeV with 137 fb^{-1}. CMS Physics Analysis Summary. CMS-PAS-HIG-19-003 (2020), http://cds.cern.ch/record/2714916
614. CMS Collaboration, Performance of deep tagging algorithms for boosted double quark jet topology in proton-proton collisions at 13 TeV with the Phase-0 CMS detector. CMS Detector Performance Summary. CMS-DP-2018-046 (2018), http://cds.cern.ch/record/2630438
615. ATLAS Collaboration, Search for boosted resonances decaying to two b-quarks and produced in association with a jet at $\sqrt{s} = 13$ TeV with the ATLAS detector. ATLAS Conference Note. ATLAS-CONF-2018-052 (2018), https://cds.cern.ch/record/2649081
616. CMS Collaboration, Evidence for Higgs boson decay to a pair of muons, arXiv:2009.04363
617. ATLAS Collaboration, Search for the decay of the Higgs boson to charm quarks with the ATLAS experiment. Phys. Rev. Lett. **120**, 211802 (2018). https://doi.org/10.1103/PhysRevLett.120.211802. arXiv:1802.04329
618. CMS Collaboration, A search for the standard model Higgs boson decaying to charm quarks. JHEP **03**, 131 (2020). https://doi.org/10.1007/JHEP03(2020)131. arXiv:1912.01662
619. N. Kidonakis, NNNLO soft-gluon corrections for the top-quark p_T and rapidity distributions. Phys. Rev. D **91**, 031501 (2015). https://doi.org/10.1103/PhysRevD.91.031501. arXiv:1411.2633
620. N. Kidonakis, NNNLO soft-gluon corrections for the top-antitop pair production cross section. Phys. Rev. D **90**, 014006 (2014). https://doi.org/10.1103/PhysRevD.90.014006. arXiv:1405.7046
621. M. Czakon, D. Heymes, A. Mitov, D. Pagani, I. Tsinikos, M. Zaro, Top-pair production at the LHC through NNLO QCD and NLO EW. JHEP **10**, 186 (2017). https://doi.org/10.1007/JHEP10(2017)186. arXiv:1705.04105
622. M. Czakon, A. Ferroglia, D. Heymes, A. Mitov, B.D. Pecjak, D.J. Scott et al., Resummation for (boosted) top-quark pair production at NNLO+NNLL' in QCD. JHEP **05**, 149 (2018). https://doi.org/10.1007/JHEP05(2018)149. arXiv:1803.07623
623. N. Kidonakis, Top-quark double-differential distributions at approximate N^3 LO. Phys. Rev. D **101**, 074006 (2020). https://doi.org/10.1103/PhysRevD.101.074006. arXiv:1912.10362
624. M. Czakon, M.L. Mangano, A. Mitov, J. Rojo, Constraints on the gluon PDF from top quark pair production at hadron colliders. JHEP **07**, 167 (2013). https://doi.org/10.1007/JHEP07(2013)167. arXiv:1303.7215
625. M. Czakon, N.P. Hartland, A. Mitov, E.R. Nocera, J. Rojo, Pinning down the large-x gluon with NNLO top-quark pair differential distributions. JHEP **04**, 044 (2017). https://doi.org/10.1007/JHEP04(2017)044. arXiv:1611.08609
626. CMS Collaboration, Measurement of $t\bar{t}$ normalised multi-differential cross sections in pp collisions at $\sqrt{s} = 13$ TeV, and simultaneous determination of the strong coupling strength, top quark pole mass, and parton distribution functions, arXiv:1904.05237
627. ATLAS Collaboration, Measurement of the $t\bar{t}$ production cross-section and lepton differential distributions in $e\mu$ dilepton events from pp collisions at $\sqrt{s} = 13$ TeV with the ATLAS detector. Eur. Phys. J. C **80**, 528 (2020). https://doi.org/10.1140/epjc/s10052-020-7907-9. arXiv:1910.08819
628. R. Frederix, F. Maltoni, Top pair invariant mass distribution: a window on new physics. JHEP **01**, 047 (2009). https://doi.org/10.1088/1126-6708/2009/01/047. arXiv:0712.2355
629. P. Ferrario, G. Rodrigo, Massive color-octet bosons and the charge asymmetries of top quarks at hadron colliders. Phys. Rev. D **78**, 094018 (2008). https://doi.org/10.1103/PhysRevD.78.094018. arXiv:0809.3354

630. C. Degrande, J.-M. Gerard, C. Grojean, F. Maltoni, G. Servant, Non-resonant new physics in top pair production at hadron colliders. JHEP **03**, 125 (2011). https://doi.org/10.1007/JHEP03(2011)125. arXiv:1010.6304
631. J.L. Hewett, J. Shelton, M. Spannowsky, T.M. Tait, M. Takeuchi, A_{FB}^{t} meets LHC. Phys. Rev. D **84**, 054005 (2011). https://doi.org/10.1103/PhysRevD.84.054005. arXiv:1103.4618
632. C. Bini, R. Contino, N. Vignaroli, Heavy-light decay topologies as a new strategy to discover a heavy gluon. JHEP **01**, 157 (2012). https://doi.org/10.1007/JHEP01(2012)157. arXiv:1110.6058
633. D. Buarque Franzosi, F. Fabbri, S. Schumann, Constraining scalar resonances with top-quark pair production at the LHC. JHEP **03**, 022 (2018). https://doi.org/10.1007/JHEP03(2018)022. arXiv:1711.00102
634. W. Bernreuther, P. Galler, C. Mellein, Z. Si, P. Uwer, Production of heavy Higgs bosons and decay into top quarks at the LHC. Phys. Rev. D **93**, 034032 (2016). https://doi.org/10.1103/PhysRevD.93.034032. arXiv:1511.05584
635. D. Buarque Franzosi, E. Vryonidou, C. Zhang, Scalar production and decay to top quarks including interference effects at NLO in QCD in an EFT approach. JHEP **10**, 096 (2017). https://doi.org/10.1007/JHEP10(2017)096. arXiv:1707.06760
636. C. Englert, P. Galler, C.D. White, Effective field theory and scalar extensions of the top quark sector. Phys. Rev. D **101**, 035035 (2020). https://doi.org/10.1103/PhysRevD.101.035035. arXiv:1908.05588
637. A. Djouadi, J. Ellis, A. Popov, J. Quevillon, Interference effects in $t\bar{t}$ production at the LHC as a window on new physics. JHEP **03**, 119 (2019). https://doi.org/10.1007/JHEP03(2019)119. arXiv:1901.03417
638. ATLAS Collaboration, Measurement of the differential cross-section of highly boosted top quarks as a function of their transverse momentum in $\sqrt{s} = 8$ TeV proton-proton collisions using the ATLAS detector. Phys. Rev. D **93**, 032009 (2016). https://doi.org/10.1103/PhysRevD.93.032009. arXiv:1510.03818
639. ATLAS Collaboration, Electron efficiency measurements with the ATLAS detector using 2012 LHC proton-proton collision data. Eur. Phys. J. C **77**, 195 (2017). https://doi.org/10.1140/epjc/s10052-017-4756-2. arXiv:1612.01456
640. ATLAS Collaboration, Electron reconstruction and identification in the ATLAS experiment using the 2015 and 2016 LHC proton-proton collision data at $\sqrt{s} = 13$ TeV. Eur. Phys. J. C **79**, 639 (2019). https://doi.org/10.1140/epjc/s10052-019-7140-6. arXiv:1902.04655
641. ATLAS Collaboration, Muon reconstruction performance of the ATLAS detector in proton-proton collision data at $\sqrt{s} = 13$ TeV. Eur. Phys. J. C **76**, 292 (2016). https://doi.org/10.1140/epjc/s10052-016-4120-y. arXiv:1603.05598
642. CMS Collaboration, Performance of electron reconstruction and selection with the CMS detector in proton-proton collisions at $\sqrt{s} = 8$ TeV. JINST **10**, P06005 (2015). https://doi.org/10.1088/1748-0221/10/06/P06005. arXiv:1502.02701
643. CMS Collaboration, Performance of the CMS muon detector and muon reconstruction with proton-proton collisions at $\sqrt{s} = 13$ TeV. JINST **13**, P06015 (2018). https://doi.org/10.1088/1748-0221/13/06/P06015. arXiv:1804.04528
644. K. Rehermann, B. Tweedie, Efficient identification of boosted semileptonic top quarks at the LHC. JHEP **03**, 059 (2011). https://doi.org/10.1007/JHEP03(2011)059. arXiv:1007.2221
645. ATLAS Collaboration, Measurement of the charge asymmetry in top quark pair production in pp collisions at $\sqrt{s} = 7$ TeV using the ATLAS detector. Eur. Phys. J. C **72**, 2039 (2012). https://doi.org/10.1140/epjc/s10052-012-2039-5. arXiv:1203.4211
646. F. Halzen, Y.S. Jeong, C. Kim, Charge asymmetry of weak boson production at the LHC and the charm content of the proton. Phys. Rev. D **88**, 073013 (2013). https://doi.org/10.1103/PhysRevD.88.073013. arXiv:1304.0322
647. ATLAS Collaboration, Measurements of normalized differential cross sections for $t\bar{t}$ production in pp collisions at $\sqrt{s} = 7$ TeV using the ATLAS detector. Phys. Rev. D **90**, 072004 (2014). https://doi.org/10.1103/PhysRevD.90.072004. arXiv:1407.0371

648. CMS Collaboration, Measurement of the differential cross section for top quark pair production in pp collisions at $\sqrt{s} = 8$ TeV. Eur. Phys. J. C **75**, 542 (2015). https://doi.org/10.1140/epjc/s10052-015-3709-x. arXiv:1505.04480
649. CMS Collaboration, Measurement of the integrated and differential $t\bar{t}$ production cross sections for high-p_T top quarks in pp collisions at $\sqrt{s} = 8$ TeV. Phys. Rev. D **94**, 072002 (2016). https://doi.org/10.1103/PhysRevD.94.072002. arXiv:1605.00116
650. CMS Collaboration, Search for $t\bar{t}$ resonances in boosted semileptonic final states in pp collisions at $\sqrt{s} = 13$ TeV. CMS Physics Analysis Summary. CMS-PAS-B2G-15-002 (2016), http://cds.cern.ch/record/2138345
651. M. Czakon, D. Heymes, A. Mitov, Dynamical scales for multi-TeV top-pair production at the LHC. JHEP **04**, 071 (2017). https://doi.org/10.1007/JHEP04(2017)071. arXiv:1606.03350
652. CMS Collaboration, Measurement of the differential $t\bar{t}$ cross section with high-p_T top-quark jets in the all-hadronic channel at $\sqrt{s} = 8$ TeV. CMS Physics Analysis Summary. CMS-PAS-TOP-16-018 (2017), http://cds.cern.ch/record/2280473
653. ATLAS Collaboration, Measurements of top-quark pair differential cross-sections in the lepton+jets channel in pp collisions at $\sqrt{s} = 13$ TeV using the ATLAS detector. JHEP **11**, 191 (2017). https://doi.org/10.1007/JHEP11(2017)191. arXiv:1708.00727
654. ATLAS Collaboration, Measurements of $t\bar{t}$ differential cross-sections of highly boosted top quarks decaying to all-hadronic final states in pp collisions at $\sqrt{s} = 13$ TeV using the ATLAS detector. Phys. Rev. D **98**, 012003 (2018). https://doi.org/10.1103/PhysRevD.98.012003. arXiv:1801.02052
655. ATLAS Collaboration, Measurements of top-quark pair differential and double-differential cross-sections in the ℓ +jets channel with pp collisions at $\sqrt{s} = 13$ TeV using the ATLAS detector. Eur. Phys. J. C **79**, 1028 (2019). https://doi.org/10.1140/epjc/s10052-019-7525-6. arXiv:1908.07305
656. M. Czakon, D. Heymes, A. Mitov, High-precision differential predictions for top-quark pairs at the LHC. Phys. Rev. Lett. **116**, 082003 (2016). https://doi.org/10.1103/PhysRevLett.116.082003. arXiv:1511.00549
657. CMS Collaboration, Measurement of differential $t\bar{t}$ production cross sections for high-p_T top quarks in proton-proton collisions at $\sqrt{s} = 13$ TeV. CMS Physics Analysis Summary. CMS-PAS-TOP-18-013 (2020), http://cds.cern.ch/record/2717638
658. M. Czakon, P. Fiedler, A. Mitov, Resolving the Tevatron top quark forward-backward asymmetry puzzle: fully differential next-to-next-to-leading-order calculation. Phys. Rev. Lett. **115**, 052001 (2015). https://doi.org/10.1103/PhysRevLett.115.052001. arXiv:1411.3007
659. Q.-H. Cao, D. McKeen, J.L. Rosner, G. Shaughnessy, C.E. Wagner, Forward-backward asymmetry of top quark pair production. Phys. Rev. D **81**, 114004 (2010). https://doi.org/10.1103/PhysRevD.81.114004. arXiv:1003.3461
660. M.I. Gresham, I.-W. Kim, K.M. Zurek, On models of new physics for the Tevatron top A_{FB}. Phys. Rev. D **83**, 114027 (2011). https://doi.org/10.1103/PhysRevD.83.114027. arXiv:1103.3501
661. D0 Collaboration, First measurement of the forward-backward charge asymmetry in top quark pair production. Phys. Rev. Lett. **100**, 142002 (2008). https://doi.org/10.1103/PhysRevLett.100.142002. arXiv:0712.0851
662. C.D.F. Collaboration, Evidence for a mass dependent forward-backward asymmetry in top quark pair production. Phys. Rev. D **83**, 12003 (2011). https://doi.org/10.1103/PhysRevD.83.112003. arXiv:1101.0034
663. D0 Collaboration, Measurement of the forward-backward asymmetry in top quark-antiquark production in $p\bar{p}$ collisions using the lepton+jets channel. Phys. Rev. D **90**, 072011 (2014). https://doi.org/10.1103/PhysRevD.90.072011. arXiv:1405.0421
664. CDF Collaboration, Measurement of the forward-backward asymmetry of top-quark and antiquark pairs using the full CDF Run II data set. Phys. Rev. D **93**, 112005 (2016). https://doi.org/10.1103/PhysRevD.93.112005. arXiv:1602.09015
665. CDF, D0 Collaborations, Combined forward-backward asymmetry measurements in top-antitop quark production at the Tevatron. Phys. Rev. Lett. **120**, 042001 (2018). https://doi.org/10.1103/PhysRevLett.120.042001. arXiv:1709.04894

666. ATLAS Collaboration, Measurement of the charge asymmetry in highly boosted top-quark pair production in $\sqrt{s} = 8$ TeV pp collision data collected by the ATLAS experiment. Phys. Lett. B **756**, 52 (2016). https://doi.org/10.1016/j.physletb.2016.02.055. arXiv:1512.06092
667. C.M.S. Collaboration, Measurement of the charge asymmetry in top-quark pair production in proton-proton collisions at $\sqrt{s} = 7$ TeV. Phys. Lett. B **709**, 28 (2012). https://doi.org/10.1016/j.physletb.2012.01.078. arXiv:1112.5100
668. C.M.S. Collaboration, Measurements of the $t\bar{t}$ charge asymmetry using the dilepton decay channel in pp collisions at $\sqrt{s} = 7$ TeV. JHEP **04**, 191 (2014). https://doi.org/10.1007/JHEP04(2014)191. arXiv:1402.3803
669. CMS Collaboration, Measurement of the charge asymmetry in top quark pair production in pp collisions at $\sqrt{(s)} = 8$ TeV using a template method. Phys. Rev. D **93**, 034014 (2016). https://doi.org/10.1103/PhysRevD.93.034014. arXiv:1508.03862
670. CMS Collaboration, Measurements of $t\bar{t}$ charge asymmetry using dilepton final states in pp collisions at $\sqrt{s} = 8$ TeV. Phys. Lett. B **760**, 365 (2016). https://doi.org/10.1016/j.physletb.2016.07.006. arXiv:1603.06221
671. CMS Collaboration, Measurement of the top quark forward-backward production asymmetry and the anomalous chromoelectric and chromomagnetic moments in pp collisions at $\sqrt{s} = 13$ TeV. JHEP **06**, 146 (2020). https://doi.org/10.1007/JHEP06(2020)146. arXiv:1912.09540
672. CMS Collaboration, Measurement of the $\mathrm{t\bar{t}}$ production cross section, the top quark mass, and the strong coupling constant using dilepton events in pp collisions at $\sqrt{s} = 13$ TeV. Eur. Phys. J. C **79**, 368 (2019). https://doi.org/10.1140/epjc/s10052-019-6863-8. arXiv:1812.10505
673. M. Czakon, A. Mitov, Top++: a program for the calculation of the top-pair cross-section at hadron colliders. Comput. Phys. Commun. **185**, 2930 (2014). https://doi.org/10.1016/j.cpc.2014.06.021. arXiv:1112.5675
674. T. Lee, A theory of spontaneous T violation. Phys. Rev. D **8**, 1226 (1973). https://doi.org/10.1103/PhysRevD.8.1226
675. R.A. Flores, M. Sher, Higgs masses in the standard, multi-Higgs and supersymmetric models. Ann. Phys. **148**, 95 (1983). https://doi.org/10.1016/0003-4916(83)90331-7
676. J. Gunion, H.E. Haber, Higgs bosons in supersymmetric models. Nucl. Phys. B **272**, 1 (1986). https://doi.org/10.1016/0550-3213(86)90340-8. Erratum: Nucl. Phys. B 402, 567 (1993)
677. H.E. Haber, Y. Nir, Multiscalar models with a high-energy scale. Nucl. Phys. B **335**, 363 (1990). https://doi.org/10.1016/0550-3213(90)90499-4
678. J.F. Gunion, H.E. Haber, The CP conserving two Higgs doublet model: the approach to the decoupling limit. Phys. Rev. D **67**, 075019 (2003). https://doi.org/10.1103/PhysRevD.67.075019. arXiv:hep-ph/0207010
679. G. Branco, P. Ferreira, L. Lavoura, M. Rebelo, M. Sher, J.P. Silva, Theory and phenomenology of two-Higgs-doublet models. Phys. Rept. **516**, 1 (2012). https://doi.org/10.1016/j.physrep.2012.02.002. arXiv:1106.0034
680. R.N. Mohapatra, J.C. Pati, Left-right gauge symmetry and an isoconjugate model of CP violation. Phys. Rev. D **11**, 566 (1975). https://doi.org/10.1103/PhysRevD.11.566
681. G. Senjanovic, R.N. Mohapatra, Exact left-right symmetry and spontaneous violation of parity. Phys. Rev. D **12**, 1502 (1975). https://doi.org/10.1103/PhysRevD.12.1502
682. M. Cvetic, P. Langacker, Implications of Abelian extended gauge structures from string models. Phys. Rev. D **54**, 3570 (1996). https://doi.org/10.1103/PhysRevD.54.3570. arXiv:hep-ph/9511378
683. M. Cvetic, P. Langacker, New gauge bosons from string models. Mod. Phys. Lett. A **11**, 1247 (1996). https://doi.org/10.1142/S0217732396001260. arXiv:hep-ph/9602424
684. P. Langacker, The physics of heavy Z' gauge bosons. Rev. Mod. Phys. **81**, 1199 (2009). https://doi.org/10.1103/RevModPhys.81.1199. arXiv:0801.1345
685. N. Arkani-Hamed, A.G. Cohen, H. Georgi, Electroweak symmetry breaking from dimensional deconstruction. Phys. Lett. B **513**, 232 (2001). https://doi.org/10.1016/S0370-2693(01)00741-9. arXiv:hep-ph/0105239

686. N. Arkani-Hamed, A.G. Cohen, T. Gregoire, J.G. Wacker, Phenomenology of electroweak symmetry breaking from theory space. JHEP **08**, 020 (2002). arXiv:hep-ph/0202089
687. H. Georgi, D.B. Kaplan, Composite Higgs and custodial SU(2). Phys. Lett. B **145**, 216 (1984). https://doi.org/10.1016/0370-2693(84)90341-1
688. H. Georgi, D.B. Kaplan, P. Galison, Calculation of the composite Higgs mass. Phys. Lett. B **143**, 152 (1984). https://doi.org/10.1016/0370-2693(84)90823-2
689. M.J. Dugan, H. Georgi, D.B. Kaplan, Anatomy of a composite Higgs model. Nucl. Phys. B **254**, 299 (1985). https://doi.org/10.1016/0550-3213(85)90221-4
690. N. Arkani-Hamed, A.G. Cohen, H. Georgi, (De)constructing dimensions. Phys. Rev. Lett. **86**, 4757 (2001). https://doi.org/10.1103/PhysRevLett.86.4757. arXiv:hep-th/0104005
691. N. Arkani-Hamed, A. Cohen, E. Katz, A. Nelson, The littlest Higgs. JHEP **07**, 034 (2002). https://doi.org/10.1088/1126-6708/2002/07/034. arXiv:hep-ph/0206021
692. N. Arkani-Hamed, A. Cohen, E. Katz, A. Nelson, T. Gregoire, J.G. Wacker, The minimal moose for a little Higgs. JHEP **08**, 021 (2002). https://doi.org/10.1088/1126-6708/2002/08/021. arXiv:hep-ph/0206020
693. T. Han, H.E. Logan, B. McElrath, L.-T. Wang, Phenomenology of the little Higgs model. Phys. Rev. D **67**, 095004 (2003). https://doi.org/10.1103/PhysRevD.67.095004. arXiv:hep-ph/0301040
694. D.E. Kaplan, M. Schmaltz, The little Higgs from a simple group. JHEP **10**, 039 (2003). https://doi.org/10.1088/1126-6708/2003/10/039. arXiv:hep-ph/0302049
695. M. Schmaltz, The simplest little Higgs. JHEP **08**, 056 (2004). https://doi.org/10.1088/1126-6708/2004/08/056. arXiv:hep-ph/0407143
696. T. Kaluza, Zum Unitätsproblem der Physik. Sitzungsber. Preuss. Akad. Wiss. Berlin (Math.Phys.) 966 (1921). https://doi.org/10.1142/S0218271818700017. arXiv:1803.08616
697. O. Klein, Quantum theory and five-dimensional theory of relativity. Z. Phys. **37**, 895 (1926). https://doi.org/10.1007/BF01397481
698. O. Klein, The atomicity of electricity as a quantum theory law. Nature **118**, 516 (1926). https://doi.org/10.1038/118516a0
699. L. Randall, R. Sundrum, A large mass hierarchy from a small extra dimension. Phys. Rev. Lett. **83**, 3370 (1999). https://doi.org/10.1103/PhysRevLett.83.3370. arXiv:hep-ph/9905221
700. L. Randall, R. Sundrum, An alternative to compactification. Phys. Rev. Lett. **83**, 4690 (1999). https://doi.org/10.1103/PhysRevLett.83.4690. arXiv:hep-th/9906064
701. A. Fitzpatrick, J. Kaplan, L. Randall, L.-T. Wang, Searching for the Kaluza-Klein graviton in bulk RS models. JHEP **09**, 013 (2007). https://doi.org/10.1088/1126-6708/2007/09/013. arXiv:hep-ph/0701150
702. N. Arkani-Hamed, S. Dimopoulos, G. Dvali (1998) The hierarchy problem and new dimensions at a millimeter. Phys. Lett. B **429**, 263 (1998). https://doi.org/10.1016/S0370-2693(98)00466-3. arXiv:hep-ph/9803315
703. I. Antoniadis, N. Arkani-Hamed, S. Dimopoulos, G. Dvali, New dimensions at a millimeter to a Fermi and superstrings at a TeV. Phys. Lett. B **436**, 257 (1998). https://doi.org/10.1016/S0370-2693(98)00860-0. arXiv:hep-ph/9804398
704. T. Dorigo, Hadron collider searches for diboson resonances. Prog. Part. Nucl. Phys. **100**, 211 (2018). https://doi.org/10.1016/j.ppnp.2018.01.009. arXiv:1802.00354
705. ATLAS Collaboration, Search for diboson resonances in hadronic final states in 139 fb^{-1} of pp collisions at $\sqrt{s} = 13$ TeV with the ATLAS detector. JHEP **09**, 091 (2019). https://doi.org/10.1007/JHEP09(2019)091. arXiv:1906.08589
706. G. Choudalakis, On hypothesis testing, trials factor, hypertests and the BumpHunter, in PHYSTAT 2011 (2011), arXiv:1101.0390
707. R.M. Harris, K. Kousouris, Searches for dijet resonances at hadron colliders. Int. J. Mod. Phys. A **26**, 5005 (2011). https://doi.org/10.1142/S0217751X11054905. arXiv:1110.5302
708. ATLAS Collaboration, Performance of missing transverse momentum reconstruction with the ATLAS detector using proton-proton collisions at $\sqrt{s} = 13$ TeV. Eur. Phys. J. C **78**, 903 (2018). https://doi.org/10.1140/epjc/s10052-018-6288-9. arXiv:1802.08168

709. CMS Collaboration, Performance of missing transverse momentum reconstruction in proton-proton collisions at $\sqrt{s} = 13$ TeV using the CMS detector. JINST **14**, P07004 (2019). https://doi.org/10.1088/1748-0221/14/07/P07004. arXiv:1903.06078
710. CMS Collaboration, Search for a heavy resonance decaying to a pair of vector bosons in the lepton plus merged jet final state at $\sqrt{s} = 13$ TeV. JHEP **05**, 088 (2018). https://doi.org/10.1007/JHEP05(2018)088. arXiv:1802.09407
711. CMS Collaboration, Search for a heavy resonance decaying into a Z boson and a vector boson in the $\nu\overline{\nu}\mathrm{q}\overline{\mathrm{q}}$ final state. JHEP **07**, 075 (2018). https://doi.org/10.1007/JHEP07(2018)075. arXiv:1803.03838
712. ATLAS Collaboration, Searches for heavy ZZ and ZW resonances in the $\ell\ell qq$ and $\nu\nu qq$ final states in pp collisions at $\sqrt{s} = 13$ TeV with the ATLAS detector. JHEP **03**, 009 (2018). https://doi.org/10.1007/JHEP03(2018)009. arXiv:1708.09638
713. CMS Collaboration, Search for a heavy resonance decaying into a Z boson and a Z or W boson in $2\ell 2q$ final states at $\sqrt{s} = 13$ TeV. JHEP **09**, 101 (2018). https://doi.org/10.1007/JHEP09(2018)101. arXiv:1803.10093
714. ATLAS Collaboration, Search for heavy diboson resonances in semileptonic final states in pp collisions at $\sqrt{s} = 13$ TeV with the ATLAS detector, arXiv:2004.14636
715. ATLAS Collaboration, Search for resonant diboson production in the $\ell\ell q\bar{q}$ final state in pp collisions at $\sqrt{s} = 8$ TeV with the ATLAS detector. Eur. Phys. J. C **75**, 69 (2015). https://doi.org/10.1140/epjc/s10052-015-3261-8. arXiv:1409.6190
716. ATLAS Collaboration, Search for production of WW/WZ resonances decaying to a lepton, neutrino and jets in pp collisions at $\sqrt{s} = 8$ TeV with the ATLAS detector. Eur. Phys. J. C **75**, 209 (2015). https://doi.org/10.1140/epjc/s10052-015-3593-4. arXiv:1503.04677. Erratum: Eur. Phys. J. C **75**, 370 (2015)
717. ATLAS Collaboration, Search for high-mass diboson resonances with boson-tagged jets in proton-proton collisions at $\sqrt{s} = 8$ TeV with the ATLAS detector. JHEP **12**, 055 (2015). https://doi.org/10.1007/JHEP12(2015)055. arXiv:1506.00962
718. C.M.S. Collaboration, Search for heavy resonances in the W/Z-tagged dijet mass spectrum in pp collisions at 7 TeV. Phys. Lett. B **723**, 280 (2013). https://doi.org/10.1016/j.physletb.2013.05.040. arXiv:1212.1910
719. C.M.S. Collaboration, Search for massive resonances in dijet systems containing jets tagged as W or Z boson decays in pp collisions at $\sqrt{s} = 8$ TeV. JHEP **08**, 173 (2014). https://doi.org/10.1007/JHEP08(2014)173. arXiv:1405.1994
720. CMS Collaboration, Search for heavy resonances that decay into a vector boson and a Higgs boson in hadronic final states at $\sqrt{s} = 13$ TeV. Eur. Phys. J. C **77**, 636 (2017). https://doi.org/10.1140/epjc/s10052-017-5192-z/ arXiv:1707.01303
721. ATLAS Collaboration, Search for resonances decaying into a weak vector boson and a Higgs boson in the fully hadronic final state produced in proton-proton collisions at $\sqrt{s} = 13$ TeV with the ATLAS detector, arXiv:2007.05293
722. ATLAS Collaboration, Search for heavy resonances decaying into a W or Z boson and a Higgs boson in final states with leptons and b-jets in 36 fb^{-1} of $\sqrt{s} = 13$ TeV pp collisions with the ATLAS detector. JHEP **03**, 174 (2018). https://doi.org/10.1007/JHEP03(2018)174. arXiv:1712.06518. Erratum: JHEP **11**, 051 (2018)
723. ATLAS Collaboration, Search for a new resonance decaying to a W or Z boson and a Higgs boson in the $\ell\ell/\ell\nu/\nu\nu + b\bar{b}$ final states with the ATLAS detector. Eur. Phys. J. C **75**, 263 (2015). https://doi.org/10.1140/epjc/s10052-015-3474-x. arXiv:1503.08089
724. ATLAS Collaboration, Search for new resonances decaying to a W or Z boson and a Higgs boson in the $\ell^+\ell^- b\bar{b}$, $\ell\nu b\bar{b}$, and $\nu\bar{\nu}b\bar{b}$ channels with pp collisions at $\sqrt{s} = 13$ TeV with the ATLAS detector. Phys. Lett. B **765**, 32 (2017). https://doi.org/10.1016/j.physletb.2016.11.045. arXiv:1607.05621
725. CMS Collaboration, Search for heavy resonances decaying into a vector boson and a Higgs boson in final states with charged leptons, neutrinos and b quarks at $\sqrt{s} = 13$ TeV. JHEP **11**, 172 (2018). https://doi.org/10.1007/JHEP11(2018)172. arXiv:1807.02826

726. A. Berlin, T. Lin, L.-T. Wang, Mono-Higgs detection of dark matter at the LHC. JHEP **06**, 078 (2014). https://doi.org/10.1007/JHEP06(2014)078. arXiv:1402.7074
727. CMS Collaboration, Search for heavy resonances decaying into a vector boson and a Higgs boson in final states with charged leptons, neutrinos, and b quarks. Phys. Lett. B **768**, 137 (2017). https://doi.org/10.1016/j.physletb.2017.02.040. arXiv:1610.08066
728. CMS Collaboration, Search for heavy resonances decaying into two Higgs bosons or into a Higgs boson and a W or Z boson in proton-proton collisions at 13 TeV. JHEP **01**, 051 (2019). https://doi.org/10.1007/JHEP01(2019)051. arXiv:1808.01365
729. CMS Collaboration, Search for narrow high-mass resonances in proton proton collisions at $\sqrt{s} = 8$ TeV decaying to a Z and a Higgs boson. Phys. Lett. B **748**, 255 (2015). https://doi.org/10.1016/j.physletb.2015.07.011. arXiv:1502.04994
730. C.M.S. Collaboration, Performance of tau-lepton reconstruction and identification in CMS. JINST **7**, P01001 (2012). https://doi.org/10.1088/1748-0221/7/01/P01001. arXiv:1109.6034
731. CMS Collaboration, Performance of reconstruction and identification of τ leptons decaying to hadrons and ν_τ in pp collisions at $\sqrt{s} = 13$ TeV. JINST **13**, P10005 (2018). https://doi.org/10.1088/1748-0221/13/10/P10005. arXiv:1809.02816
732. V. Branchina, H. Faivre, V. Pangon, Effective potential and vacuum stability. J. Phys. G **36**, 015006 (2009). https://doi.org/10.1088/0954-3899/36/1/015006. arXiv:0802.4423
733. D. Buttazzo, G. Degrassi, P.P. Giardino, G.F. Giudice, F. Sala, A. Salvio et al., Investigating the near-criticality of the Higgs boson. JHEP **12**, 089 (2013). https://doi.org/10.1007/JHEP12(2013)089. arXiv:1307.3536
734. B. Di Micco, M. Gouzevitch, J. Mazzitelli, C. Vernieri et al., Higgs boson pair production at colliders: status and perspectives, arXiv:1910.00012
735. ATLAS Collaboration, Search for Higgs boson pair production in the $\gamma\gamma b\bar{b}$ final state using pp collision data at $\sqrt{s} = 8$ TeV from the ATLAS detector. Phys. Rev. Lett. **114**, 081802 (2015). https://doi.org/10.1103/PhysRevLett.114.081802. arXiv:1406.5053
736. ATLAS Collaboration, Search for Higgs boson pair production in the $b\bar{b}b\bar{b}$ final state from pp collisions at $\sqrt{s} = 8$ TeV with the ATLAS detector. Eur. Phys. J. C **75**, 412 (2015). https://doi.org/10.1140/epjc/s10052-015-3628-x. arXiv:1506.00285
737. ATLAS Collaboration, Searches for Higgs boson pair production in the $hh \to bb\tau\tau, \gamma\gamma WW^*, \gamma\gamma bb, bbbb$ channels with the ATLAS detector. Phys. Rev. D **92**, 092004 (2015). https://doi.org/10.1103/PhysRevD.92.092004. arXiv:1509.04670
738. ATLAS Collaboration, Search for pair production of Higgs bosons in the $b\bar{b}b\bar{b}$ final state using proton-proton collisions at $\sqrt{s} = 13$ TeV with the ATLAS detector. Phys. Rev. D **94**, 052002 (2016). https://doi.org/10.1103/PhysRevD.94.052002. arXiv:1606.04782
739. CMS Collaboration, Search for resonant pair production of Higgs bosons decaying to two bottom quark-antiquark pairs in proton-proton collisions at 8 TeV. Phys. Lett. B **749**, 560 (2015). https://doi.org/10.1016/j.physletb.2015.08.047. arXiv:1503.04114
740. CMS Collaboration, Search for heavy resonances decaying to two Higgs bosons in final states containing four b quarks. Eur. Phys. J. C **76**, 371 (2016). https://doi.org/10.1140/epjc/s10052-016-4206-6. arXiv:1602.08762
741. CMS Collaboration, Search for two Higgs bosons in final states containing two photons and two bottom quarks in proton-proton collisions at 8 TeV. Phys. Rev. D **94**, 052012 (2016). https://doi.org/10.1103/PhysRevD.94.052012. arXiv:1603.06896
742. B. Cooper, N. Konstantinidis, L. Lambourne, D. Wardrope, Boosted $HH \to b\bar{b}b\bar{b}$: a new topology in searches for TeV-scale resonances at the LHC. Phys. Rev. D **88**, 114005 (2013). https://doi.org/10.1103/PhysRevD.88.114005. arXiv:1307.0407
743. ATLAS Collaboration, Search for pair production of Higgs bosons in the $b\bar{b}b\bar{b}$ final state using proton-proton collisions at $\sqrt{s} = 13$ TeV with the ATLAS detector. JHEP **01**, 030 (2019). https://doi.org/10.1007/JHEP01(2019)030. arXiv:1804.06174
744. CMS Collaboration, Search for a massive resonance decaying to a pair of Higgs bosons in the four b quark final state in proton-proton collisions at $\sqrt{s} = 13$ TeV. Phys. Lett. B **781**, 244 (2018). https://doi.org/10.1016/j.physletb.2018.03.084. arXiv:1710.04960

745. CMS Collaboration, Search for resonant pair production of Higgs bosons decaying to bottom quark-antiquark pairs in proton-proton collisions at 13 TeV. JHEP **08**, 152 (2018). https://doi.org/10.1007/JHEP08(2018)152. arXiv:1806.03548
746. CMS Collaboration, Search for production of Higgs boson pairs in the four b quark final state using large-area jets in proton-proton collisions at $\sqrt{s} = 13$ TeV. JHEP **01**, 040 (2019). https://doi.org/10.1007/JHEP01(2019)040. arXiv:1808.01473
747. ATLAS Collaboration, Reconstruction and identification of boosted di-τ systems in a search for Higgs boson pairs using 13 TeV proton-proton collision data in ATLAS, arXiv:2007.14811
748. CMS Collaboration, Search for resonances decaying to a pair of Higgs bosons in the $b\bar{b}q\bar{q}'\ell\nu$ final state in proton-proton collisions at $\sqrt{s} = 13$ TeV. JHEP **10**, 125 (2019). https://doi.org/10.1007/JHEP10(2019)125. arXiv:1904.04193
749. K.S. Cranmer, Kernel estimation in high-energy physics. Comput. Phys. Commun. **136**, 198 (2001). https://doi.org/10.1016/S0010-4655(00)00243-5. arXiv:hep-ex/0011057
750. J. de Blas, J. Lizana, M. Perez-Victoria, Combining searches of Z′ and W′ bosons. JHEP **01**, 166 (2013). https://doi.org/10.1007/JHEP01(2013)166. arXiv:1211.2229
751. D. Pappadopulo, A. Thamm, R. Torre, A. Wulzer, Heavy vector triplets: bridging theory and data. JHEP **09**, 060 (2014). https://doi.org/10.1007/JHEP09(2014)060. arXiv:1402.4431
752. ATLAS Collaboration, Combination of searches for WW, WZ, and ZZ resonances in pp collisions at $\sqrt{s} = 8$ TeV with the ATLAS detector. Phys. Lett. B **755**, 285 (2016). https://doi.org/10.1016/j.physletb.2016.02.015. arXiv:1512.05099
753. ATLAS Collaboration, Searches for heavy diboson resonances in pp collisions at $\sqrt{s} = 13$ TeV with the ATLAS detector. JHEP **09**, 173 (2016). https://doi.org/10.1007/JHEP09(2016)173. arXiv:1606.04833
754. CMS Collaboration, Combination of searches for heavy resonances decaying to WW, WZ, ZZ, WH, and ZH boson pairs in proton-proton collisions at $\sqrt{s} = 8$ and 13 TeV. Phys. Lett. B **774**, 533 (2017). https://doi.org/10.1016/j.physletb.2017.09.083. arXiv:1705.09171
755. ATLAS Collaboration, Combination of searches for heavy resonances decaying into bosonic and leptonic final states using 36 fb^{-1} of proton-proton collision data at $\sqrt{s} = 13$ TeV with the ATLAS detector. Phys. Rev. D **98**, 052008 (2018). https://doi.org/10.1103/PhysRevD.98.052008. arXiv:1808.02380
756. CMS Collaboration, Combination of CMS searches for heavy resonances decaying to pairs of bosons or leptons. Phys. Lett. B **798**, 134952 (2019). https://doi.org/10.1016/j.physletb.2019.134952. arXiv:1906.00057
757. ATLAS Collaboration, Search for a new heavy gauge boson resonance decaying into a lepton and missing transverse momentum in 36 fb^{-1} of pp collisions at $\sqrt{s} = 13$ TeV with the ATLAS experiment. Eur. Phys. J. C **78**, 401 (2018). https://doi.org/10.1140/epjc/s10052-018-5877-y. arXiv:1706.04786
758. CMS Collaboration, Search for high-mass resonances in final states with a lepton and missing transverse momentum at $\sqrt{s} = 13$ TeV. JHEP **06**, 128 (2018). https://doi.org/10.1007/JHEP06(2018)128. arXiv:1803.11133
759. ATLAS Collaboration, Search for new high-mass phenomena in the dilepton final state using 36 fb^{-1} of proton-proton collision data at $\sqrt{s} = 13$ TeV with the ATLAS detector. JHEP **10**, 182 (2017). https://doi.org/10.1007/JHEP10(2017)182. arXiv:1707.02424
760. CMS Collaboration, Search for high-mass resonances in dilepton final states in proton-proton collisions at $\sqrt{s} = 13$ TeV. JHEP **06**, 120 (2018). https://doi.org/10.1007/JHEP06(2018)120. arXiv:1803.06292
761. ATLAS Collaboration, Search for heavy resonances decaying into WW in the $e\nu\mu\nu$ final state in pp collisions at $\sqrt{s} = 13$ TeV with the ATLAS detector. Eur. Phys. J. C **78**, 24 (2018). https://doi.org/10.1140/epjc/s10052-017-5491-4. arXiv:1710.01123
762. ATLAS Collaboration, Search for resonant WZ production in the fully leptonic final state in proton-proton collisions at $\sqrt{s} = 13$ TeV with the ATLAS detector. Phys. Lett. B **787**, 68 (2018). https://doi.org/10.1016/j.physletb.2018.10.021. arXiv:1806.01532

763. ATLAS Collaboration, Search for heavy ZZ resonances in the $\ell^+\ell^-\ell^+\ell^-$ and $\ell^+\ell^-\nu\bar{\nu}$ final states using proton-proton collisions at $\sqrt{s} = 13$ TeV with the ATLAS detector. Eur. Phys. J. C **78**, 293 (2018). https://doi.org/10.1140/epjc/s10052-018-5686-3. arXiv:1712.06386
764. ATLAS Collaboration, Combination of searches for Higgs boson pairs in *pp* collisions at $\sqrt{s} = 13$ TeV with the ATLAS detector. Phys. Lett. B **800**, 135103 (2020). https://doi.org/10.1016/j.physletb.2019.135103. arXiv:1906.02025
765. CMS Collaboration, Combination of searches for Higgs boson pair production in proton-proton collisions at $\sqrt{s} = 13$ TeV. Phys. Rev. Lett. **122**, 121803 (2019). https://doi.org/10.1103/PhysRevLett.122.121803. arXiv:1811.09689
766. ATLAS Collaboration, Search for Higgs boson pair production in the $\gamma\gamma b\bar{b}$ final state with 13 TeV *pp* collision data collected by the ATLAS experiment. JHEP **11**, 040 (2018). https://doi.org/10.1007/JHEP11(2018)040. arXiv:1807.04873
767. CMS Collaboration, Search for Higgs boson pair production in the $\gamma\gamma \mathrm{b}\bar{b}$ final state in pp collisions at $\sqrt{s} = 13$ TeV. Phys. Lett. B **788**, 7 (2019). https://doi.org/10.1016/j.physletb.2018.10.056. arXiv:1806.00408
768. ATLAS Collaboration, A search for the $Z\gamma$ decay mode of the Higgs boson in *pp* collisions at $\sqrt{s} = 13$ TeV with the ATLAS detector. Phys. Lett. B **809**, 135754 (2020). https://doi.org/10.1016/j.physletb.2020.135754 arXiv:2005.05382
769. A. Freitas, P. Schwaller, Multi-photon signals from composite models at LHC. JHEP **01**, 022 (2011). https://doi.org/10.1007/JHEP01(2011)022. arXiv:1010.2528
770. H. Davoudiasl, J. Hewett, T. Rizzo, Experimental probes of localized gravity: on and off the wall. Phys. Rev. D **63**, 075004 (2001). https://doi.org/10.1103/PhysRevD.63.075004. arXiv:hep-ph/0006041
771. B.C. Allanach, J.P. Skittrall, K. Sridhar, Z boson decay to photon plus Kaluza-Klein graviton in large extra dimensions. JHEP **11**, 089 (2007). https://doi.org/10.1088/1126-6708/2007/11/089. arXiv:0705.1953
772. I. Low, J. Lykken, G. Shaughnessy, Singlet scalars as Higgs imposters at the Large Hadron Collider. Phys. Rev. D **84**, 035027 (2011). https://doi.org/10.1103/PhysRevD.84.035027. arXiv:1105.4587
773. K.D. Lane, E. Eichten, Two scale technicolor. Phys. Lett. B **222**, 274 (1989). https://doi.org/10.1016/0370-2693(89)91265-3
774. B.A. Dobrescu, P.J. Fox, J. Kearney, Higgs-photon resonances. Eur. Phys. J. C **77**, 704 (2017). https://doi.org/10.1140/epjc/s10052-017-5276-9. arXiv:1705.08433
775. C.M.S. Collaboration, Search for a Higgs boson decaying into a Z and a photon in *pp* Collisions at $\sqrt{s} = 7$ and 8 TeV. Phys. Lett. B **726**, 587 (2013). https://doi.org/10.1016/j.physletb.2013.09.057. arXiv:1307.5515
776. ATLAS Collaboration, Search for new resonances in $W\gamma$ and $Z\gamma$ final states in *pp* collisions at $\sqrt{s} = 8$ TeV with the ATLAS detector. Phys. Lett. B **738**, 428 (2014). https://doi.org/10.1016/j.physletb.2014.10.002. arXiv:1407.8150
777. CMS Collaboration, Search for high-mass Zγ resonances in $e^+e^-\ \gamma$ and $\mu^+\mu^-\ \gamma$ final states in proton-proton collisions at $\sqrt{s} = 8$ and 13 TeV. JHEP **01**, 076 (2017). https://doi.org/10.1007/JHEP01(2017)076. arXiv:1610.02960
778. ATLAS Collaboration, Searches for the $Z\gamma$ decay mode of the Higgs boson and for new high-mass resonances in *pp* collisions at $\sqrt{s} = 13$ TeV with the ATLAS detector. JHEP **10**, 112 (2017). https://doi.org/10.1007/JHEP10(2017)112. arXiv:1708.00212
779. ATLAS Collaboration, Search for dark matter at $\sqrt{s} = 13$ TeV in final states containing an energetic photon and large missing transverse momentum with the ATLAS detector. Eur. Phys. J. C **77**, 393 (2017). https://doi.org/10.1140/epjc/s10052-017-4965-8. arXiv:1704.03848
780. ATLAS Collaboration, Search for heavy resonances decaying to a Z boson and a photon in *pp* collisions at $\sqrt{s} = 13$ TeV with the ATLAS detector. Phys. Lett. B **764**, 11 (2017). https://doi.org/10.1016/j.physletb.2016.11.005. arXiv:1607.06363
781. CMS Collaboration, Search for $Z\gamma$ resonances using leptonic and hadronic final states in proton-proton collisions at $\sqrt{s} = 13$ TeV. JHEP **09**, 148 (2018). https://doi.org/10.1007/JHEP09(2018)148. arXiv:1712.03143

782. CMS Collaboration, Search for high-mass $Z\gamma$ resonances in proton-proton collisions at $\sqrt{s} = 8$ and 13 TeV using jet substructure techniques. Phys. Lett. B **772**, 363 (2017). https://doi.org/10.1016/j.physletb.2017.06.062. arXiv:1612.09516
783. ATLAS Collaboration, Search for heavy resonances decaying to a photon and a hadronically decaying $Z/W/H$ boson in pp collisions at $\sqrt{s} = 13$ TeV with the ATLAS detector. Phys. Rev. D **98**, 032015 (2018). https://doi.org/10.1103/PhysRevD.98.032015. arXiv:1805.01908
784. CMS Collaboration, Search for narrow $h\gamma$ resonances in proton-proton collisions at $\sqrt{s} = 13$ TeV. Phys. Rev. Lett. **122**, 081804 (2019). https://doi.org/10.1103/PhysRevLett.122.081804. arXiv:1808.01257
785. ATLAS Collaboration, Search for heavy resonances decaying into a photon and a hadronically decaying Higgs boson in pp collisions at $\sqrt{s} = 13$ TeV with the ATLAS detector. Phys. Rev. Lett. **125**, 251802 (2020). https://doi.org/10.1103/PhysRevLett.125.251802 arXiv:2008.05928
786. B. Lillie, L. Randall, L.-T. Wang, The bulk RS KK-gluon at the LHC. JHEP **09**, 074 (2007). https://doi.org/10.1088/1126-6708/2007/09/074. arXiv:hep-ph/0701166
787. H. Davoudiasl, J. Hewett, T. Rizzo, Phenomenology of the Randall-Sundrum gauge hierarchy model. Phys. Rev. Lett. **84**, 2080 (2000). https://doi.org/10.1103/PhysRevLett.84.2080. arXiv:hep-ph/9909255
788. A. Goyal, R. Islam, M. Kumar, Dark matter in the Randall-Sundrum model with non-universal coupling. JHEP **10**, 050 (2019). https://doi.org/10.1007/JHEP10(2019)050 arXiv:1905.10583
789. J.L. Rosner, Prominent decay modes of a leptophobic Z'. Phys. Lett. B **387**, 113 (1996). https://doi.org/10.1016/0370-2693(96)01022-2. arXiv:hep-ph/9607207
790. D.J. Muller, S. Nandi, Top flavor: a Separate SU(2) for the third family. Phys. Lett. B **383**, 345 (1996). https://doi.org/10.1016/0370-2693(96)00745-9. arXiv:hep-ph/9602390
791. E. Malkawi, T.M. Tait, C. Yuan, A Model of strong flavor dynamics for the top quark. Phys. Lett. B **385**, 304 (1996). https://doi.org/10.1016/0370-2693(96)00859-3. arXiv:hep-ph/9603349
792. K.R. Lynch, E.H. Simmons, M. Narain, S. Mrenna, Finding Z' bosons coupled preferentially to the third family at LEP and the Tevatron. Phys. Rev. D **63**, 035006 (2001). https://doi.org/10.1103/PhysRevD.63.035006. arXiv:hep-ph/0007286
793. M. Carena, A. Daleo, B.A. Dobrescu, T.M. Tait, Z' gauge bosons at the Tevatron. Phys. Rev. D **70**, 093009 (2004). https://doi.org/10.1103/PhysRevD.70.093009. arXiv:hep-ph/0408098
794. C.T. Hill, Topcolor: Top quark condensation in a gauge extension of the standard model. Phys. Lett. B **266**, 419 (1991). https://doi.org/10.1016/0370-2693(91)91061-Y
795. C.T. Hill, S.J. Parke, Top production: sensitivity to new physics. Phys. Rev. D **49**, 4454 (1994). https://doi.org/10.1103/PhysRevD.49.4454. arXiv:hep-ph/9312324
796. C.T. Hill, Topcolor assisted technicolor. Phys. Lett. B **345**, 483 (1995). https://doi.org/10.1016/0370-2693(94)01660-5. arXiv:hep-ph/9411426
797. P.H. Frampton, S.L. Glashow, Chiral color: an alternative to the standard model. Phys. Lett. B **190**, 157 (1987). https://doi.org/10.1016/0370-2693(87)90859-8
798. D. Choudhury, R.M. Godbole, R.K. Singh, K. Wagh, Top production at the Tevatron/LHC and nonstandard, strongly interacting spin one particles. Phys. Lett. B **657**, 69 (2007). https://doi.org/10.1016/j.physletb.2007.09.057. arXiv:0705.1499
799. D. Dicus, A. Stange, S. Willenbrock, Higgs decay to top quarks at hadron colliders. Phys. Lett. B **333**, 126 (1994). https://doi.org/10.1016/0370-2693(94)91017-0. arXiv:hep-ph/9404359
800. K. Agashe, H. Davoudiasl, S. Gopalakrishna, T. Han, G.-Y. Huang, G. Perez et al., LHC signals for warped electroweak neutral gauge bosons. Phys. Rev. D **76**, 115015 (2007). https://doi.org/10.1103/PhysRevD.76.115015. arXiv:0709.0007
801. ATLAS Collaboration, Search for resonances in the mass distribution of jet pairs with one or two jets identified as b-jets in proton-proton collisions at $\sqrt{s} = 13$ TeV with the ATLAS detector. Phys. Rev. D **98**, 032016 (2018). https://doi.org/10.1103/PhysRevD.98.032016arXiv:1805.09299

802. M. Nemevsek, F. Nesti, G. Senjanovic, Y. Zhang, First limits on left-right symmetry scale from LHC data. Phys. Rev. D **83**, 115014 (2011). https://doi.org/10.1103/PhysRevD.83.115014. arXiv:1103.1627
803. CMS Collaboration, Search for anomalous $t\bar{t}$ production in the highly-boosted all-hadronic final state. JHEP **09**, 029 (2012). https://doi.org/10.1007/JHEP09(2012)029. arXiv:1204.2488. Erratum: JHEP **03**, 132 (2014)
804. ATLAS Collaboration, Search for $t\bar{t}$ resonances in the lepton plus jets final state with ATLAS using 4.7 fb^{-1} of pp collisions at $\sqrt{s} = 7$ TeV. Phys. Rev. D **88**, 012004 (2013). https://doi.org/10.1103/PhysRevD.88.012004. arXiv:1305.2756
805. CMS Collaboration, Searches for new physics using the $t\bar{t}$ invariant mass distribution in pp collisions a $\sqrt{s} = 8$ TeV. Phys. Rev. Lett. **111**, 211804 (2013). https://doi.org/10.1103/PhysRevLett.111.211804. arXiv:1309.2030. Erratum: Phys. Rev. Lett. **112**, 119903 (2014)
806. CMS Collaboration, Search for resonant $t\bar{t}$ production in proton-proton collisions at $\sqrt{s} =$ 8 TeV. Phys. Rev. D **93**, 012001 (2016). https://doi.org/10.1103/PhysRevD.93.012001. arXiv:1506.03062
807. CMS Collaboration, Search for $t\bar{t}$ resonances in highly boosted lepton+jets and fully hadronic final states in proton-proton collisions at $\sqrt{s} = 13$ TeV. JHEP **07**, 001 (2017). https://doi.org/10.1007/JHEP07(2017)001. arXiv:1704.03366
808. CMS Collaboration, Search for resonant $t\bar{t}$ production in proton-proton collisions at $\sqrt{s} = 13$ TeV. JHEP **04**, 031, (2019). https://doi.org/10.1007/JHEP04(2019)031. arXiv:1810.05905
809. ATLAS Collaboration, Search for heavy particles decaying into a top-quark pair in the fully hadronic final state in pp collisions at $\sqrt{s} = 13$ TeV with the ATLAS detector. Phys. Rev. D **99**, 092004 (2019). https://doi.org/10.1103/PhysRevD.99.092004. arXiv:1902.10077
810. M.R. Buckley, T. Plehn, M. Takeuchi, Buckets of tops. JHEP **08**, 086 (2013). https://doi.org/10.1007/JHEP08(2013)086. arXiv:1302.6238
811. ATLAS Collaboration, Search for $t\bar{t}$ resonances in fully hadronic final states in pp collisions at $\sqrt{s} = 13$ TeV with the ATLAS detector. JHEP **10**, 061 (2020). https://doi.org/10.1007/JHEP10(2020)061. arXiv:2005.05138
812. ATLAS Collaboration, A search for $t\bar{t}$ resonances with the ATLAS detector in 2.05 fb^{-1} of proton-proton collisions at $\sqrt{s} = 7$ TeV. Eur. Phys. J. C **72**, 2083 (2012). https://doi.org/10.1140/epjc/s10052-012-2083-1. arXiv:1205.5371
813. ATLAS Collaboration, A search for $t\bar{t}$ resonances in lepton+jets events with highly boosted top quarks collected in pp collisions at $\sqrt{s} = 7$ TeV with the ATLAS detector. JHEP **09**, 041 (2012). https://doi.org/10.1007/JHEP09(2012)041. arXiv:1207.2409
814. C.M.S. Collaboration, Search for resonant $t\bar{t}$ production in lepton+jets events in pp collisions at $\sqrt{s} = 7$ TeV. JHEP **12**, 015 (2012). https://doi.org/10.1007/JHEP12(2012)015. arXiv:1209.4397
815. ATLAS Collaboration, Search for heavy Higgs bosons A/H decaying to a top quark pair in pp collisions at $\sqrt{s} = 8$ TeV with the ATLAS Detector. Phys. Rev. Lett. **119**, 191803 (2017). https://doi.org/10.1103/PhysRevLett.119.191803. arXiv:1707.06025
816. CMS Collaboration, Search for heavy Higgs bosons decaying to a top quark pair in proton-proton collisions at $\sqrt{s} = 13$ TeV. JHEP **04**, 171 (2020). https://doi.org/10.1007/JHEP04(2020)171. arXiv:1908.01115
817. ATLAS Collaboration, Search for heavy particles decaying into top-quark pairs using lepton-plus-jets events in proton-proton collisions at $\sqrt{s} = 13$ TeV with the ATLAS detector. Eur. Phys. J. C **78**, 565 (2018). https://doi.org/10.1140/epjc/s10052-018-5995-6. arXiv:1804.10823
818. E. Megías, M. Quirós, Gapped continuum Kaluza-Klein spectrum. JHEP **08**, 166 (2019). https://doi.org/10.1007/JHEP08(2019)166. arXiv:1905.07364
819. M. Gavela, J. No, V. Sanz, J. de Trocóniz, Nonresonant searches for axionlike particles at the LHC. Phys. Rev. Lett. **124**, 051802 (2020). https://doi.org/10.1103/PhysRevLett.124.051802. arXiv:1905.12953
820. ATLAS Collaboration, Search for tb resonances in proton-proton collisions at $\sqrt{s} = 7$ TeV with the ATLAS detector. Phys. Rev. Lett. **109**, 081801 (2012). https://doi.org/10.1103/PhysRevLett.109.081801. arXiv:1205.1016

821. C.M.S. Collaboration, Search for a W' boson decaying to a bottom quark and a top quark in pp collisions at $\sqrt{s} = 7$ TeV. Phys. Lett. B **718**, 1229 (2013). https://doi.org/10.1016/j.physletb.2012.12.008. arXiv:1208.0956
822. ATLAS Collaboration, Search for $W' \to t\bar{b}$ in the lepton plus jets final state in proton-proton collisions at a centre-of-mass energy of $\sqrt{s} = 8$ TeV with the ATLAS detector. Phys. Lett. B **743**, 235 (2015). https://doi.org/10.1016/j.physletb.2015.02.051. arXiv:1410.4103
823. C.M.S. Collaboration, S. Chatrchyan et al., Search for $W' \to tb$ decays in the lepton+jets final state in pp collisions at $\sqrt{s} = 8$ TeV. JHEP **05**, 108 (2014). https://doi.org/10.1007/JHEP05(2014)108. arXiv:1402.2176
824. ATLAS Collaboration, Search for $W' \to tb \to qqbb$ decays in pp collisions at $\sqrt{s} = 8$ TeV with the ATLAS detector. Eur. Phys. J. C **75**, 165 (2015). https://doi.org/10.1140/epjc/s10052-015-3372-2. arXiv:1408.0886
825. CMS Collaboration, Search for $W' \to tb$ in proton-proton collisions at $\sqrt{s} = 8$ TeV. JHEP **02**, 122 (2016). https://doi.org/10.1007/JHEP02(2016)122. arXiv:1509.06051
826. ATLAS Collaboration, Search for vector-boson resonances decaying to a top quark and bottom quark in the lepton plus jets final state in pp collisions at $\sqrt{s} = 13$ TeV with the ATLAS detector. Phys. Lett. B **788**, 347 (2019). https://doi.org/10.1016/j.physletb.2018.11.032. arXiv:1807.10473
827. CMS Collaboration, Search for heavy resonances decaying to a top quark and a bottom quark in the lepton+jets final state in proton-proton collisions at 13 TeV. Phys. Lett. B **777**, 39 (2018). https://doi.org/10.1016/j.physletb.2017.12.006. arXiv:1708.08539
828. X.-G. He, G. Valencia, The $Z \to b\bar{b}$ decay asymmetry and left-right models. Phys. Rev. D **66**, 013004 (2002). https://doi.org/10.1103/PhysRevD.66.079901. arXiv:hep-ph/0203036. Erratum: Phys. Rev. D **66**, 079901 (2002)
829. X.-G. He, G. Valencia, $A_{\rm FB}^b$ and R_b at LEP and new right-handed gauge bosons. Phys. Rev. D **68**, 033011 (2003). https://doi.org/10.1103/PhysRevD.68.033011. arXiv:hep-ph/0304215
830. C.-W. Chiang, X.-G. He, G. Valencia, Z' model for $b \to s\ell\bar{\ell}$ flavor anomalies. Phys. Rev. D **93**, 074003 (2016). https://doi.org/10.1103/PhysRevD.93.074003. arXiv:1601.07328
831. X.-G. He, G. Valencia, Lepton universality violation and right-handed currents in $b \to c\tau\nu$. Phys. Lett. B **779**, 52 (2018). https://doi.org/10.1016/j.physletb.2018.01.073. arXiv:1711.09525
832. A. Hayreter, X.-G. He, G. Valencia, LHC constraints on W', Z' that couple mainly to third generation fermions, arXiv:1912.06344
833. A. Djouadi, A. Lenz, Sealing the fate of a fourth generation of fermions. Phys. Lett. B **715**, 310–314 (2012). https://doi.org/10.1016/j.physletb.2012.07.060. arXiv:1204.1252
834. O. Eberhardt, G. Herbert, H. Lacker, A. Lenz, A. Menzel, U. Nierste et al., Impact of a Higgs boson at a mass of 126 GeV on the standard model with three and four fermion generations. Phys. Rev. Lett. **109**, 241802 (2012). https://doi.org/10.1103/PhysRevLett.109.241802. arXiv:1209.1101
835. J. Bulava, K. Jansen, A. Nagy, Constraining a fourth generation of quarks: non-perturbative Higgs boson mass bounds. Phys. Lett. B **723**, 95 (2013). https://doi.org/10.1016/j.physletb.2013.04.041. arXiv:1301.3416
836. J. Berger, J. Hubisz, M. Perelstein, A fermionic top partner: naturalness and the LHC. JHEP **07**, 016 (2012). https://doi.org/10.1007/JHEP07(2012)016. arXiv:1205.0013
837. J. Aguilar-Saavedra, R. Benbrik, S. Heinemeyer, M. Pérez-Victoria, Handbook of vectorlike quarks: mixing and single production. Phys. Rev. D **88**, 094010 (2013). https://doi.org/10.1103/PhysRevD.88.094010. arXiv:1306.0572
838. J. Aguilar-Saavedra, Identifying top partners at LHC. JHEP **11**, 030 (2009). https://doi.org/10.1088/1126-6708/2009/11/030. arXiv:0907.3155
839. J. Aguilar-Saavedra, Mixing with vector-like quarks: constraints and expectations, in *1st Large Hadron Collider Physics Conference (LHCP). Proceedings, EPJ Web Conference*, Barcelona, Spain, 13–18 May 2013, vol. 60 (2013), p. 16012. https://doi.org/10.1051/epjconf/20136016012. arXiv:1306.4432

840. F. del Aguila, M. Perez-Victoria, J. Santiago, Effective description of quark mixing. Phys. Lett. B **492**, 98 (2000). https://doi.org/10.1016/S0370-2693(00)01071-6. arXiv:hep-ph/0007160
841. F. del Aguila, M. Perez-Victoria, J. Santiago, Observable contributions of new exotic quarks to quark mixing. JHEP **09**, 011 (2000). https://doi.org/10.1088/1126-6708/2000/09/011. arXiv:hep-ph/0007316
842. G. Azuelos et al., Exploring little Higgs models with ATLAS at the LHC. Eur. Phys. J. C **39S2**, 13 (2005). https://doi.org/10.1140/epjcd/s2004-02-002-x. arXiv:hep-ph/0402037
843. Y. Okada, L. Panizzi, LHC signatures of vector-like quarks. Adv. High Energy Phys. **2013**, 364936 (2013). https://doi.org/10.1155/2013/364936. arXiv:1207.5607
844. A. De Simone, O. Matsedonskyi, R. Rattazzi, A. Wulzer, A first top partner hunter's guide. JHEP **04**, 004 (2013). https://doi.org/10.1007/JHEP04(2013)004. arXiv:1211.5663
845. O. Matsedonskyi, G. Panico, A. Wulzer, Top partners searches and composite Higgs models. JHEP **04**, 003 (2016). https://doi.org/10.1007/JHEP04(2016)003. arXiv:1512.04356
846. A. Angelescu, A. Djouadi, G. Moreau, Vector-like top/bottom quark partners and Higgs physics at the LHC. Eur. Phys. J. C **76**, 99 (2016). https://doi.org/10.1140/epjc/s10052-016-3950-y. arXiv:1510.07527
847. M.J. Dolan, J. Hewett, M. Krämer, T. Rizzo, Simplified models for Higgs physics: Singlet scalar and vector-like quark phenomenology. JHEP **07**, 039 (2016). https://doi.org/10.1007/JHEP07(2016)039. arXiv:1601.07208
848. B. Fuks, H.-S. Shao, QCD next-to-leading-order predictions matched to parton showers for vector-like quark models. Eur. Phys. J. C **77**, 135 (2017). https://doi.org/10.1140/epjc/s10052-017-4686-z. arXiv:1610.04622
849. A. Carvalho, S. Moretti, D. O'Brien, L. Panizzi, H. Prager, Single production of vectorlike quarks with large width at the Large Hadron Collider. Phys. Rev. D **98**, 015029 (2018). https://doi.org/10.1103/PhysRevD.98.015029. arXiv:1805.06402
850. M. Perelstein, M.E. Peskin, A. Pierce, Top quarks and electroweak symmetry breaking in little Higgs models. Phys. Rev. D **69**, 075002 (2004). https://doi.org/10.1103/PhysRevD.69.075002. arXiv:hep-ph/0310039
851. G. Dissertori, E. Furlan, F. Moortgat, P. Nef, Discovery potential of top-partners in a realistic composite Higgs model with early LHC data. JHEP **09**, 019 (2010). https://doi.org/10.1007/JHEP09(2010)019. arXiv:1005.4414
852. K. Agashe, R. Contino, A. Pomarol, The minimal composite Higgs model. Nucl. Phys. B **719**, 165 (2005). https://doi.org/10.1016/j.nuclphysb.2005.04.035. arXiv:hep-ph/0412089
853. O. Matsedonskyi, G. Panico, A. Wulzer, Light top partners for a light composite Higgs. JHEP **01**, 164 (2013). https://doi.org/10.1007/JHEP01(2013)164. arXiv:1204.6333
854. B. Bellazzini, C. Csáki, J. Serra, Composite Higgses. Eur. Phys. J. C **74**, 2766 (2014). https://doi.org/10.1140/epjc/s10052-014-2766-x. arXiv:1401.2457
855. D.B. Kaplan, Flavor at SSC energies: a new mechanism for dynamically generated fermion masses. Nucl. Phys. B **365**, 259 (1991). https://doi.org/10.1016/S0550-3213(05)80021-5
856. I. Antoniadis, K. Benakli, M. Quiros, Finite Higgs mass without supersymmetry. New J. Phys. **3**, 20 (2001). https://doi.org/10.1088/1367-2630/3/1/320. arXiv:hep-th/0108005
857. R. Contino, L. Da Rold, A. Pomarol, Light custodians in natural composite Higgs models. Phys. Rev. D **75**, 055014 (2007). https://doi.org/10.1103/PhysRevD.75.055014. arXiv:hep-ph/0612048
858. J.L. Hewett, T.G. Rizzo, Low-energy phenomenology of superstring inspired $E(6)$ models. Phys. Rept. **183**, 193 (1989). https://doi.org/10.1016/0370-1573(89)90071-9
859. M. Buchkremer, G. Cacciapaglia, A. Deandrea, L. Panizzi, Model independent framework for searches of top partners. Nucl. Phys. B **876**, 376 (2013). https://doi.org/10.1016/j.nuclphysb.2013.08.010. arXiv:1305.4172
860. CMS Collaboration, Search for vector-like T and B quark pairs in final states with leptons at $\sqrt{s} = 13$ TeV. JHEP **08**, 177 (2018). https://doi.org/10.1007/JHEP08(2018)177. arXiv:1805.04758

861. G. Cacciapaglia, S. Choudhury, A. Deandrea, N. Gaur, M. Klasen, Dileptonic signatures of T-odd quarks at the LHC. JHEP **03**, 059 (2010). https://doi.org/10.1007/JHEP03(2010)059. arXiv:0911.4630
862. A. Buckley, J. Butterworth, L. Corpe, D. Huang, P. Sun, New sensitivity of current LHC measurements to vector-like quarks, arXiv:2006.07172
863. ATLAS Collaboration, Search for pair and single production of new heavy quarks that decay to a Z boson and a third-generation quark in pp collisions at $\sqrt{s} = 8$ TeV with the ATLAS detector. JHEP **11**, 104 (2014). https://doi.org/10.1007/JHEP11(2014)104. arXiv:1409.5500
864. ATLAS Collaboration, Analysis of events with b-jets and a pair of leptons of the same charge in pp collisions at $\sqrt{s} = 8$ TeV with the ATLAS detector. JHEP **10**, 150 (2015). https://doi.org/10.1007/JHEP10(2015)150. arXiv:1504.04605
865. ATLAS Collaboration, Search for production of vector-like quark pairs and of four top quarks in the lepton-plus-jets final state in pp collisions at $\sqrt{s} = 8$ TeV with the ATLAS detector. JHEP **08**, 105 (2015). https://doi.org/10.1007/JHEP08(2015)105. arXiv:1505.04306
866. ATLAS Collaboration, Search for vector-like B quarks in events with one isolated lepton, missing transverse momentum and jets at $\sqrt{s} = 8$ TeV with the ATLAS detector. Phys. Rev. D **91**, 112011 (2015). https://doi.org/10.1103/PhysRevD.91.112011. arXiv:1503.05425
867. C.M.S. Collaboration, Inclusive Search for a vector-like T quark with charge $\frac{2}{3}$ in pp collisions at $\sqrt{s} = 8$ TeV. Phys. Lett. B **729**, 149 (2014). https://doi.org/10.1016/j.physletb.2014.01.006. arXiv:1311.7667
868. CMS Collaboration, Search for vector-like T quarks decaying to top quarks and Higgs bosons in the all-hadronic channel using jet substructure. JHEP **06**, 080 (2015). https://doi.org/10.1007/JHEP06(2015)080. arXiv:1503.01952
869. CMS Collaboration, Search for vector-like charge 2/3 T quarks in proton-proton collisions at $\sqrt{s} = 8$ TeV. Phys. Rev. D **93**, 012003 (2016). https://doi.org/10.1103/PhysRevD.93.012003. arXiv:1509.04177
870. CMS Collaboration, Search for pair-produced vectorlike B quarks in proton-proton collisions at $\sqrt{s} = 8$ TeV. Phys. Rev. D **93**, 112009 (2016). https://doi.org/10.1103/PhysRevD.93.112009. arXiv:1507.07129
871. C.M.S. Collaboration, Search for top-quark partners with charge 5/3 in the same-sign dilepton final state. Phys. Rev. Lett. **112**, 171801 (2014). https://doi.org/10.1103/PhysRevLett.112.171801. arXiv:1312.2391
872. CMS Collaboration, Search for top quark partners with charge 5/3 in proton-proton collisions at $\sqrt{s} = 13$ TeV. JHEP **08**, 073 (2017). https://doi.org/10.1007/JHEP08(2017)073. arXiv:1705.10967
873. CMS Collaboration, Search for top quark partners with charge 5/3 in the same-sign dilepton and single-lepton final states in proton-proton collisions at $\sqrt{s} = 13$ TeV. JHEP **03**, 082 (2019). https://doi.org/10.1007/JHEP03(2019)082. arXiv:1810.03188
874. ATLAS Collaboration, Search for new phenomena in events with same-charge leptons and b-jets in pp collisions at $\sqrt{s} = 13$ TeV with the ATLAS detector. JHEP **12**, 039 (2018). https://doi.org/10.1007/JHEP12(2018)039. arXiv:1807.11883
875. ATLAS Collaboration, Search for pair production of heavy vector-like quarks decaying into high-p_T W bosons and top quarks in the lepton-plus-jets final state in pp collisions at $\sqrt{s} = 13$ TeV with the ATLAS detector. JHEP **08**, 048 (2018). https://doi.org/10.1007/JHEP08(2018)048. arXiv:1806.01762
876. CMS Collaboration, Search for pair production of vector-like T and B quarks in single-lepton final states using boosted jet substructure in proton-proton collisions at $\sqrt{s} = 13$ TeV. JHEP **11**, 085 (2017). https://doi.org/10.1007/JHEP11(2017)085. arXiv:1706.03408
877. ATLAS Collaboration, Search for pair production of heavy vector-like quarks decaying to high-p_T W bosons and b quarks in the lepton-plus-jets final state in pp collisions at $\sqrt{s} = 13$ TeV with the ATLAS detector. JHEP **10**, 141 (2017). https://doi.org/10.1007/JHEP10(2017)141. arXiv:1707.03347
878. ATLAS Collaboration, Search for pair production of vector-like top quarks in events with one lepton, jets, and missing transverse momentum in $\sqrt{s} = 13$ TeV pp collisions with

the ATLAS detector. JHEP **08**, 052 (2017). https://doi.org/10.1007/JHEP08(2017)052. arXiv:1705.10751

879. ATLAS Collaboration, Search for pair production of up-type vector-like quarks and for four-top-quark events in final states with multiple b-jets with the ATLAS detector. JHEP **07**, 089 (2018). https://doi.org/10.1007/JHEP07(2018)089. arXiv:1803.09678
880. CMS Collaboration, Search for pair production of vector-like quarks in the bW$\overline{b}$ W channel from proton-proton collisions at $\sqrt{s} = 13$ TeV. Phys. Lett. B **779**, 82 (2018). https://doi.org/10.1016/j.physletb.2018.01.077. arXiv:1710.01539
881. D0 Collaboration, Direct measurement of the top quark mass at D0. Phys. Rev. D **58**, 052001 (1998). https://doi.org/10.1103/PhysRevD.58.052001. arXiv:hep-ex/9801025
882. D0 Collaboration, Measurement of the top quark mass in the lepton+jets channel using the ideogram method. Phys. Rev. D **75**, 092001 (2007). https://doi.org/10.1103/PhysRevD.75.092001. arXiv:hep-ex/0702018
883. ATLAS Collaboration, Search for pair- and single-production of vector-like quarks in final states with at least one Z boson decaying into a pair of electrons or muons in pp collision data collected with the ATLAS detector at $\sqrt{s} = 13$ TeV. Phys. Rev. D **98**, 112010 (2018). https://doi.org/10.1103/PhysRevD.98.112010. arXiv:1806.10555
884. CMS Collaboration, Search for vector-like quarks in events with two oppositely charged leptons and jets in proton-proton collisions at $\sqrt{s} = 13$ TeV. Eur. Phys. J. C **79**, 364 (2019). https://doi.org/10.1140/epjc/s10052-019-6855-8. arXiv:1812.09768
885. ATLAS Collaboration, Search for pair production of heavy vector-like quarks decaying into hadronic final states in pp collisions at $\sqrt{s} = 13$ TeV with the ATLAS detector. Phys. Rev. D **98**, 092005 (2018). https://doi.org/10.1103/PhysRevD.98.092005. arXiv:1808.01771
886. K. Kondo, Dynamical likelihood method for reconstruction of events with missing momentum. 1: Method and toy models. J. Phys. Soc. Jap. **57**, 4126 (1988). https://doi.org/10.1143/JPSJ.57.4126
887. CMS Collaboration, Search for pair production of vector-like quarks in the fully hadronic final state. Phys. Rev. D **100**, 072001 (2019). https://doi.org/10.1103/PhysRevD.100.072001. arXiv:1906.11903
888. ATLAS Collaboration, Combination of the searches for pair-produced vector-like partners of the third-generation quarks at $\sqrt{s} = 13$ TeV with the ATLAS detector. Phys. Rev. Lett. **121**, 211801 (2018). https://doi.org/10.1103/PhysRevLett.121.211801. arXiv:1808.02343
889. CMS Collaboration, A search for bottom-type, vector-like quark pair production in a fully hadronic final state in proton-proton collisions at $\sqrt{s} = 13$ TeV, arXiv:2008.09835
890. J. Serra, Beyond the minimal top partner decay. JHEP **09**, 176 (2015). https://doi.org/10.1007/JHEP09(2015)176. arXiv:1506.05110
891. G. Cacciapaglia, T. Flacke, M. Park, M. Zhang, Exotic decays of top partners: mind the search gap. Phys. Lett. B **798**, 135015 (2019). https://doi.org/10.1016/j.physletb.2019.135015. arXiv:1908.07524
892. R. Benbrik et al., Signatures of vector-like top partners decaying into new neutral scalar or pseudoscalar bosons. JHEP **05**, 028 (2020). https://doi.org/10.1007/JHEP05(2020)028. arXiv:1907.05929
893. A. Anandakrishnan, J. H. Collins, M. Farina, E. Kuflik, M. Perelstein, Odd top partners at the LHC. Phys. Rev. D **93**, 075009 (2016). https://doi.org/10.1103/PhysRevD.93.075009. arXiv:1506.05130
894. J.H. Kim, S.D. Lane, H.-S. Lee, I.M. Lewis, M. Sullivan, Searching for dark photons with maverick top partners. Phys. Rev. D **101**, 035041 (2020). https://doi.org/10.1103/PhysRevD.101.035041. arXiv:1904.05893
895. R. Dermíšek, E. Lunghi, S. Shin, Hunting for vector-like quarks. JHEP **04**, 019 (2019). https://doi.org/10.1007/JHEP04(2019)019. arXiv:1901.03709
896. N. Vignaroli, Early discovery of top partners and test of the Higgs nature. Phys. Rev. D **86**, 075017 (2012). https://doi.org/10.1103/PhysRevD.86.075017. arXiv:1207.0830
897. C.-X. Yue, H.-D. Yang, W. Ma, Single production of the top partners at high energy colliders. Nucl. Phys. B **818**, 1 (2009). https://doi.org/10.1016/j.nuclphysb.2009.03.014. arXiv:0903.3720

898. N. Vignaroli, Discovering the composite Higgs through the decay of a heavy fermion. JHEP **07**, 158 (2012). https://doi.org/10.1007/JHEP07(2012)158. arXiv:1204.0468
899. S. Gopalakrishna, T. Mandal, S. Mitra, G. Moreau, LHC signatures of warped-space vector-like quarks. JHEP **08**, 079 (2014). https://doi.org/10.1007/JHEP08(2014)079. arXiv:1306.2656
900. A. Azatov, M. Salvarezza, M. Son, M. Spannowsky, Boosting top partner searches in composite Higgs models. Phys. Rev. D **89**, 075001 (2014). https://doi.org/10.1103/PhysRevD.89.075001. arXiv:1308.6601
901. M. Backovic, T. Flacke, J.H. Kim, S.J. Lee, Search strategies for TeV scale fermionic top partners with charge 2/3. JHEP **04**, 014 (2016). https://doi.org/10.1007/JHEP04(2016)014. arXiv:1507.06568
902. Y.-B. Liu, Search for single production of the heavy vectorlike T quark with $T \to th$ and $h \to \gamma\gamma$ at the high-luminosity LHC. Phys. Rev. D **95**, 035013 (2017). https://doi.org/10.1103/PhysRevD.95.035013. arXiv:1612.05851
903. Y.-B. Liu, Y.-Q. Li, Search for single production of the vector-like top partner at the 14 TeV LHC. Eur. Phys. J. C **77**, 654 (2017). https://doi.org/10.1140/epjc/s10052-017-5228-4. arXiv:1709.06427
904. A. Deandrea, A. Iyer, Vectorlike quarks and heavy colored bosons at the LHC. Phys. Rev. D **97**, 055002 (2018). https://doi.org/10.1103/PhysRevD.97.055002. arXiv:1710.01515
905. ATLAS Collaboration, Search for the production of single vector-like and excited quarks in the Wt final state in pp collisions at $\sqrt{s} = 8$ TeV with the ATLAS detector. JHEP **02**, 110 (2016). https://doi.org/10.1007/JHEP02(2016)110. arXiv:1510.02664
906. ATLAS Collaboration, Search for single production of vector-like quarks decaying into Wb in pp collisions at $\sqrt{s} = 8$ TeV with the ATLAS detector. Eur. Phys. J. C **76**, 442 (2016). https://doi.org/10.1140/epjc/s10052-016-4281-8. arXiv:1602.05606
907. CMS Collaboration, Search for electroweak production of a vector-like T quark using fully hadronic final states. JHEP **01**, 036 (2020). https://doi.org/10.1007/JHEP01(2020)036. arXiv:1909.04721
908. ATLAS Collaboration, Search for large missing transverse momentum in association with one top-quark in proton-proton collisions at $\sqrt{s} = 13$ TeV with the ATLAS detector. JHEP **05**, 041 (2019). https://doi.org/10.1007/JHEP05(2019)041. arXiv:1812.09743
909. CMS Collaboration, Search for single production of vector-like quarks decaying to a Z boson and a top or a bottom quark in proton-proton collisions at $\sqrt{s} = 13$ TeV. JHEP **05**, 029 (2017). https://doi.org/10.1007/JHEP05(2017)029. arXiv:1701.07409
910. CMS Collaboration, Search for single production of a vector-like T quark decaying to a Z boson and a top quark in proton-proton collisions at $\sqrt{s} = 13$ TeV. Phys. Lett. B **781**, 574 (2018). https://doi.org/10.1016/j.physletb.2018.04.036. arXiv:1708.01062
911. ATLAS Collaboration, Search for single production of vector-like quarks decaying into Wb in pp collisions at $\sqrt{s} = 13$ TeV with the ATLAS detector. JHEP **05**, 164 (2019). https://doi.org/10.1007/JHEP05(2019)164. arXiv:1812.07343
912. CMS Collaboration, Search for single production of vector-like quarks decaying into a b quark and a W boson in proton-proton collisions at $\sqrt{s} = 13$ TeV. Phys. Lett. B **772**, 634 (2017). https://doi.org/10.1016/j.physletb.2017.07.022. arXiv:1701.08328
913. CMS Collaboration, Search for single production of vector-like quarks decaying to ab quark and a Higgs boson. JHEP **06** 031 (2018). https://doi.org/10.1007/JHEP06(2018)031. arXiv:1802.01486
914. CMS Collaboration, Search for single production of vector-like quarks decaying to a top quark and a W boson in proton-proton collisions at $\sqrt{s} = 13$ TeV. Eur. Phys. J. C **79**, 90 (2019). https://doi.org/10.1140/epjc/s10052-019-6556-3. arXiv:1809.08597
915. X. Gong, C.-X. Yue, Y.-C. Guo, Search for vector-like bottom quark via Zb production at the LHC. Phys. Lett. B **793**, 175 (2019). https://doi.org/10.1016/j.physletb.2019.04.051. arXiv:1904.08206
916. CMS Collaboration, Search for vector-like light-flavor quark partners in proton-proton collisions at $\sqrt{s} = 8$ TeV. Phys. Rev. D **97**, 072008 (2018). https://doi.org/10.1103/PhysRevD.97.072008. arXiv:1708.02510

917. ATLAS Collaboration, Search for heavy vector-like quarks coupling to light quarks in proton-proton collisions at $\sqrt{s} = 7$ TeV with the ATLAS detector. Phys. Lett. B **712**, 22 (2012). https://doi.org/10.1016/j.physletb.2012.03.082. arXiv:1112.5755
918. ATLAS Collaboration, Search for pair production of a new heavy quark that decays into a W boson and a light quark in pp collisions at $\sqrt{s} = 8$ TeV with the ATLAS detector. Phys. Rev. D **92**, 112007 (2015). https://doi.org/10.1103/PhysRevD.92.112007. arXiv:1509.04261
919. O. Matsedonskyi, G. Panico, A. Wulzer, On the interpretation of top partners searches. JHEP **12**, 097 (2014). https://doi.org/10.1007/JHEP12(2014)097. arXiv:1409.0100
920. A. Roy, N. Nikiforou, N. Castro, T. Andeen, Novel interpretation strategy for searches of singly produced vectorlike quarks at the LHC. Phys. Rev. D **101**, 115027 (2020). https://doi.org/10.1103/PhysRevD.101.115027. arXiv:2003.00640
921. R. Contino, D. Marzocca, D. Pappadopulo, R. Rattazzi, On the effect of resonances in composite Higgs phenomenology. JHEP **10**, 081 (2011). https://doi.org/10.1007/JHEP10(2011)081. arXiv:1109.1570
922. D. Greco, D. Liu, Hunting composite vector resonances at the LHC: naturalness facing data. JHEP **12**, 126 (2014). https://doi.org/10.1007/JHEP12(2014)126. arXiv:1410.2883
923. G. Giudice, C. Grojean, A. Pomarol, R. Rattazzi, The strongly-interacting light Higgs. JHEP **06**, 045 (2007). https://doi.org/10.1088/1126-6708/2007/06/045. arXiv:hep-ph/0703164
924. R. Contino, T. Kramer, M. Son, R. Sundrum, Warped/composite phenomenology simplified. JHEP **05**, 074 (2007). https://doi.org/10.1088/1126-6708/2007/05/074. arXiv:hep-ph/0612180
925. M. Chala, J. Juknevich, G. Perez, J. Santiago, The elusive gluon. JHEP **01**, 092 (2015). https://doi.org/10.1007/JHEP01(2015)092. arXiv:1411.1771
926. D. Marzocca, M. Serone, J. Shu, General composite Higgs models. JHEP **08**, 013 (2012). https://doi.org/10.1007/JHEP08(2012)013. arXiv:1205.0770
927. A. Pomarol, F. Riva, The composite Higgs and light resonance connection. JHEP **08**, 135 (2012). https://doi.org/10.1007/JHEP08(2012)135. arXiv:1205.6434
928. J.P. Araque, N.F. Castro, J. Santiago, Interpretation of vector-like quark searches: Heavy gluons in composite Higgs models. JHEP **11**, 120 (2015). https://doi.org/10.1007/JHEP11(2015)120. arXiv:1507.05628
929. D. Liu, L.-T. Wang, K.-P. Xie, Prospects of searching for composite resonances at the LHC and beyond. JHEP **01**, 157 (2019). https://doi.org/10.1007/JHEP01(2019)157. arXiv:1810.08954
930. ATLAS Collaboration, Search for single production of a vector-like quark via a heavy gluon in the $4b$ final state with the ATLAS detector in pp collisions at $\sqrt{s} = 8$ TeV. Phys. Lett. B **758**, 249 (2016). https://doi.org/10.1016/j.physletb.2016.04.061. arXiv:1602.06034
931. M. Chala, J. Santiago, $Hb\bar{b}$ production in composite Higgs models. Phys. Rev. D **88**, 035010 (2013). https://doi.org/10.1103/PhysRevD.88.035010. arXiv:1305.1940
932. CMS Collaboration, Search for a heavy resonance decaying to a top quark and a vector-like top quark at $\sqrt{s} = 13$ TeV. JHEP **09**, 053 (2017). https://doi.org/10.1007/JHEP09(2017)053. arXiv:1703.06352
933. CMS Collaboration, Search for a heavy resonance decaying to a top quark and a vector-like top quark in the lepton+jets final state in pp collisions at $\sqrt{s} = 13$ TeV. Eur. Phys. J. C **79**, 208 (2019). https://doi.org/10.1140/epjc/s10052-019-6688-5. arXiv:1812.06489
934. CMS Collaboration, Search for a W' boson decaying to a vector-like quark and a top or bottom quark in the all-jets final state. JHEP **03**, 127 (2019). https://doi.org/10.1007/JHEP03(2019)127. arXiv:1811.07010
935. N. Vignaroli, New W' signals at the LHC. Phys. Rev. D **89**, 095027 (2014). https://doi.org/10.1103/PhysRevD.89.095027. arXiv:1404.5558
936. CMS Collaboration, Measurement of the cross section for top quark pair production in association with a W or Z boson in proton-proton collisions at $\sqrt{s} = 13$ TeV. JHEP **08**, 011 (2018). https://doi.org/10.1007/JHEP08(2018)011. arXiv:1711.02547
937. ATLAS Collaboration, Measurement of the $t\bar{t}Z$ and $t\bar{t}W$ cross sections in proton-proton collisions at $\sqrt{s} = 13$ TeV with the ATLAS detector. Phys. Rev. D **99**, 072009 (2019). https://doi.org/10.1103/PhysRevD.99.072009. arXiv:1901.03584

938. C.M.S. Collaboration, Measurement of the production cross sections for a Z boson and one or more b jets in pp collisions at $\sqrt{s} = 7$ TeV. JHEP **06**, 120 (2014). https://doi.org/10.1007/JHEP06(2014)120. arXiv:1402.1521
939. ATLAS Collaboration, Measurement of differential production cross-sections for a Z boson in association with b-jets in 7 TeV proton-proton collisions with the ATLAS detector. JHEP **10**, 141 (2014). https://doi.org/10.1007/JHEP10(2014)141. arXiv:1407.3643
940. LHCb Collaboration, Measurement of the Z +b-jet cross-section in pp collisions at $\sqrt{s} = 7$ TeV in the forward region. JHEP **01**, 064 (2015). https://doi.org/10.1007/JHEP01(2015)064. arXiv:1411.1264
941. ATLAS Collaboration, Measurement of the production cross-section of a single top quark in association with a Z boson in proton-proton collisions at 13 TeV with the ATLAS detector. Phys. Lett. B **780**, 557 (2018). https://doi.org/10.1016/j.physletb.2018.03.023. arXiv:1710.03659
942. CMS Collaboration, Measurement of the associated production of a single top quark and a Z boson in pp collisions at $\sqrt{s} = 13$ TeV. Phys. Lett. B **779**, 358 (2018). https://doi.org/10.1016/j.physletb.2018.02.025. arXiv:1712.02825
943. C.M.S. Collaboration, Observation of single top quark production in association with a Z boson in proton-proton collisions at $\sqrt{s} = 13$ TeV. Phys. Rev. Lett. **122**, (2019). https://doi.org/10.1103/PhysRevLett.122.132003. arXiv:1812.05900
944. E. Eichten, K.D. Lane, M.E. Peskin, New tests for quark and lepton substructure. Phys. Rev. Lett. **50**, 811 (1983). https://doi.org/10.1103/PhysRevLett.50.811
945. E. Eichten, I. Hinchliffe, K.D. Lane, C. Quigg, Super collider physics. Rev. Mod. Phys. **56**, 579 (1984). https://doi.org/10.1103/RevModPhys.56.579. Addendum: Rev. Mod. Phys. 58, 1065 (1986)
946. J. Gao, C.S. Li, J. Wang, H.X. Zhu, C.-P. Yuan, Next-to-leading QCD effect to the quark compositeness search at the LHC. Phys. Rev. Lett. **106**, 142001 (2011). https://doi.org/10.1103/PhysRevLett.106.142001. arXiv:1101.4611
947. ATLAS Collaboration, Search for quark contact interactions in dijet angular distributions in pp collisions at $\sqrt{s} = 7$ TeV measured with the ATLAS Detector. Phys. Lett. B **694**, 327 (2011). https://doi.org/10.1016/j.physletb.2010.10.021. arXiv:1009.5069
948. ATLAS Collaboration, Search for new physics in dijet mass and angular distributions in pp collisions at $\sqrt{s} = 7$ TeV measured with the ATLAS detector. New J. Phys. **13**, 053044 (2011). https://doi.org/10.1088/1367-2630/13/5/053044. arXiv:1103.3864
949. ATLAS Collaboration, ATLAS search for new phenomena in dijet mass and angular distributions using pp collisions at $\sqrt{s} = 7$ TeV. JHEP **01**, 029 (2013). https://doi.org/10.1007/JHEP01(2013)029. arXiv:1210.1718
950. ATLAS Collaboration, Search for new phenomena in dijet angular distributions in proton-proton collisions at $\sqrt{s} = 8$ TeV measured with the ATLAS detector. Phys. Rev. Lett. **114**, 221802 (2015). https://doi.org/10.1103/PhysRevLett.114.221802. arXiv:1504.00357
951. ATLAS Collaboration, Search for new phenomena in dijet events using 37 fb^{-1} of pp collision data collected at $\sqrt{s} = 13$ TeV with the ATLAS detector. Phys. Rev. D **96**, 052004 (2017). https://doi.org/10.1103/PhysRevD.96.052004. arXiv:1703.09127
952. C.M.S. Collaboration, Measurement of dijet angular distributions and search for quark compositeness in pp collisions at $sqrts = 7$ TeV. Phys. Rev. Lett. **106**, 201804 (2011). https://doi.org/10.1103/PhysRevLett.106.201804. arXiv:1102.2020
953. C.M.S. Collaboration, Search for quark compositeness in dijet angular distributions from pp collisions at $\sqrt{s} = 7$ TeV. JHEP **05**, 055 (2012). https://doi.org/10.1007/JHEP05(2012)055. arXiv:1202.5535
954. C.M.S. Collaboration, Search for quark contact interactions and extra spatial dimensions using dijet angular distributions in proton-proton collisions at $\sqrt{s} = 8$ TeV. Phys. Lett. B **746**, 79 (2015). https://doi.org/10.1016/j.physletb.2015.04.042. arXiv:1411.2646
955. CMS Collaboration, Search for new physics with dijet angular distributions in proton-proton collisions at $\sqrt{s} = 13$ TeV. JHEP **07**, 013 (2017). https://doi.org/10.1007/JHEP07(2017)013. arXiv:1703.09986

956. CMS Collaboration, Search for new physics in dijet angular distributions using proton-proton collisions at $\sqrt{s} = 13$ TeV and constraints on dark matter and other models. Eur. Phys. J. C **78**, 789 (2018). https://doi.org/10.1140/epjc/s10052-018-6242-x. arXiv:1803.08030
957. C. Cheung, A. Fitzpatrick, L. Randall, Sequestering CP violation and GIM-violation with warped extra dimensions. JHEP **01**, 069 (2008). https://doi.org/10.1088/1126-6708/2008/01/069. arXiv:0711.4421
958. A. Fitzpatrick, G. Perez, L. Randall, Flavor anarchy in a Randall-Sundrum model with 5D minimal flavor violation and a low Kaluza-Klein scale. Phys. Rev. Lett. **100**, 171604 (2008). https://doi.org/10.1103/PhysRevLett.100.171604. arXiv:0710.1869
959. N. Vignaroli, $\Delta F = 1$ constraints on composite Higgs models with LR parity. Phys. Rev. D **86**, 115011 (2012). https://doi.org/10.1103/PhysRevD.86.115011. arXiv:1204.0478
960. H. Georgi, L. Kaplan, D. Morin, A. Schenk, Effects of top compositeness. Phys. Rev. D **51**, 3888 (1995). https://doi.org/10.1103/PhysRevD.51.3888. arXiv:hep-ph/9410307
961. B. Lillie, J. Shu, T.M. Tait, Top compositeness at the Tevatron and LHC. JHEP **04**, 087 (2008). https://doi.org/10.1088/1126-6708/2008/04/087. arXiv:0712.3057
962. A. Pomarol, J. Serra, Top quark compositeness: Feasibility and implications. Phys. Rev. D **78**, 074026 (2008). https://doi.org/10.1103/PhysRevD.78.074026. arXiv:0806.3247
963. K. Kumar, T.M. Tait, R. Vega-Morales, Manifestations of top compositeness at colliders. JHEP **05**, 022 (2009). https://doi.org/10.1088/1126-6708/2009/05/022. arXiv:0901.3808
964. B. Moussallam, V. Soni, Production of heavy spin 3/2 fermions in colliders. Phys. Rev. D **39**, 1883 (1989). https://doi.org/10.1103/PhysRevD.39.1883
965. B. Hassanain, J. March-Russell, J. Rosa, On the possibility of light string resonances at the LHC and Tevatron from Randall-Sundrum throats. JHEP **07**, 077 (2009). https://doi.org/10.1088/1126-6708/2009/07/077. arXiv:0904.4108
966. W. Stirling, E. Vryonidou, Effect of spin-3/2 top quark excitation on $t\bar{t}$ production at the LHC. JHEP **01**, 055 (2012). https://doi.org/10.1007/JHEP01(2012)055. arXiv:1110.1565
967. D.A. Dicus, D. Karabacak, S. Nandi, S.K. Rai, Search for spin-3/2 quarks at the Large Hadron Collider. Phys. Rev. D **87**, 015023 (2013). https://doi.org/10.1103/PhysRevD.87.015023. arXiv:1208.5811
968. J. H. Kim, I.M. Lewis, Loop induced single top partner production and decay at the LHC. JHEP **05**, 095 (2018). https://doi.org/10.1007/JHEP05(2018)095. arXiv:1803.06351
969. H. Alhazmi, J.H. Kim, K. Kong, I.M. Lewis, Shedding light on top partner at the LHC. JHEP **01**, 139 (2019). https://doi.org/10.1007/JHEP01(2019)139. arXiv:1808.03649
970. U. Baur, I. Hinchliffe, D. Zeppenfeld, Excited quark production at hadron colliders. Int. J. Mod. Phys. A **2**, 1285 (1987). https://doi.org/10.1142/S0217751X87000661
971. ATLAS Collaboration, Search for new resonances in mass distributions of jet pairs using 139 fb^{-1} of pp collisions at $\sqrt{s} = 13$ TeV with the ATLAS detector. JHEP **03**, 145 (2020). https://doi.org/10.1007/JHEP03(2020)145. arXiv:1910.08447
972. CMS Collaboration, Search for high mass dijet resonances with a new background prediction method in proton-proton collisions at $\sqrt{s} = 13$ TeV. JHEP **05**, 033 (2020). https://doi.org/10.1007/JHEP05(2020)033. arXiv:1911.03947
973. J. Nutter, R. Schwienhorst, D.G. Walker, J.-H. Yu, Single top production as a probe of B' quarks. Phys. Rev. D **86**, 094006 (2012). https://doi.org/10.1103/PhysRevD.86.094006. arXiv:1207.5179
974. T.M. Tait, C.-P. Yuan, Single top quark production as a window to physics beyond the standard model. Phys. Rev. D **63**, 014018 (2000). https://doi.org/10.1103/PhysRevD.63.014018. arXiv:hep-ph/0007298
975. C.T. Hill, E.H. Simmons, Strong dynamics and electroweak symmetry breaking. Phys. Rept. **381**, 235 (2003). https://doi.org/10.1016/S0370-1573(03)00140-6. arXiv:hep-ph/0203079. Erratum: Phys. Rept. **390**, 553 (2004)
976. D. Choudhury, T.M. Tait, C. Wagner, Beautiful mirrors and precision electroweak data. Phys. Rev. D **65**, 053002 (2002). https://doi.org/10.1103/PhysRevD.65.053002. arXiv:hep-ph/0109097

977. K. Kumar, W. Shepherd, T.M. Tait, R. Vega-Morales, Beautiful mirrors at the LHC. JHEP **08**, 052 (2010). https://doi.org/10.1007/JHEP08(2010)052. arXiv:1004.4895
978. A. de Rújula, L. Maiani, R. Petronzio, Search for excited quarks. Phys. Lett. B **140**, 253 (1984). https://doi.org/10.1016/0370-2693(84)90930-4
979. ATLAS Collaboration, Search for single b^*-quark production with the ATLAS detector at $\sqrt{s} = 7$ TeV. Phys. Lett. B **721**, 171 (2013). https://doi.org/10.1016/j.physletb.2013.03.016. arXiv:1301.1583
980. CMS Collaboration, Search for the production of an excited bottom quark decaying to tW in proton-proton collisions at $\sqrt{s} = 8$ TeV. JHEP **01**, 166 (2016). https://doi.org/10.1007/JHEP01(2016)166. arXiv:1509.08141
981. CMS Collaboration, Search for a heavy resonance decaying to a top quark and a W boson at $\sqrt{s} = 13$ TeV in the fully hadronic final state. CMS Physics Analysis Summary.CMS-PAS-B2G-19-003 (2021), http://cds.cern.ch/record/2749405
982. C.M.S. Collaboration, Search for pair production of excited top quarks in the lepton+jets final state. JHEP **06**, 125 (2014). https://doi.org/10.1007/JHEP06(2014)125. arXiv:1311.5357
983. CMS Collaboration, Measurements of $t\bar{t}$ differential cross sections in proton-proton collisions at $\sqrt{s} = 13$ TeV using events containing two leptons. JHEP **02**, 149 (2019). https://doi.org/10.1007/JHEP02(2019)149. arXiv:1811.06625
984. ATLAS Collaboration, Measurement of jet activity produced in top-quark events with an electron, a muon and two b-tagged jets in the final state in pp collisions at $\sqrt{s} = 13$ TeV with the ATLAS detector. Eur. Phys. J. C **77**, 220 (2017). https://doi.org/10.1140/epjc/s10052-017-4766-0. arXiv:1610.09978
985. ATLAS Collaboration, Measurements of differential cross sections of top quark pair production in association with jets in pp collisions at $\sqrt{s} = 13$ TeV using the ATLAS detector.JHEP **10**, 159 (2018). https://doi.org/10.1007/JHEP10(2018)159 arXiv:1802.06572
986. CMS Collaboration, Search for pair production of excited top quarks in the lepton+jets final state. Phys. Lett. B **778**, 349 (2018). https://doi.org/10.1016/j.physletb.2018.01.049. arXiv:1711.10949
987. G. Bertone, D. Hooper, History of dark matter. Rev. Mod. Phys. **90**, 045002 (2018). https://doi.org/10.1103/RevModPhys.90.045002. arXiv:1605.04909
988. J. de Swart, G. Bertone, J. van Dongen, How dark matter came to matter. Nature Astron. **1**, 0059 (2017). https://doi.org/10.1038/s41550017-0059. arXiv:1703.00013
989. V. Trimble, Existence and nature of dark matter in the universe. Ann. Rev. Astron. Astrophys. **25**, 425 (1987). https://doi.org/10.1146/annurev.aa.25.090187.002233
990. R. Peccei, H.R. Quinn, Constraints imposed by CP conservation in the presence of instantons. Phys. Rev. D **16**, 1791 (1977). https://doi.org/10.1103/PhysRevD.16.1791
991. R. Peccei, H.R. Quinn, CP conservation in the presence of instantons. Phys. Rev. Lett. **38**, 1440 (1977). https://doi.org/10.1103/PhysRevLett.38.1440
992. L. Visinelli, P. Gondolo, Dark matter axions revisited. Phys. Rev. D **80**, 035024 (2009). https://doi.org/10.1103/PhysRevD.80.035024. arXiv:0903.4377
993. G. Bertone, D. Hooper, J. Silk, Particle dark matter: evidence, candidates and constraints. Phys. Rept. **405**, 279 (2005). https://doi.org/10.1016/j.physrep.2004.08.031. arXiv:hep-ph/0404175
994. S. Dodelson, L.M. Widrow, Sterile-neutrinos as dark matter. Phys. Rev. Lett. **72**, 17 (1994). https://doi.org/10.1103/PhysRevLett.72.17. arXiv:hep-ph/9303287
995. A. Boyarsky, O. Ruchayskiy, M. Shaposhnikov, The role of sterile neutrinos in cosmology and astrophysics. Ann. Rev. Nucl. Part. Sci. **59**, 191 (2009). https://doi.org/10.1146/annurev.nucl.010909.083654. arXiv:0901.0011
996. G. Jungman, M. Kamionkowski, K. Griest, Supersymmetric dark matter. Phys. Rept. **267**, 195 (1996). https://doi.org/10.1016/0370-1573(95)00058-5. arXiv:hep-ph/9506380
997. H.-C. Cheng, I. Low, TeV symmetry and the little hierarchy problem. JHEP **09**, 051 (2003). arXiv:hep-ph/0308199
998. K. Agashe, G. Servant, Warped unification, proton stability and dark matter. Phys. Rev. Lett. **93**, 231805 (2004). https://doi.org/10.1103/PhysRevLett.93.231805. arXiv:hep-ph/0403143

999. ATLAS Collaboration, Searching for dark matter with the ATLAS detector, ATLAS Feature, March 5, 2019, https://atlas.cern/updates/atlas-feature/dark-matter
1000. H.E.S.S. Collaborations, The status of the H.E.S.S. project. New Astron. Rev. **48**, 331 (2004). https://doi.org/10.1016/j.newar.2003.12.004. arXiv:astro-ph/0403052
1001. MAGIC Collaboration, Commissioning and first tests of the MAGIC telescope. Nucl. Instrum. Meth. A **518**, 188 (2004). https://doi.org/10.1016/j.nima.2003.10.057
1002. VERITAS Collaborations, VERITAS: the very energetic radiation imaging telescope array system. Astropart. Phys. **17**, 221 (2002). https://doi.org/10.1016/S0927-6505(01)00152-9. arXiv:astro-ph/0108478
1003. AMANDA Collaboration, The AMANDA neutrino telescope: principle of operation and first results. Astropart. Phys. **13**, 1 (2000). https://doi.org/10.1016/S0927-6505(99)00092-4. arXiv:astro-ph/9906203
1004. Super-Kamiokande Collaborations, The Super-Kamiokande detector. Nucl. Instrum. Meth. A **501**, 418 (2003). https://doi.org/10.1016/S0168-9002(03)00425-X
1005. IceCube Collaborations, First year performance of the IceCube neutrino telescope. Astropart. Phys. **26**, 155 (2006). https://doi.org/10.1016/j.astropartphys.2006.06.007. arXiv:astro-ph/0604450
1006. IceCube Collaborations, The design and performance of IceCube DeepCore. Astropart. Phys. **35**, 615 (2012). https://doi.org/10.1016/j.astropartphys.2012.01.004. arXiv:1109.6096
1007. AMS Collaboration, The Alpha Magnetic Spectrometer (AMS) on the International Space Station. I: Results from the test flight on the space shuttle. Phys. Rept. **366**, 331 (2002). https://doi.org/10.1016/S0370-1573(02)00013-3. Erratum: Phys. Rept. **380**, 97 (2003)
1008. PAMELA Collaboration, An anomalous positron abundance in cosmic rays with energies 1.5–100 GeV. Nature **458**, 607 (2009). https://doi.org/10.1038/nature07942. arXiv:0810.4995
1009. PAMELA Collaboration, Pamela results on the cosmic-ray antiproton flux from 60 mev to 180 gev in kinetic energy. Phys. Rev. Lett. **105** (September 2010). https://doi.org/10.1103/physrevlett.105.121101. arXiv:1007.0821
1010. AMS Collaboration, First result from the Alpha Magnetic Spectrometer on the International Space Station: precision measurement of the positron fraction in primary cosmic rays of 0.5–350 GeV. Phys. Rev. Lett. **110**, 141102 (2013). https://doi.org/10.1103/PhysRevLett.110.141102
1011. L. Bergström, Nonbaryonic dark matter: observational evidence and detection methods. Rept. Prog. Phys. **63**, 793 (2000). https://doi.org/10.1088/0034-4885/63/5/2r3. arXiv:hep-ph/0002126
1012. XENON Collaboration, The XENON1T dark matter experiment. Eur. Phys. J. C **77**, 881 (2017). https://doi.org/10.1140/epjc/s10052-017-5326-3. arXiv:1708.07051
1013. L.U.X. Collaboration, The Large Underground Xenon (LUX) experiment. Nucl. Instrum. Meth. A **704**, 111 (2013). https://doi.org/10.1016/j.nima.2012.11.135. arXiv:1211.3788
1014. PandaX Collaboration, Dark matter search results from the commissioning run of PandaX-II. Phys. Rev. D **93**, 122009 (2016). https://doi.org/10.1103/PhysRevD.93.122009. arXiv:1602.06563
1015. SuperCDMS Collaborations, Search for low-mass weakly interacting massive particles using voltage-assisted calorimetric ionization detection in the SuperCDMS experiment. Phys. Rev. Lett. **112**, 041302 (2014). https://doi.org/10.1103/PhysRevLett.112.041302. arXiv:1309.3259
1016. SuperCDMS Collaborations, Projected sensitivity of the SuperCDMS SNOLAB experiment. Phys. Rev. D **95**, 082002 (2017). https://doi.org/10.1103/PhysRevD.95.082002. arXiv:1610.00006
1017. Q.-H. Cao, C.-R. Chen, C.S. Li, H. Zhang, Effective dark matter model: Relic density, CDMS II. Fermi LAT and LHC. JHEP **08**, 018 (2011). https://doi.org/10.1007/JHEP08(2011)018. arXiv:0912.4511
1018. J. Goodman, M. Ibe, A. Rajaraman, W. Shepherd, T.M. Tait, H.-B. Yu, Constraints on light Majorana dark matter from colliders. Phys. Lett. B **695**, 185 (2011). https://doi.org/10.1016/j.physletb.2010.11.009. arXiv:1005.1286

1019. J. Goodman, M. Ibe, A. Rajaraman, W. Shepherd, T.M. Tait, H.-B. Yu, Constraints on dark matter from colliders. Phys. Rev. D **82**, 116010116010 (2010). https://doi.org/10.1103/PhysRevD.82.116010. arXiv:1008.1783
1020. P.J. Fox, R. Harnik, J. Kopp, Y. Tsai, Missing energy signatures of dark matter at the LHC. Phys. Rev. D **85**, 056011 (2012). https://doi.org/10.1103/PhysRevD.85.056011. arXiv:1109.4398
1021. G. Busoni, A. De Simone, E. Morgante, A. Riotto, On the validity of the effective field theory for dark matter searches at the LHC. Phys. Lett. B **728**, 412 (2014). https://doi.org/10.1016/j.physletb.2013.11.069. arXiv:1307.2253
1022. G. Busoni, A. De Simone, J. Gramling, E. Morgante, A. Riotto, On the validity of the effective field theory for dark matter searches at the LHC, part II: complete analysis for the s-channel. JCAP **06**, 060 (2014). https://doi.org/10.1088/1475-7516/2014/06/060. arXiv:1402.1275
1023. G. Busoni, A. De Simone, T. Jacques, E. Morgante, A. Riotto, On the validity of the effective field theory for dark matter searches at the LHC part III: analysis for the t-channel. JCAP **09**, 022 (2014). https://doi.org/10.1088/1475-7516/2014/09/022. arXiv:1405.3101
1024. O. Buchmueller, M.J. Dolan, C. McCabe, Beyond effective field theory for dark matter searches at the LHC. JHEP **01**, 025 (2014). https://doi.org/10.1007/JHEP01(2014)025. arXiv:1308.6799
1025. J. Abdallah et al., Simplified models for dark matter searches at the LHC. Phys. Dark Univ. **9**, 8 (2015). https://doi.org/10.1016/j.dark.2015.08.001. arXiv:1506.03116
1026. D. Abercrombie et al., Dark matter benchmark models for early LHC Run-2 searches: report of the ATLAS/CMS dark matter forum. Phys. Dark Univ. **27**, 100371 (2020). https://doi.org/10.1016/j.dark.2019.100371. arXiv:1507.00966
1027. LHC Dark Matter Working Group, LHC Dark Matter Working Group: Next-generation spin-0 dark matter models. Phys. Dark Univ. **27**, 100351 (2020). https://doi.org/10.1016/j.dark.2019.100351. arXiv:1810.09420
1028. G. Bertone, The moment of truth for WIMP dark matter. Nature **468**, 389 (2010). https://doi.org/10.1038/nature09509. arXiv:1011.3532
1029. ATLAS Collaboration, Search for new phenomena with the monojet and missing transverse momentum signature using the ATLAS detector in $\sqrt{s} = 7$ TeV proton-proton collisions. Phys. Lett. B **705**, 294 (2011). https://doi.org/10.1016/j.physletb.2011.10.006. arXiv:1106.5327
1030. ATLAS Collaboration, Search for dark matter candidates and large extra dimensions in events with a jet and missing transverse momentum with the ATLAS detector. JHEP **04**, 075 (2013). https://doi.org/10.1007/JHEP04(2013)075. arXiv:1210.4491
1031. C.M.S. Collaboration, Search for dark matter and large extra dimensions in monojet events in pp collisions at $\sqrt{s} = 7$ TeV. JHEP **09**, 094 (2012). https://doi.org/10.1007/JHEP09(2012)094. arXiv:1206.5663
1032. ATLAS Collaboration, Search for dark matter candidates and large extra dimensions in events with a photon and missing transverse momentum in pp collision data at $\sqrt{s} = 7$ TeV with the ATLAS detector. Phys. Rev. Lett. **110**, 011802 (2013). https://doi.org/10.1103/PhysRevLett.110.011802. arXiv:1209.4625
1033. C.M.S. Collaboration, Search for dark matter and large extra dimensions in pp collisions yielding a photon and missing transverse energy. Phys. Rev. Lett. **108**, 261803 (2012). https://doi.org/10.1103/PhysRevLett.108.261803. arXiv:1204.0821
1034. ATLAS Collaboration, Search for new phenomena in final states with an energetic jet and large missing transverse momentum in pp collisions at $\sqrt{s} = 8$ TeV with the ATLAS detector. Eur. Phys. J. C **75**, 299 (2015). https://doi.org/10.1140/epjc/s10052-015-3517-3. arXiv:1502.01518. Erratum: Eur. Phys. J. C **75**, 408 (2015)
1035. ATLAS Collaboration, Search for new phenomena in events with a photon and missing transverse momentum in pp collisions at $\sqrt{s} = 8$ TeV with the ATLAS detector. Phys. Rev. D **91**, 012008 (2015). https://doi.org/10.1103/PhysRevD.91.012008. arXiv:1411.1559. Erratum: Phys. Rev. D **92**, 059903 (2015)

1036. C.M.S. Collaboration, Search for dark matter, extra dimensions, and unparticles in monojet events in proton-proton collisions at $\sqrt{s} = 8$ TeV. Eur. Phys. J. C **75**, 235 (2015). https://doi.org/10.1140/epjc/s10052-015-3451-4. arXiv:1408.3583
1037. C.M.S. Collaboration, Search for new phenomena in monophoton final states in proton-proton collisions at $\sqrt{s} = 8$ TeV. Phys. Lett. B **755**, 102 (2016). https://doi.org/10.1016/j.physletb.2016.01.057. arXiv:1410.8812
1038. Y. Bai, T.M. Tait, Searches with mono-leptons. Phys. Lett. B **723**, 384 (2013). https://doi.org/10.1016/j.physletb.2013.05.057. arXiv:1208.4361
1039. ATLAS Collaboration, Search for dark matter in events with a Z boson and missing transverse momentum in *pp* collisions at $\sqrt{s} = 8$ TeV with the ATLAS detector. Phys. Rev. D **90**, 012004 (2014). https://doi.org/10.1103/PhysRevD.90.012004. arXiv:1404.0051
1040. CMS Collaboration, Search for dark matter and unparticles produced in association with a Z boson in proton-proton collisions at $\sqrt{s} = 8$ TeV. Phys. Rev. D **93**, 052011 (2016). https://doi.org/10.1103/PhysRevD.93.052011. arXiv:1511.09375. Erratum: Phys. Rev. D **97**, 099903 (2018)
1041. C.M.S. Collaboration, Search for physics beyond the standard model in final states with a lepton and missing transverse energy in proton-proton collisions at $\sqrt{s} = 8$ TeV. Phys. Rev. D **91**, 041802 (2015). https://doi.org/10.1103/PhysRevD.91.092005. arXiv:1408.2745
1042. ATLAS Collaboration, Search for dark matter in events with a hadronically decaying W or Z boson and missing transverse momentum in *pp* collisions at $\sqrt{s} = 8$ TeV with the ATLAS detector. Phys. Rev. Lett. **112**, 041802 (2014). https://doi.org/10.1103/PhysRevLett.112.041802. arXiv:1309.4017
1043. CMS Collaboration, Search for dark matter in proton-proton collisions at 8 TeV with missing transverse momentum and vector boson tagged jets. JHEP **12**, 083 (2016). https://doi.org/10.1007/JHEP12(2016)083. arXiv:1607.05764. Erratum: JHEP **08**, 035 (2017)
1044. L. Carpenter, A. DiFranzo, M. Mulhearn, C. Shimmin, S. Tulin, D. Whiteson, Mono-Higgs-boson: a new collider probe of dark matter. Phys. Rev. D **89**, 075017 (2014). https://doi.org/10.1103/PhysRevD.89.075017. arXiv:1312.2592
1045. A.A. Petrov, W. Shepherd, Searching for dark matter at LHC with mono-Higgs production. Phys. Lett. B **730**, 178 (2014). https://doi.org/10.1016/j.physletb.2014.01.051. arXiv:1311.1511
1046. ATLAS Collaboration, Search for invisible Higgs boson decays with vector boson fusion signatures with the ATLAS detector using an integrated luminosity of 139 fb^{-1}. ATLAS Conference Note. ATLAS-CONF-2020-008 (2020), http://cds.cern.ch/record/2715447
1047. ATLAS Collaboration, Search for dark matter produced in association with a Higgs boson decaying to two bottom quarks in *pp* collisions at $\sqrt{s} = 8$ TeV with the ATLAS detector. Phys. Rev. D **93**, 072007 (2016). https://doi.org/10.1103/PhysRevD.93.072007. arXiv:1510.06218
1048. T. Lin, E.W. Kolb, L.-T. Wang, Probing dark matter couplings to top and bottom quarks at the LHC. Phys. Rev. D **88**, 063510 (2013). https://doi.org/10.1103/PhysRevD.88.063510. arXiv:1303.6638
1049. J. Andrea, B. Fuks, F. Maltoni, Monotops at the LHC. Phys. Rev. D **84**, 074025 (2011). https://doi.org/10.1103/PhysRevD.84.074025. arXiv:1106.6199
1050. J.-L. Agram, J. Andrea, M. Buttignol, E. Conte, B. Fuks, Monotop phenomenology at the Large Hadron Collider. Phys. Rev. D **89**, 014028 (2014). https://doi.org/10.1103/PhysRevD.89.014028. arXiv:1311.6478
1051. G. Cacciapaglia, E. Conte, A. Deandrea, B. Fuks, H.-S. Shao, LHC constraints and potential on resonant monotop production. Eur. Phys. J. C **79**, 174 (2019). https://doi.org/10.1140/epjc/s10052-019-6675-x. arXiv:1811.03626
1052. ATLAS Collaboration, Search for dark matter in events with heavy quarks and missing transverse momentum in *pp* collisions with the ATLAS detector. Eur. Phys. J. C **75**, 92 (2015). https://doi.org/10.1140/epjc/s10052-015-3306-z. arXiv:1410.4031
1053. ATLAS Collaboration, Search for invisible particles produced in association with single-top-quarks in proton-proton collisions at $\sqrt{s} = 8$ TeV with the ATLAS detector. Eur. Phys. J. C **75**, 79 (2015). https://doi.org/10.1140/epjc/s10052-014-3233-4. arXiv:1410.5404

1054. C.M.S. Collaboration, Search for monotop signatures in proton-proton collisions at $\sqrt{s} = 8$ TeV. Phys. Rev. Lett. **114**, 101801 (2015). https://doi.org/10.1103/PhysRevLett.114.101801. arXiv:1410.1149
1055. ATLAS Collaboration, Search for new phenomena in events with a photon and missing transverse momentum in pp collisions at $\sqrt{s} = 13$ TeV with the ATLAS detector. JHEP **06**, 059 (2016). https://doi.org/10.1007/JHEP06(2016)059. arXiv:1604.01306
1056. ATLAS Collaboration, Search for dark matter in association with an energetic photon in pp collisions at $\sqrt{s} = 13$ TeV with the ATLAS detector. ATLAS Conference Note. ATLAS-CONF-2020-020 (2020), http://cds.cern.ch/record/2720250
1057. CMS Collaboration, Search for new physics in the monophoton final state in proton-proton collisions at $\sqrt{s} = 13$ TeV. JHEP **10**, 073 (2017). https://doi.org/10.1007/JHEP10(2017)073. arXiv:1706.03794
1058. ATLAS Collaboration, Search for new phenomena in final states with an energetic jet and large missing transverse momentum in pp collisions at $\sqrt{s} = 13$ TeV using the ATLAS detector. Phys. Rev. D **94**, 032005 (2016). https://doi.org/10.1103/PhysRevD.94.032005. arXiv:1604.07773
1059. ATLAS Collaboration, Search for new phenomena in events with jets and missing transverse momentum in pp collisions at $\sqrt{s} = 13$ TeV with the ATLAS detector. ATLAS Conference Note. ATLAS-CONF-2020-048 (2020), http://cds.cern.ch/record/2728058
1060. ATLAS Collaboration, Search for dark matter and other new phenomena in events with an energetic jet and large missing transverse momentum using the ATLAS detector. JHEP **01**, 126 (2018). https://doi.org/10.1007/JHEP01(2018)126. arXiv:1711.03301
1061. ATLAS Collaboration, Search for an invisibly decaying Higgs boson or dark matter candidates produced in association with a Z boson in pp collisions at $\sqrt{s} = 13$ TeV with the ATLAS detector. Phys. Lett. B **776**, 318 (2018). https://doi.org/10.1016/j.physletb.2017.11.049. arXiv:1708.09624
1062. CMS Collaboration, Search for dark matter and unparticles in events with a Z boson and missing transverse momentum in proton-proton collisions at $\sqrt{s} = 13$ TeV. JHEP **03**, 061 (2017). https://doi.org/10.1007/JHEP03(2017)061. arXiv:1701.02042. Erratum: JHEP **09**, 106 (2017)
1063. CMS Collaboration, Search for new physics in events with a leptonically decaying Z boson and a large transverse momentum imbalance in proton-proton collisions at $\sqrt{s} = 13$ TeV. Eur. Phys. J. C **78**, 291 (2018). https://doi.org/10.1140/epjc/s10052-018-5740-1. arXiv:1711.00431
1064. CMS Collaboration, Search for dark matter produced in association with a leptonically decaying Z boson in proton-proton collisions at $\sqrt{s} = 13$ TeV, arXiv:2008.04735
1065. CMS Collaboration, Search for dark matter produced with an energetic jet or a hadronically decaying W or Z boson at $\sqrt{s} = 13$ TeV. JHEP **07**, 014 (2017). https://doi.org/10.1007/JHEP07(2017)014. arXiv:1703.01651
1066. CMS Collaboration, Search for new physics in final states with an energetic jet or a hadronically decaying W or Z boson and transverse momentum imbalance at $\sqrt{s} = 13$ TeV. Phys. Rev. D **97**, 092005 (2018). https://doi.org/10.1103/PhysRevD.97.092005. arXiv:1712.02345
1067. ATLAS Collaboration, Search for dark matter produced in association with a hadronically decaying vector boson in pp collisions at $\sqrt{s} = 13$ TeV with the ATLAS detector. Phys. Lett. B **763**, 251 (2016). https://doi.org/10.1016/j.physletb.2016.10.042. arXiv:1608.02372
1068. ATLAS Collaboration, Search for dark matter in events with a hadronically decaying vector boson and missing transverse momentum in pp collisions at $\sqrt{s} = 13$ TeV with the ATLAS detector. JHEP **10**, 180 (2018). https://doi.org/10.1007/JHEP10(2018)180. arXiv:1807.11471
1069. ATLAS Collaboration, Search for dark matter in association with a Higgs boson decaying to two photons at $\sqrt{s} = 13$ TeV with the ATLAS detector. Phys. Rev. D **96**, 112004 (2017). https://doi.org/10.1103/PhysRevD.96.112004. arXiv:1706.03948
1070. CMS Collaboration, Search for associated production of dark matter with a Higgs boson decaying to $b\bar{b}$ or $\gamma\gamma$ at $\sqrt{s} = 13$ TeV. JHEP **10**, 180 (2017). https://doi.org/10.1007/JHEP10(2017)180. arXiv:1703.05236

1071. CMS Collaboration, Search for dark matter produced in association with a Higgs boson decaying to $\gamma\gamma$ or $\tau^+\tau^-$ at $\sqrt{s} = 13$ TeV. JHEP **09**, 046 (2018). https://doi.org/10.1007/JHEP09(2018)046. arXiv:1806.04771
1072. ATLAS Collaboration, Search for dark matter in association with a Higgs boson decaying to b-quarks in pp collisions at $\sqrt{s} = 13$ TeV with the ATLAS detector. Phys. Lett. B **765**, 11 (2017). https://doi.org/10.1016/j.physletb.2016.11.035. arXiv:1609.04572
1073. ATLAS Collaboration, Search for dark matter produced in association with a Higgs boson decaying to $b\bar{b}$ using 36 fb^{-1} of pp collisions at $\sqrt{s} = 13$ TeV with the ATLAS detector. Phys. Rev. Lett. **119**, 181804 (2017). https://doi.org/10.1103/PhysRevLett.119.181804. arXiv:1707.01302
1074. ATLAS Collaboration, Search for dark matter produced in association with a Higgs boson decaying to $b\bar{b}$ at $\sqrt{s} = 13$ TeV with the ATLAS detector using 79.8 fb^{-1} of proton-proton collision data. ATLAS Conference Note. ATLAS-CONF-2018-039 (2018), http://cds.cern.ch/record/2632344
1075. CMS Collaboration, Search for dark matter produced in association with a Higgs boson decaying to a pair of bottom quarks in proton-proton collisions at $\sqrt{s} = 13$ TeV. Eur. Phys. J. C **79**, 280 (2019). https://doi.org/10.1140/epjc/s10052-019-6730-7. arXiv:1811.06562
1076. M. Bauer, U. Haisch, F. Kahlhoefer, Simplified dark matter models with two Higgs doublets: I. Pseudoscalar mediators. JHEP **05**, 138 (2017). https://doi.org/10.1007/JHEP05(2017)138. arXiv:1701.07427
1077. CMS Collaboration, Search for dark matter particles produced in association with a Higgs boson in proton-proton collisions at $\sqrt{s} = 13$ TeV. JHEP **03**, 025 (2020). https://doi.org/10.1007/JHEP03(2020)025. arXiv:1908.01713
1078. ATLAS Collaboration, Search for dark matter produced in association with a dark Higgs boson decaying to $W^\pm W^\mp$ or ZZ in fully hadronic final states from $\sqrt{s} = 13$ TeV pp collisions recorded with the ATLAS detector, arXiv:2010.06548
1079. M. Duerr, A. Grohsjean, F. Kahlhoefer, B. Penning, K. Schmidt-Hoberg, C. Schwanenberger, Hunting the dark Higgs. JHEP **04**, 143 (2017). https://doi.org/10.1007/JHEP04(2017)143. arXiv:1701.08780
1080. M. Duerr, F. Kahlhoefer, K. Schmidt-Hoberg, T. Schwetz, S. Vogl, How to save the WIMP: global analysis of a dark matter model with two s-channel mediators. JHEP **09**, 042 (2016). https://doi.org/10.1007/JHEP09(2016)042. arXiv:1606.07609
1081. ATLAS Collaboration, Search for dark matter produced in association with bottom or top quarks in $\sqrt{s} = 13$ TeV pp collisions with the ATLAS detector. Eur. Phys. J. C **78**, 18 (2018). https://doi.org/10.1140/epjc/s10052-017-5486-1. arXiv:1710.11412
1082. ATLAS Collaboration, Search for new phenomena in events with two opposite-charge leptons, jets and missing transverse momentum in pp collisions at $\sqrt{s} = 13$ TeV with the ATLAS detector. ATLAS Conference Note. ATLAS-CONF-2020-046 (2020), http://cds.cern.ch/record/2728056
1083. CMS Collaboration, Search for dark matter produced in association with heavy-flavor quark pairs in proton-proton collisions at $\sqrt{s} = 13$ TeV. Eur. Phys. J. C **77**, 845 (2017). https://doi.org/10.1140/epjc/s10052-017-5317-4. arXiv:1706.02581
1084. CMS Collaboration, Search for dark matter particles produced in association with a top quark pair at $\sqrt{s} = 13$ TeV. Phys. Rev. Lett. **122**, 011803 (2019). https://doi.org/10.1103/PhysRevLett.122.011803. arXiv:1807.06522
1085. CMS Collaboration, Search for dark matter in events with energetic, hadronically decaying top quarks and missing transverse momentum at $\sqrt{s} = 13$ TeV. JHEP **06**, 027 (2018). https://doi.org/10.1007/JHEP06(2018)027. arXiv:1801.08427
1086. I. Boucheneb, G. Cacciapaglia, A. Deandrea, B. Fuks, Revisiting monotop production at the LHC. JHEP **01**, 017 (2015). https://doi.org/10.1007/JHEP01(2015)017. arXiv:1407.7529
1087. D. Pinna, A. Zucchetta, M. R. Buckley, F. Canelli, Single top quarks and dark matter. Phys. Rev. D **96**, 035031 (2017). https://doi.org/10.1103/PhysRevD.96.035031. arXiv:1701.05195
1088. P. Pani, G. Polesello, Dark matter production in association with a single top-quark at the LHC in a two-Higgs-doublet model with a pseudoscalar mediator. Phys. Dark Univ. **21**, 8 (2018). https://doi.org/10.1016/j.dark.2018.04.006. arXiv:1712.03874

1089. T. Plehn, J. Thompson, S. Westhoff, Dark matter from electroweak single top production. Phys. Rev. D **98**, 015012 (2018). https://doi.org/10.1103/PhysRevD.98.015012. arXiv:1712.08065
1090. ATLAS Collaboration, Search for dark matter produced in association with a single top quark in $\sqrt{s} = 13$ TeV pp collisions with the ATLAS detector, arXiv:2011.09308
1091. CMS Collaboration, Search for dark matter produced in association with a single top quark or a top quark pair in proton-proton collisions at $\sqrt{s} = 13$ TeV. JHEP **03**, 141 (2019). https://doi.org/10.1007/JHEP03(2019)141. arXiv:1901.01553
1092. ATLAS Collaboration, Constraints on mediator-based dark matter and scalar dark energy models using $\sqrt{s} = 13$ TeV pp collision data collected by the ATLAS detector. JHEP **05**, 142 (2019). https://doi.org/10.1007/JHEP05(2019)142. arXiv:1903.01400
1093. G. Busoni et al., Recommendations on presenting LHC searches for missing transverse energy signals using simplified s-channel models of dark matter. Phys. Dark Univ. **27**, 100365 (2020). https://doi.org/10.1016/j.dark.2019.100365. arXiv:1603.04156
1094. L. Roszkowski, E.M. Sessolo, S. Trojanowski, WIMP dark matter candidates and searches—current status and future prospects. Rept. Prog. Phys. **81**, 066201 (2018). https://doi.org/10.1088/1361-6633/aab913. arXiv:1707.06277
1095. S. Kraml, U. Laa, K. Mawatari, K. Yamashita, Simplified dark matter models with a spin-2 mediator at the LHC. Eur. Phys. J. C **77**, 326 (2017). https://doi.org/10.1140/epjc/s10052-017-4871-0. arXiv:1701.07008
1096. S. Banerjee, D. Barducci, G. Bélanger, B. Fuks, A. Goudelis, B. Zaldivar, Cornering pseudoscalar-mediated dark matter with the LHC and cosmology. JHEP **07**, 080 (2017). https://doi.org/10.1007/JHEP07(2017)080. arXiv:1705.02327
1097. E.H. Simmons, Coloron phenomenology. Phys. Rev. D **55**, 1678 (1997). https://doi.org/10.1103/PhysRevD.55.1678. arXiv:hep-ph/9608269
1098. S. Cullen, M. Perelstein, M.E. Peskin, TeV strings and collider probes of large extra dimensions. Phys. Rev. D **62**, 055012 (2000). https://doi.org/10.1103/PhysRevD.62.055012. arXiv:hep-ph/0001166
1099. L.A. Anchordoqui, H. Goldberg, D. Lust, S. Nawata, S. Stieberger, T.R. Taylor, Dijet signals for low mass strings at the LHC. Phys. Rev. Lett. **101**, 241803 (2008). https://doi.org/10.1103/PhysRevLett.101.241803. arXiv:0808.0497
1100. T. Han, I. Lewis, Z. Liu, Colored resonant signals at the LHC: largest rate and simplest topology. JHEP **12**, 085 (2010). https://doi.org/10.1007/JHEP12(2010)085. arXiv:1010.4309
1101. R. Chivukula, A. Farzinnia, E.H. Simmons, R. Foadi, Production of massive color-octet vector bosons at next-to-leading order. Phys. Rev. D **85**, 054005 (2012). https://doi.org/10.1103/PhysRevD.85.054005. arXiv:1111.7261
1102. U. Baur, M. Spira, P. Zerwas, Excited quark and lepton production at hadron colliders. Phys. Rev. D **42**, 815 (1990). https://doi.org/10.1103/PhysRevD.42.815
1103. UA1 Collaboration, Two jet mass distributions at the CERN proton-anti-proton collider. Phys. Lett. B **209**, 127 (1988). https://doi.org/10.1016/0370-2693(88)91843-6
1104. UA2 Collaboration, A measurement of two jet decays of the W and Z bosons at the CERN $\bar{p}p$ collider. Z. Phys. C **49**, 17 (1991). https://doi.org/10.1007/BF01570793
1105. UA2 Collaboration, A search for new intermediate vector mesons and excited quarks decaying to two jets at the CERN $\bar{p}p$ collider. Nucl. Phys. B **400**, 3 (1993). https://doi.org/10.1016/0550-3213(93)90395-6
1106. CDF Collaboration, The two jet invariant mass distribution at $\sqrt{s} = 1.8$ TeV. Phys. Rev. D **41**, 1722 (1990). https://doi.org/10.1103/PhysRevD.41.1722
1107. C.D.F. Collaboration, Search for quark compositeness, axigluons and heavy particles using the dijet invariant mass spectrum observed in $p\bar{p}$ collisions. Phys. Rev. Lett. **71**, 2542 (1993). https://doi.org/10.1103/PhysRevLett.71.2542
1108. CDF Collaboration, Search for new particles decaying to dijets in $p\bar{p}$ collisions at $\sqrt{s} = 1.8$ TeV. Phys. Rev. Lett. **74**, 3538 (1995). https://doi.org/10.1103/PhysRevLett.74.3538. arXiv:hep-ex/9501001

1109. C.D.F. Collaboration, Search for new particles decaying to dijets at CDF. Phys. Rev. D **55**, 5263 (1997). https://doi.org/10.1103/PhysRevD.55.R5263. arXiv:hep-ex/9702004
1110. CDF Collaboration, Search for new particles decaying into dijets in proton-antiproton collisions at $\sqrt{s} = 1.96$ TeV. Phys. Rev. D **79**, 112002 (2009). https://doi.org/10.1103/PhysRevD.79.112002. arXiv:0812.4036
1111. D0 Collaboration, Measurement of dijet angular distributions and search for quark compositeness. Phys. Rev. Lett. **80**, 666 (1998). https://doi.org/10.1103/PhysRevLett.80.666. arXiv:hep-ex/9707016
1112. D0 Collaboration, The dijet mass spectrum and a search for quark compositeness in $\bar{p}p$ collisions at $\sqrt{s} = 1.8$ TeV. Phys. Rev. Lett. **82**, 2457 (1999). https://doi.org/10.1103/PhysRevLett.82.2457. arXiv:hep-ex/9807014
1113. D0 Collaboration, Search for new particles in the two jet decay channel with the D0 detector. Phys. Rev. D **69**, 111101 (2004). https://doi.org/10.1103/PhysRevD.69.111101. arXiv:hep-ex/0308033
1114. ATLAS Collaboration, Search for new particles in two-jet final states in 7 TeV proton-proton collisions with the ATLAS detector at the LHC. Phys. Rev. Lett. **105**, 161801 (2010). https://doi.org/10.1103/PhysRevLett.105.161801. arXiv:1008.2461
1115. ATLAS Collaboration, Search for quark contact interactions in dijet angular distributions in pp collisions at $\sqrt{s} = 7$ TeV measured with the ATLAS detector. Phys. Lett. B **694**, 327 (2011). https://doi.org/10.1016/j.physletb.2010.10.021. arXiv:1009.5069
1116. ATLAS Collaboration, Search for new physics in the dijet mass distribution using 1 fb^{-1} of pp collision data at $\sqrt{s} = 7$ TeV collected by the ATLAS detector. Phys. Lett. B **708**, 37 (2012). https://doi.org/10.1016/j.physletb.2012.01.035. arXiv:1108.6311
1117. C.M.S. Collaboration, Search for dijet resonances in 7 TeV pp collisions at CMS. Phys. Rev. Lett. **105**, 111101 (2010). https://doi.org/10.1103/PhysRevLett.105.211801. arXiv:1010.0203
1118. C.M.S. Collaboration, Search for quark compositeness with the dijet centrality ratio in pp collisions at $\sqrt{s} = 7$ TeV. Phys. Rev. Lett. **105**, 262001 (2010). https://doi.org/10.1103/PhysRevLett.105.262001. arXiv:1010.4439
1119. C.M.S. Collaboration, Search for resonances in the dijet mass spectrum from 7 TeV pp collisions at CMS. Phys. Lett. B **704**, 123 (2011). https://doi.org/10.1016/j.physletb.2011.09.015. arXiv:1107.4771
1120. C.M.S. Collaboration, Search for narrow resonances and quantum black holes in inclusive and b-tagged dijet mass spectra from pp collisions at $\sqrt{s} = 7$ TeV. JHEP **01**, 013 (2013). https://doi.org/10.1007/JHEP01(2013)013. arXiv:1210.2387
1121. ATLAS Collaboration, Search for new phenomena in the dijet mass distribution using pp collision data at $\sqrt{s} = 8$ TeV with the ATLAS detector. Phys. Rev. D **91**, 052007 (2015). https://doi.org/10.1103/PhysRevD.91.052007. arXiv:1407.1376
1122. C.M.S. Collaboration, Search for narrow resonances using the dijet mass spectrum in pp collisions at $\sqrt{s} = 8$ TeV. Phys. Rev. D **87**, 114015 (2013). https://doi.org/10.1103/PhysRevD.87.114015. arXiv:1302.4794
1123. CMS Collaboration, Search for resonances and quantum black holes using dijet mass spectra in proton-proton collisions at $\sqrt{s} = 8$ TeV. Phys. Rev. D **91**, 052009 (2015). https://doi.org/10.1103/PhysRevD.91.052009. arXiv:1501.04198
1124. CMS Collaboration, Search for narrow resonances in the b-tagged dijet mass spectrum in proton-proton collisions at $\sqrt{s} = 8$ TeV. Phys. Rev. Lett. **120**, 201801 (2018). https://doi.org/10.1103/PhysRevLett.120.201801. arXiv:1802.06149
1125. ATLAS Collaboration, Search for new phenomena in dijet mass and angular distributions from pp collisions at $\sqrt{s} = 13$ TeV with the ATLAS detector. Phys. Lett. B **754**, 302 (2016). https://doi.org/10.1016/j.physletb.2016.01.032. arXiv:1512.01530
1126. ATLAS Collaboration, Search for resonances in the mass distribution of jet pairs with one or two jets identified as b-jets in proton-proton collisions at $\sqrt{s} = 13$ TeV with the ATLAS detector. Phys. Lett. B **759**, 229 (2016). https://doi.org/10.1016/j.physletb.2016.05.064. arXiv:1603.08791

1127. CMS Collaboration, Search for narrow resonances decaying to dijets in proton-proton collisions at $(\sqrt{s}) = 13$ TeV. Phys. Rev. Lett. **116**, 071801 (2016). https://doi.org/10.1103/PhysRevLett.116.071801. arXiv:1512.01224
1128. CMS Collaboration, Search for dijet resonances in proton-proton collisions at $\sqrt{s} = 13$ TeV and constraints on dark matter and other models. Phys. Lett. B **769**, 520 (2017). https://doi.org/10.1016/j.physletb.2017.02.012. arXiv:1611.03568. Erratum: Phys. Lett. B **772**, 882 (2017)
1129. CMS Collaboration, Search for narrow and broad dijet resonances in proton-proton collisions at $\sqrt{s} = 13$ TeV and constraints on dark matter mediators and other new particles. JHEP **08**, 130 (2018). https://doi.org/10.1007/JHEP08(2018)130. arXiv:1806.00843
1130. CMS Collaboration, Data parking and data scouting at the CMS experiment. CMS Detector Performance Summary. CMS-DP-2012-022 (September 2012), https://cds.cern.ch/record/1480607
1131. R. Aaij et al., Tesla: an application for real-time data analysis in high energy physics. Comput. Phys. Commun. **208**, 35 (2016). https://doi.org/10.1016/j.cpc.2016.07.022. arXiv:1604.05596
1132. CMS Collaboration, Search for narrow resonances in dijet final states at $\sqrt{s} = 8$ TeV with the novel CMS technique of data scouting. Phys. Rev. Lett. **117**, 031802 (2016). https://doi.org/10.1103/PhysRevLett.117.031802. arXiv:1604.08907
1133. ATLAS Collaboration, Search for low-mass dijet resonances using trigger-level jets with the ATLAS detector in pp collisions at $\sqrt{s} = 13$ TeV. Phys. Rev. Lett. **121**, 081801 (2018). https://doi.org/10.1103/PhysRevLett.121.081801. arXiv:1804.03496
1134. CMS Collaboration, Search for dijet resonances using events with three jets in proton-proton collisions at $\sqrt{s} = 13$ TeV. Phys. Lett. B **805**, 135448 (2020). https://doi.org/10.1016/j.physletb.2020.135448. arXiv:1911.03761
1135. ATLAS Collaboration, Search for low-mass resonances decaying into two jets and produced in association with a photon using pp collisions at $\sqrt{s} = 13$ TeV with the ATLAS detector. Phys. Lett. B **795**, 56 (2019). https://doi.org/10.1016/j.physletb.2019.03.067. arXiv:1901.10917
1136. E. Eichten, K.D. Lane, J. Womersley, Finding low scale technicolor at hadron colliders. Phys. Lett. B **405**, 305 (1997). https://doi.org/10.1016/S0370-2693(97)00637-0. arXiv:hep-ph/9704455
1137. G. Altarelli, B. Mele, M. Ruiz-Altaba, Searching for new heavy vector bosons in $p\bar{p}$ colliders. Z. Phys. C **45**, 109 (1989). https://doi.org/10.1007/BF01556677. Erratum: Z. Phys. C **47**, 676 (1990)
1138. ATLAS Collaboration, Search for dijet resonances in events with an isolated charged lepton using $\sqrt{s} = 13$ TeV proton-proton collision data collected by the ATLAS detector. JHEP **06**, 151 (2020). https://doi.org/10.1007/JHEP06(2020)151. arXiv:2002.11325
1139. H. An, R. Huo, L.-T. Wang, Searching for low mass dark portal at the LHC. Phys. Dark Univ. **2**, 50 (2013). https://doi.org/10.1016/j.dark.2013.03.002. arXiv:1212.2221
1140. C. Shimmin, D. Whiteson, Boosting low-mass hadronic resonances. Phys. Rev. D **94**, 055001 (2016). https://doi.org/10.1103/PhysRevD.94.055001. arXiv:1602.07727
1141. CMS Collaboration, Search for low mass vector resonances decaying to quark-antiquark pairs in proton-proton collisions at $\sqrt{s} = 13$ TeV. Phys. Rev. Lett. **119**, 111802 (2017). https://doi.org/10.1103/PhysRevLett.119.111802. arXiv:1705.10532
1142. CMS Collaboration, Search for low mass vector resonances decaying into quark-antiquark pairs in proton-proton collisions at $\sqrt{s} = 13$ TeV. Phys. Rev. D **100**, 112007 (2019). https://doi.org/10.1103/PhysRevD.100.112007. arXiv:1909.04114
1143. CMS Collaboration, Search for low-mass quark-antiquark resonances produced in association with a photon at $\sqrt{s} = 13$ TeV. Phys. Rev. Lett. **123**, 231803 (2019). https://doi.org/10.1103/PhysRevLett.123.231803. arXiv:1905.10331
1144. ATLAS Collaboration, Search for light resonances decaying to boosted quark pairs and produced in association with a photon or a jet in proton-proton collisions at $\sqrt{s} = 13$ TeV with the ATLAS detector. Phys. Lett. B **788**, 316 (2019). https://doi.org/10.1016/j.physletb.2018.09.062. arXiv:1801.08769

1145. B.A. Dobrescu, F. Yu, Coupling-mass mapping of dijet peak searches. Phys. Rev. D **88**, 035021 (2013). https://doi.org/10.1103/PhysRevD.88.035021. arXiv:1306.2629. Erratum: Phys. Rev. D **90**, 079901 (2014)
1146. B.A. Dobrescu, C. Frugiuele, Hidden GeV-scale interactions of quarks. Phys. Rev. Lett. **113**, 061801 (2014). https://doi.org/10.1103/PhysRevLett.113.061801. arXiv:1404.3947
1147. CMS Collaboration, CMS Exotica summary plots for 13 TeV data. Results prepared for LHCP2020 (2020), https://twiki.cern.ch/twiki/bin/view/CMSPublic/SummaryPlotsEXO13TeV
1148. D. Lopez-Fogliani, C. Munoz, Proposal for a supersymmetric standard model. Phys. Rev. Lett. **97**, 041801 (2006). https://doi.org/10.1103/PhysRevLett.97.041801. arXiv:hep-ph/0508297
1149. S.P. Martin, A supersymmetry primer. Adv. Ser. Direct. High Energy Phys. **21**, 1 (2010). https://doi.org/10.1142/9789812839657_0001. arXiv:hep-ph/9709356
1150. H.E. Haber, L. Stephenson Haskins, Supersymmetric Theory and Models, in *Theoretical Advanced Study Institute in Elementary Particle Physics: Anticipating the Next Discoveries in Particle Physics* (WSP, 2018), p. 355. https://doi.org/10.1142/9789813233348_0006. arXiv:1712.05926
1151. Particle Data Group, B.C. Allanach, H.E. Haber, Supersymmetry part I (Theory), in *Review of Particle Physics*. Prog. Theo. Exp. Phys. 083C01 (2020). https://doi.org/10.1093/ptep/ptaa104DOI
1152. A. Canepa, Searches for supersymmetry at the Large Hadron Collider. Rev. Phys. **4**, 100033 (2019). https://doi.org/10.1016/j.revip.2019.100033
1153. J. Alwall, M.-P. Le, M. Lisanti, J.G. Wacker, Searching for directly decaying gluinos at the Tevatron. Phys. Lett. B **666**, 34 (2008). https://doi.org/10.1016/j.physletb.2008.06.065. arXiv:0803.0019
1154. J. Alwall, P. Schuster, N. Toro, Simplified models for a first characterization of new physics at the LHC. Phys. Rev. D **79**, 075020 (2009). https://doi.org/10.1103/PhysRevD.79.075020. arXiv:0810.3921
1155. LHC New Physics Working Group, D. Alves, Simplified models for LHC new physics searches. J. Phys. G **39**, 105005 (2012). https://doi.org/10.1088/0954-3899/39/10/105005. arXiv:1105.2838
1156. ATLAS Collaboration, Search for new phenomena in final states with large jet multiplicities and missing transverse momentum using $\sqrt{s} = 13$ TeV proton-proton collisions recorded by ATLAS in Run 2 of the LHC. JHEP **10**, 062 (2020). https://doi.org/10.1007/JHEP10(2020)062. arXiv:2008.06032
1157. S.P. Martin, Compressed supersymmetry and natural neutralino dark matter from top squark-mediated annihilation to top quarks. Phys. Rev. D **75**, 115005 (2007). https://doi.org/10.1103/PhysRevD.75.115005. arXiv:hep-ph/0703097
1158. S.P. Martin, The Top squark-mediated annihilation scenario and direct detection of dark matter in compressed supersymmetry. Phys. Rev. D **76**, 095005 (2007). https://doi.org/10.1103/PhysRevD.76.095005. arXiv:0707.2812
1159. M. Carena, A. Freitas, C. Wagner, Light stop searches at the LHC in events with one hard photon or jet and missing energy. JHEP **10**, 109 (2008). https://doi.org/10.1088/1126-6708/2008/10/109. arXiv:0808.2298
1160. ATLAS Collaboration, Search for new phenomena with top quark pairs in final states with one lepton, jets, and missing transverse momentum in *pp* collisions at $\sqrt{s} = 13$ TeV with the ATLAS detector, arXiv:2012.03799
1161. CMS Collaboration, Inclusive search for supersymmetry in pp collisions at $\sqrt{s} = 13$ TeV using razor variables and boosted object identification in zero and one lepton final states. JHEP **03**, 031 (2019). https://doi.org/10.1007/JHEP03(2019)031. arXiv:1812.06302
1162. C. Rogan, Kinematical variables towards new dynamics at the LHC, arXiv:1006.2727
1163. C.M.S. Collaboration, Inclusive search for supersymmetry using the razor variables in *pp* collisions at $\sqrt{s} = 7$ TeV. Phys. Rev. Lett. **111**, 081802 (2013). https://doi.org/10.1103/PhysRevLett.111.081802. arXiv:1212.6961

1164. CMS Collaboration, Search for direct top squark pair production in events with one lepton, jets, and missing transverse momentum at 13 TeV with the CMS experiment. JHEP **05**, 032 (2020). https://doi.org/10.1007/JHEP05(2020)032. arXiv:1912.08887
1165. ATLAS Collaboration, Search for a scalar partner of the top quark in the all-hadronic $t\bar{t}$ plus missing transverse momentum final state at $\sqrt{s} = 13$ TeV with the ATLAS detector. Eur. Phys. J. C **80**, 737 (2020). https://doi.org/10.1140/epjc/s10052-020-8102-8. arXiv:2004.14060
1166. CMS Collaboration, Search for supersymmetry in pp collisions at $\sqrt{s} = 8$ TeV in final states with boosted W bosons and b jets using razor variables. Phys. Rev. D **93**, 092009 (2016). https://doi.org/10.1103/PhysRevD.93.092009. arXiv:1602.02917
1167. CMS Collaboration, Search for supersymmetry in the all-hadronic final state using top quark tagging in pp collisions at $\sqrt{s} = 13$ TeV. Phys. Rev. D **96**, 012004 (2017). https://doi.org/10.1103/PhysRevD.96.012004. arXiv:1701.01954
1168. CMS Collaboration, Searches for third-generation squark production in fully hadronic final states in proton-proton collisions at $\sqrt{s} = 8$ TeV. JHEP **06**, 116 (2015). https://doi.org/10.1007/JHEP06(2015)116. arXiv:1503.08037
1169. CMS Collaboration, Search for supersymmetry in multijet events with missing transverse momentum in proton-proton collisions at 13 TeV. Phys. Rev. D **96**, 032003 (2017). https://doi.org/10.1103/PhysRevD.96.032003. arXiv:1704.07781
1170. CMS Collaboration, Search for new phenomena with the M_{T2} variable in the all-hadronic final state produced in proton-proton collisions at $\sqrt{s} = 13$ TeV. Eur. Phys. J. C **77**, 710 (2017). https://doi.org/10.1140/epjc/s10052-017-5267-x. arXiv:1705.04650
1171. M. Dine, A.E. Nelson, Dynamical supersymmetry breaking at low-energies. Phys. Rev. D **48**, 1277 (1993). https://doi.org/10.1103/PhysRevD.48.1277. arXiv:hep-ph/9303230
1172. M. Dine, A.E. Nelson, Y. Shirman, Low-energy dynamical supersymmetry breaking simplified. Phys. Rev. D **51**, 1362 (1995). https://doi.org/10.1103/PhysRevD.51.1362. arXiv:hep-ph/9408384
1173. M. Dine, A.E. Nelson, Y. Nir, Y. Shirman, New tools for low-energy dynamical supersymmetry breaking. Phys. Rev. D **53**, 2658 (1996). https://doi.org/10.1103/PhysRevD.53.2658. arXiv:hep-ph/9507378
1174. H.K. Dreiner, An introduction to explicit R-parity violation. Adv. Ser. Direct. High Energy Phys. **21**, 565 (2010). https://doi.org/10.1142/9789814307505_0017. arXiv:hep-ph/9707435
1175. R. Barbier et al., R-parity violating supersymmetry. Phys. Rept. **420**, 1 (2005). https://doi.org/10.1016/j.physrep.2005.08.006. arXiv:hep-ph/0406039
1176. A. Hook, E. Izaguirre, M. Lisanti, J.G. Wacker, High multiplicity searches at the LHC using jet masses. Phys. Rev. D **85**, 055029 (2012). https://doi.org/10.1103/PhysRevD.85.055029. arXiv:1202.0558
1177. S. El Hedri, A. Hook, M. Jankowiak, J.G. Wacker, Learning how to count: a high multiplicity search for the LHC. JHEP **08**, 136 (2013). https://doi.org/10.1007/JHEP08(2013)136. arXiv:1302.1870
1178. T. Cohen, E. Izaguirre, M. Lisanti, H.K. Lou, Jet substructure by accident. JHEP **03**, 161 (2013). https://doi.org/10.1007/JHEP03(2013)161. arXiv:1212.1456
1179. ATLAS Collaboration, Search for R-parity-violating supersymmetric particles in multi-jet final states produced in pp collisions at $\sqrt{s} = 13$ TeV using the ATLAS detector at the LHC. Phys. Lett. B **785**, 136 (2018). https://doi.org/10.1016/j.physletb.2018.08.021. arXiv:1804.03568
1180. ATLAS Collaboration, Search for massive supersymmetric particles decaying to many jets using the ATLAS detector in pp collisions at $\sqrt{s} = 8$ TeV. Phys. Rev. D **91**, 112016 (2015). https://doi.org/10.1103/PhysRevD.91.112016. arXiv:1502.05686. Erratum: Phys. Rev. D **93**, 039901 (2016)
1181. CMS Collaboration, Search for supersymmetry in pp collisions at $\sqrt{s} = 13$ TeV with 137 fb^{-1} in final states with a single lepton using the sum of masses of large-radius jets. Phys. Rev. D **101**, 052010 (2020). https://doi.org/10.1103/PhysRevD.101.052010. arXiv:1911.07558

1182. H. Baer, V. Barger, P. Huang, D. Mickelson, M. Padeffke-Kirkland, X. Tata, Natural SUSY with a bino- or wino-like LSP. Phys. Rev. D **91**, 075005 (2015). https://doi.org/10.1103/PhysRevD.91.075005. arXiv:1501.06357

1183. CMS Collaboration, Search for physics beyond the standard model in events with high-momentum Higgs bosons and missing transverse momentum in proton-proton collisions at 13 TeV. Phys. Rev. Lett. **120**, 241801 (2018). https://doi.org/10.1103/PhysRevLett.120.241801. arXiv:1712.08501

1184. CMS Collaboration, Search for supersymmetry in proton-proton collisions at $\sqrt{s} = 13$ TeV in events with high-momentum Z bosons and missing transverse momentum. JHEP **09**, 149 (2020). https://doi.org/10.1007/JHEP09(2020)149. arXiv:2008.04422

1185. J. Fan, M. Reece, J.T. Ruderman, Stealth supersymmetry. JHEP **11**, 012 (2011). https://doi.org/10.1007/JHEP11(2011)012. arXiv:1105.5135

1186. J. Fan, M. Reece, J.T. Ruderman, A stealth supersymmetry sampler. JHEP **07**, 196 (2012). https://doi.org/10.1007/JHEP07(2012)196. arXiv:1201.4875

1187. J. Fan, R. Krall, D. Pinner, M. Reece, J.T. Ruderman, Stealth supersymmetry simplified.JHEP **07**, 016 (2016). https://doi.org/10.1007/JHEP07(2016)016. arXiv:1512.05781

1188. CMS Collaboration, Search for physics beyond the standard model in events with overlapping photons and jets. Phys. Rev. Lett. **123**, 241801 (2019). https://doi.org/10.1103/PhysRevLett.123.241801. arXiv:1907.06275

1189. P. Fayet, Supergauge invariant extension of the Higgs mechanism and a model for the electron and its neutrino. Nucl. Phys. B **90**, 104 (1975). https://doi.org/10.1016/0550-3213(75)90636-7

1190. P. Fayet, Spontaneously broken supersymmetric theories of weak, electromagnetic and strong interactions. Phys. Lett. B **69**, 489 (1977). https://doi.org/10.1016/0370-2693(77)90852-8

1191. B.A. Dobrescu, K.T. Matchev, Light axion within the next-to-minimal supersymmetric standard model. JHEP **09**, 031 (2000). https://doi.org/10.1088/1126-6708/2000/09/031. arXiv:hep-ph/0008192

1192. U. Ellwanger, J.F. Gunion, C. Hugonie, S. Moretti, Towards a no lose theorem for NMSSM Higgs discovery at the LHC, arXiv:hep-ph/0305109

1193. R. Dermisek, J.F. Gunion, Escaping the large fine tuning and little hierarchy problems in the next to minimal supersymmetric model and $h \to aa$ decays. Phys. Rev. Lett. **95**, 04180 (2005). https://doi.org/10.1103/PhysRevLett.95.041801. arXiv:hep-ph/0502105

1194. D.E. Morrissey, A. Pierce, Modified Higgs boson phenomenology from gauge or gaugino mediation in the NMSSM. Phys. Rev. D **78**, 075029 (2008). https://doi.org/10.1103/PhysRevD.78.075029. arXiv:0807.2259

1195. CMS Collaboration, Search for invisible decays of a Higgs boson produced through vector boson fusion in proton-proton collisions at $\sqrt{s} = 13$ TeV. Phys. Lett. B **793**, 520 (2019). https://doi.org/10.1016/j.physletb.2019.04.025. arXiv:1809.05937

1196. ATLAS Collaboration, Combination of searches for invisible Higgs boson decays with the ATLAS experiment. Phys. Rev. Lett. **122**, 231801 (2019). https://doi.org/10.1103/PhysRevLett.122.231801. arXiv:1904.05105

1197. D. Curtin et al., Exotic decays of the 125 GeV Higgs boson. Phys. Rev. D **90**, 075004 (2014). https://doi.org/10.1103/PhysRevD.90.075004. arXiv:1312.4992

1198. A. Martin, Higgs cascade decays to $\gamma\gamma$ + jet jet at the LHC, arXiv:hep-ph/0703247

1199. CMS Collaboration, Search for a very light NMSSM Higgs boson produced in decays of the 125 GeV scalar boson and decaying into τ leptons in pp collisions at $\sqrt{s} = 8$ TeV. JHEP **01**, 079 (2016). https://doi.org/10.1007/JHEP01(2016)079. arXiv:1510.06534

1200. CMS Collaboration, Search for light bosons in decays of the 125 GeV Higgs boson in proton-proton collisions at $\sqrt{s} = 8$ TeV. JHEP **10**, 076 (2017). https://doi.org/10.1007/JHEP10(2017)076. arXiv:1701.02032

1201. CMS Collaboration, A search for pair production of new light bosons decaying into muons. Phys. Lett. B **752**, 146 (2016). https://doi.org/10.1016/j.physletb.2015.10.067. arXiv:1506.00424

1202. ATLAS Collaboration, Search for Higgs bosons decaying to aa in the $\mu\mu\tau\tau$ final state in pp collisions at $\sqrt{s} = 8$ TeV with the ATLAS experiment. Phys. Rev. D **92**, 052002 (2015). https://doi.org/10.1103/PhysRevD.92.052002. arXiv:1505.01609
1203. ATLAS Collaboration, Search for Higgs boson decays to beyond-the-standard-model light bosons in four-lepton events with the ATLAS detector at $\sqrt{s} = 13$ TeV. JHEP **06**, 166 (2018). https://doi.org/10.1007/JHEP06(2018)166. arXiv:1802.03388
1204. CMS Collaboration, Search for light pseudoscalar boson pairs produced from decays of the 125 GeV Higgs boson in final states with two muons and two nearby tracks in pp collisions at $\sqrt{s} = 13$ TeV. Phys. Lett. B **800**, 135087 (2020). https://doi.org/10.1016/j.physletb.2019.135087. arXiv:1907.07235
1205. CMS Collaboration, Search for a light pseudoscalar Higgs boson in the boosted $\mu\mu\tau\tau$ final state in proton-proton collisions at $\sqrt{s} = 13$ TeV. JHEP **08**, 139 (2020). https://doi.org/10.1007/JHEP08(2020)139. arXiv:2005.08694
1206. ATLAS Collaboration, Search for Higgs boson decays into pairs of light (pseudo)scalar particles in the $\gamma\gamma jj$ final state in pp collisions at $\sqrt{s} = 13$ TeV with the ATLAS detector. Phys. Lett. B **782**, 750 (2018). https://doi.org/10.1016/j.physletb.2018.06.011. arXiv:1803.11145
1207. ATLAS Collaboration, Search for Higgs boson decays into a pair of light bosons in the $bb\mu\mu$ final state in pp collision at $\sqrt{s} = 13$ TeV with the ATLAS detector. Phys. Lett. B **790**, 1 (2019). https://doi.org/10.1016/j.physletb.2018.10.073. arXiv:1807.00539
1208. CMS Collaboration, Search for an exotic decay of the Higgs boson to a pair of light pseudoscalars in the final state with two muons and two b quarks in pp collisions at 13 TeV. Phys. Lett. B **795**, 398 (2019). https://doi.org/10.1016/j.physletb.2019.06.021. arXiv:1812.06359
1209. CMS Collaboration, Search for an exotic decay of the Higgs boson to a pair of light pseudoscalars in the final state with two b quarks and two τ leptons in proton-proton collisions at $\sqrt{s} = 13$ TeV. Phys. Lett. B **785**, 462 (2018). https://doi.org/10.1016/j.physletb.2018.08.057. arXiv:1805.10191
1210. ATLAS Collaboration, Search for the Higgs boson produced in association with a vector boson and decaying into two spin-zero particles in the $H \to aa \to 4b$ channel in pp collisions at $\sqrt{s} = 13$ TeV with the ATLAS detector. JHEP **10**, 031 (2018). https://doi.org/10.1007/JHEP10(2018)031. arXiv:1806.07355
1211. ATLAS Collaboration, Search for Higgs boson decays into two new low-mass spin-0 particles in the $4b$ channel with the ATLAS detector using pp collisions at $\sqrt{s} = 13$ TeV, arXiv:2005.12236
1212. CMS Collaboration, Search for leptoquarks coupled to third-generation quarks in proton-proton collisions at $\sqrt{s} = 13$ TeV. Phys. Rev. Lett. **121**, 241802 (2018). https://doi.org/10.1103/PhysRevLett.121.241802. arXiv:1809.05558
1213. J.C. Pati, A. Salam, Lepton number as the fourth color. Phys. Rev. D **10**, 275 (1974). https://doi.org/10.1103/PhysRevD.10.275. Erratum: Phys. Rev. D **11**, 703 (1975)
1214. H. Georgi, S.L. Glashow, Unity of all elementary-particle forces. Phys. Rev. Lett. **32**, 438 (1974). https://doi.org/10.1103/PhysRevLett.32.438
1215. H. Fritzsch, P. Minkowski, Unified interactions of leptons and hadrons. Ann. Phys. **93**, 193 (1975). https://doi.org/10.1016/0003-4916(75)90211-0
1216. E. Farhi, L. Susskind, Technicolor. Phys. Rept. **74**, 277 (1981). https://doi.org/10.1016/0370-1573(81)90173-3
1217. K. Lane, M.V. Ramana, Walking technicolor signatures at hadron colliders. Phys. Rev. D **44**, 2678 (1991). https://doi.org/10.1103/PhysRevD.44.2678
1218. B. Schrempp, F. Schrempp, Light leptoquarks. Phys. Lett. B **153**, 101 (1985). https://doi.org/10.1016/0370-2693(85)91450-9
1219. B. Gripaios, Composite leptoquarks at the LHC. JHEP **02**, 045 (2010). https://doi.org/10.1007/JHEP02(2010)045. arXiv:0910.1789
1220. W. Buchmüller, R. Rückl, D. Wyler, Leptoquarks in Lepton-Quark Collisions. Phys. Lett. B **191**, 442 (1987). https://doi.org/10.1016/0370-2693(87)90637-X. Erratum: Phys. Lett. B **448**, 320 (1999)

1221. D.E. Acosta, S.K. Blessing, Leptoquark searches at HERA and the Tevatron. Ann. Rev. Nucl. Part. Sci. **49**, 389 (1999). https://doi.org/10.1146/annurev.nucl.49.1.389
1222. I. Doršner, S. Fajfer, A. Greljo, J.F. Kamenik, N. Košnik, Physics of leptoquarks in precision experiments and at particle colliders. Phys. Rept. **641**, 1 (2016). https://doi.org/10.1016/j.physrep.2016.06.001. arXiv:1603.04993
1223. D. Toback, L. ŽIvković, Review of physics results from the Tevatron: searches for new particles and interactions. Int. J. Mod. Phys. A **30**, 1541007 (2015). https://doi.org/10.1142/S0217751X15410079. arXiv:1409.4910
1224. D.M. South, M. Turcato, Review of searches for rare processes and physics beyond the standard model at HERA. Eur. Phys. J. C **76**, 336 (2016). https://doi.org/10.1140/epjc/s10052-016-4152-3. arXiv:1605.03459
1225. R. Mandal, A. Pich, Constraints on scalar leptoquarks from lepton and kaon physics. JHEP **12**, 089 (2019). https://doi.org/10.1007/JHEP12(2019)089. arXiv:1908.11155
1226. A. Crivellin, D. Müller, T. Ota, Simultaneous explanation of $R(D^{(*)})$ and $\mathrm{b} \to \mathrm{s}\mu^+\mu^-$: the last scalar leptoquarks standing. JHEP **09**, 040 (2017). https://doi.org/10.1007/JHEP09(2017)040. arXiv:1703.09226
1227. G. Hiller, I. Niš, žić, R_K and R_{K^*} beyond the standard model. Phys. Rev. D **96**, 035003 (2017). https://doi.org/10.1103/PhysRevD.96.035003. arXiv:1704.05444
1228. B. Diaz, M. Schmaltz, Y.-M. Zhong, The leptoquark hunter's guide: pair production. JHEP **10**, 097 (2017). https://doi.org/10.1007/JHEP10(2017)097. arXiv:1706.05033
1229. BABAR Collaboration, Evidence for an excess of $\bar{B} \to D^{(*)}\tau^-\bar{\nu}_\tau$ decays. Phys. Rev. Lett. **109**, 101802 (2012). https://doi.org/10.1103/PhysRevLett.109.101802. arXiv:1205.5442
1230. BABAR Collaboration, Measurement of an excess of $\bar{B} \to D^{(*)}\tau^-\bar{\nu}_\tau$ decays and implications for charged Higgs bosons. Phys. Rev. D **88**, 072012 (2013). https://doi.org/10.1103/PhysRevD.88.072012. arXiv:1303.0571
1231. Belle Collaboration, Observation of $B^0 \to D^{*-}\tau^+\nu_\tau$ decay at Belle. Phys. Rev. Lett. **99**, 191807 (2007). https://doi.org/10.1103/PhysRevLett.99.191807. arXiv:0706.4429
1232. Belle Collaboration, Observation of $B^+ \to \bar{D}^{*0}\tau^+\nu_\tau$ and evidence for $B^+ \to \bar{D}^0\tau^+\nu_\tau$ at Belle. Phys. Rev. D **82**, 072005 (2010). https://doi.org/10.1103/PhysRevD.82.072005. arXiv:1005.2302
1233. Belle Collaboration, Measurement of the branching ratio of $\bar{B} \to D^{(*)}\tau^-\bar{\nu}_\tau$ relative to $\bar{B} \to D^{(*)}\ell^-\bar{\nu}_\ell$ decays with hadronic tagging at Belle. Phys. Rev. D **92**, 072014 (2015). https://doi.org/10.1103/PhysRevD.92.072014. arXiv:1507.03233
1234. Belle Collaboration, Measurement of $\mathcal{R}(D)$ and $\mathcal{R}(D^*)$ with a semileptonic tagging method. Phys. Rev. Lett. **124**, 161803 (2020). https://doi.org/10.1103/PhysRevLett.124.161803. arXiv:1910.05864
1235. LHCb Collaboration, Measurement of the ratio of branching fractions $\mathcal{B}(\bar{B}^0 \to D^{*+}\tau^-\bar{\nu}_\tau)/\mathcal{B}(\bar{B}^0 \to D^{*+}\mu^-\bar{\nu}_\mu)$. Phys. Rev. Lett. **115**, 111803 (2015). https://doi.org/10.1103/PhysRevLett.115.159901. arXiv:1506.08614. Erratum: Phys. Rev. Lett. **115**, 159901 (2015)
1236. LHCb Collaboration, Test of lepton flavor universality by the measurement of the $B^0 \to D^{*-}\tau^+\nu_\tau$ branching fraction using three-prong τ decays. Phys. Rev. D **97**, 072013 (2018). https://doi.org/10.1103/PhysRevD.97.072013. arXiv:1711.02505
1237. HFLAV Collaboration, Averages of b-hadron, c-hadron, and τ-lepton properties as of summer 2016. Eur. Phys. J. C **77**, 895 (2017). https://doi.org/10.1140/epjc/s10052-017-5058-4. arXiv:1612.07233
1238. For updated results, see the HFLAV average of $R_{D^{(*)}}$ for Spring 2019. https://hflav-eos.web.cern.ch/hflav-eos/semi/spring19/html/RDsDsstar/RDRDs.html
1239. LHCb Collaboration, Test of lepton universality using $B^+ \to K^+\ell^+\ell^-$ decays. Phys. Rev. Lett. **113**, 151601 (2014). https://doi.org/10.1103/PhysRevLett.113.151601. arXiv:1406.6482
1240. LHCb Collaboration, Differential branching fractions and isospin asymmetries of $B \to K^{(*)}\mu^+\mu^-$ decays. JHEP **06**, 133 (2014). https://doi.org/10.1007/JHEP06(2014)133. arXiv:1403.8044

1241. LHCb Collaboration, Angular analysis of the $B^0 \to K^{*0}\mu^+\mu^-$ decay using 3 fb^{-1} of integrated luminosity. JHEP **02**, 104 (2016). https://doi.org/10.1007/JHEP02(2016)104. arXiv:1512.04442
1242. LHCb Collaboration, Test of lepton universality with $B^0 \to K^{*0}\ell^+\ell^-$ decays. JHEP **08**, 055 (2017). https://doi.org/10.1007/JHEP08(2017)055. arXiv:1705.05802
1243. LHCb Collaboration, Search for lepton-universality violation in $B^+ \to K^+\ell^+\ell^-$ decays. Phys. Rev. Lett. **122**, 191801 (2019). https://doi.org/10.1103/PhysRevLett.122.191801. arXiv:1903.09252
1244. Muon g-2 Collaborations, G.W. Bennett et al., Final report of the muon E821 anomalous magnetic moment measurement at BNL. Phys. Rev. D **73**, 072003 (2006). https://doi.org/10.1103/PhysRevD.73.072003. arXiv:hep-ex/0602035
1245. M. Davier, A. Hoecker, B. Malaescu, Z. Zhang, Reevaluation of the hadronic vacuum polarisation contributions to the Standard Model predictions of the muon $g-2$ and $\alpha(m_Z^2)$ using newest hadronic cross-section data. Eur. Phys. J. C **77**, 827 (2017). https://doi.org/10.1140/epjc/s10052-017-5161-6. arXiv:1706.09436
1246. M. Tanaka, R. Watanabe, New physics in the weak interaction of $\bar{B} \to D^{(*)}\tau\bar{\nu}$. Phys. Rev. D **87**, 034028 (2013). https://doi.org/10.1103/PhysRevD.87.034028. arXiv:1212.1878
1247. Y. Sakaki, M. Tanaka, A. Tayduganov, R. Watanabe, Testing leptoquark models in $\bar{B} \to D^{(*)}\tau\bar{\nu}$. Phys. Rev. D **88**, 094012 (2013). https://doi.org/10.1103/PhysRevD.88.094012. arXiv:1309.0301
1248. I. Doršner, S. Fajfer, N. Košnik, I. Niš, žić, Minimally flavored colored scalar in $\bar{B} \to D^{(*)}\tau\bar{\nu}$ and the mass matrices constraints. JHEP **11**, 084 (2013). https://doi.org/10.1007/JHEP11(2013)084. arXiv:1306.6493
1249. B. Dumont, K. Nishiwaki, R. Watanabe, LHC constraints and prospects for S_1 scalar leptoquark explaining the $\bar{B} \to D^{(*)}\tau\bar{\nu}$ anomaly. Phys. Rev. D **94**, 034001 (2016). https://doi.org/10.1103/PhysRevD.94.034001. arXiv:1603.05248
1250. B. Gripaios, M. Nardecchia, S.A. Renner, Composite leptoquarks and anomalies in b-meson decays. JHEP **05**, 006 (2015). https://doi.org/10.1007/JHEP05(2015)006. arXiv:1412.1791
1251. M. Bauer, M. Neubert, Minimal leptoquark explanation for the $R_{D^{(*)}}$, R_K, and $(g-2)_\mu$ anomalies. Phys. Rev. Lett. **116**, 141802 (2016). https://doi.org/10.1103/PhysRevLett.116.141802. arXiv:1511.01900
1252. E. Coluccio Leskow, G. D'Ambrosio, A. Crivellin, D. Müller, $(g-2)_\mu$, lepton flavor violation, and Z decays with leptoquarks: correlations and future prospects. Phys. Rev. D **95**, 055018 (2017). https://doi.org/10.1103/PhysRevD.95.055018. arXiv:1612.06858
1253. D. Bečirević, O. Sumensari, A leptoquark model to accommodate $R_K^{\mathrm{exp}} < R_K^{\mathrm{SM}}$ and $R_{K^*}^{\mathrm{exp}} < R_{K^*}^{\mathrm{SM}}$. JHEP **08**, 104 (2017). https://doi.org/10.1007/JHEP08(2017)104. arXiv:1704.05835
1254. G. Hiller, D. Loose, I. Nišandžić, Flavorful leptoquarks at hadron colliders. Phys. Rev. D **97**, 075004 (2018). https://doi.org/10.1103/PhysRevD.97.075004. arXiv:1801.09399
1255. E. Alvarez, L. Da Rold, A. Juste, M. Szewc, T. Vazquez Schroeder, A composite pNGB leptoquark at the LHC. JHEP **12**, 027 (2018). https://doi.org/10.1007/JHEP12(2018)027. arXiv:1808.02063
1256. D. Buttazzo, A. Greljo, G. Isidori, D. Marzocca, B-physics anomalies: a guide to combined explanations. JHEP **11**, 044 (2017). https://doi.org/10.1007/JHEP11(2017)044. arXiv:1706.07808
1257. M. Kramer, T. Plehn, M. Spira, P.M. Zerwas, Pair production of scalar leptoquarks at the CERN LHC. Phys. Rev. D **71**, 057503 (2005). https://doi.org/10.1103/PhysRevD.71.057503. arXiv:hep-ph/0411038
1258. I. Doršner, A. Greljo, Leptoquark toolbox for precision collider studies. JHEP **05**, 126 (2018). https://doi.org/10.1007/JHEP05(2018)126. arXiv:1801.07641
1259. M.J. Baker, J. Fuentes-Martín, G. Isidori, M. König, High-p_T signatures in vector-leptoquark models. Eur. Phys. J. C **79**, 334 (2019). https://doi.org/10.1140/epjc/s10052-019-6853-x. arXiv:1901.10480
1260. D0 Collaboration, Search for third-generation leptoquarks in $p\bar{p}$ collisions at $\sqrt{s} = 1.96$ TeV. Phys. Rev. Lett. **99**, 061801 (2007). https://doi.org/10.1103/PhysRevLett.99.061801. arXiv:0705.0812

1261. CDF Collaboration, Search for third generation vector leptoquarks in $p\bar{p}$ collisions at $\sqrt{s} =$ 1.96 TeV. Phys. Rev. D **77**, 091105 (2008). https://doi.org/10.1103/PhysRevD.77.091105. arXiv:0706.2832
1262. ATLAS Collaboration, Search for third generation scalar leptoquarks in pp collisions at $\sqrt{s} = 7$ TeV with the ATLAS detector. JHEP **06**, 033 (2013). https://doi.org/10.1007/JHEP06(2013)033. arXiv:1303.0526
1263. C.M.S. Collaboration, Search for third-generation leptoquarks and scalar bottom quarks in pp collisions at $\sqrt{s} = 7$ TeV. JHEP **12**, 055 (2012). https://doi.org/10.1007/JHEP12(2012)055. arXiv:1210.5627
1264. C.M.S. Collaboration, Search for pair production of third-generation scalar leptoquarks and top squarks in proton-proton collisions at $\sqrt{s} = 8$ TeV. Phys. Lett. B **739**, 229 (2014). https://doi.org/10.1016/j.physletb.2014.10.063. arXiv:1408.0806
1265. ATLAS Collaboration, Searches for scalar leptoquarks in pp collisions at $\sqrt{s} = 8$ TeV with the ATLAS detector. Eur. Phys. J. C **76**, 5 (2016). https://doi.org/10.1140/epjc/s10052-015-3823-9. arXiv:1508.04735
1266. CMS Collaboration, Search for heavy neutrinos or third-generation leptoquarks in final states with two hadronically decaying τ leptons and two jets in proton-proton collisions at $\sqrt{s} = 13$ TeV. JHEP **03**, 077 (2017). https://doi.org/10.1007/JHEP03(2017)077. arXiv:1612.01190
1267. CMS Collaboration, Search for third-generation scalar leptoquarks and heavy right-handed neutrinos in final states with two tau leptons and two jets in proton-proton collisions at $\sqrt{s} =$ 13 TeV. JHEP **07**, 121 (2017). https://doi.org/10.1007/JHEP07(2017)121. arXiv:1703.03995
1268. CMS Collaboration, Search for third-generation scalar leptoquarks in the tτ channel in proton-proton collisions at $\sqrt{s} = 8$ TeV. JHEP **07**, 042 (2015). https://doi.org/10.1007/JHEP07(2015)042. arXiv:1503.09049. Erratum: JHEP **11**, 056 (2016)
1269. CMS Collaboration, Search for third-generation scalar leptoquarks decaying to a top quark and a τ lepton at $\sqrt{s} = 13$ TeV. Eur. Phys. J. C **78**, 707 (2018). https://doi.org/10.1140/epjc/s10052-018-6143-z. arXiv:1803.02864
1270. ATLAS Collaboration, Search for pair production of third-generation scalar leptoquarks decaying into a top quark and a τ-lepton in pp collisions at $\sqrt{s} = 13$ TeV with the ATLAS detector, arXiv:2101.11582
1271. CMS Collaboration, Constraints on models of scalar and vector leptoquarks decaying to a quark and a neutrino at $\sqrt{s} = 13$ TeV. Phys. Rev. D **98**, 032005 (2018). https://doi.org/10.1103/PhysRevD.98.032005. arXiv:1805.10228
1272. ATLAS Collaboration, Searches for third-generation scalar leptoquarks in $\sqrt{s} = 13$ TeV pp collisions with the ATLAS detector. JHEP **06**, 144 (2019). https://doi.org/10.1007/JHEP06(2019)144. arXiv:1902.08103
1273. ATLAS Collaboration, Search for a scalar partner of the top quark in the jets plus missing transverse momentum final state at $\sqrt{s} = 13$ TeV with the ATLAS detector. JHEP **12**, 085 (2017). https://doi.org/10.1007/JHEP12(2017)085. arXiv:1709.04183
1274. ATLAS Collaboration, Search for supersymmetry in events with b-tagged jets and missing transverse momentum in pp collisions at $\sqrt{s} = 13$ TeV with the ATLAS detector. JHEP **11**, 195 (2017). https://doi.org/10.1007/JHEP11(2017)195. arXiv:1708.09266
1275. ATLAS Collaboration, Search for top squarks decaying to tau sleptons in pp collisions at $\sqrt{s} = 13$ TeV with the ATLAS detector. Phys. Rev. D **98**, 032008 (2018). https://doi.org/10.1103/PhysRevD.98.032008. arXiv:1803.10178
1276. K. Chandak, T. Mandal, S. Mitra, Hunting for scalar leptoquarks with boosted tops and light leptons. Phys. Rev. D **100**, 075019 (2019). https://doi.org/10.1103/PhysRevD.100.075019. arXiv:1907.11194
1277. A. Bhaskar, T. Mandal, S. Mitra, Boosting vector leptoquark searches with boosted tops. Phys. Rev. D **101**, 115015. (2020). https://doi.org/10.1103/PhysRevD.101.115015. arXiv:2004.01096
1278. CMS Collaboration, Search for singly and pair-produced leptoquarks coupling to third-generation fermions in proton-proton collisions at $\sqrt{s} = 13$ TeV, arXiv:2012.04178

1279. ATLAS Collaboration, G. Aad et al., Search for pair production of scalar leptoquarks decaying into first- or second-generation leptons and top quarks in proton-proton collisions at $\sqrt{s} = 13$ TeV with the ATLAS detector, arXiv:2010.02098
1280. P. Jackson, C. Rogan, Recursive jigsaw reconstruction: HEP event analysis in the presence of kinematic and combinatoric ambiguities. Phys. Rev. D **96**, 112007 (2017). https://doi.org/10.1103/PhysRevD.96.112007. arXiv:1705.10733
1281. P. Baldi, K. Cranmer, T. Faucett, P. Sadowski, D. Whiteson, Parameterized neural networks for high-energy physics. Eur. Phys. J. C **76**, 235 (2016). https://doi.org/10.1140/epjc/s10052-016-4099-4. arXiv:1601.07913
1282. ATLAS Collaboration, Search for pairs of scalar leptoquarks decaying into quarks and electrons or muons in $\sqrt{s} = 13$ TeV pp collisions with the ATLAS detector, arXiv:2006.05872
1283. N. Craig, P. Draper, K. Kong, Y. Ng, D. Whiteson, The unexplored landscape of two-body resonances. Acta Phys. Polon. B **50**, 837 (2019). https://doi.org/10.5506/APhysPolB.50.837. arXiv:1610.09392
1284. J.H. Kim, K. Kong, B. Nachman, D. Whiteson, The motivation and status of two-body resonance decays after the LHC Run 2 and beyond. JHEP **04**, 030 (2020). https://doi.org/10.1007/JHEP04(2020)030. arXiv:1907.06659
1285. K. Agashe, M. Ekhterachian, D. Kim, D. Sathyan, LHC signals for KK graviton from an extended warped extra dimension, arXiv:2008.06480
1286. R. Baier, Y.L. Dokshitzer, A.H. Mueller, S. Peigne, D. Schiff, Radiative energy loss and p_T broadening of high-energy partons in nuclei. Nucl. Phys. B **484**, 265 (1997). https://doi.org/10.1016/S0550-3213(96)00581-0. arXiv:hep-ph/9608322
1287. ALICE Collaboration, Exploration of jet substructure using iterative declustering in pp and Pb-Pb collisions at LHC energies. Phys. Lett. B **802**, 135227 (2020). https://doi.org/10.1016/j.physletb.2020.135227. arXiv:1905.02512
1288. CMS Collaboration, Measurement of the groomed jet mass in $PbPb$ and pp collisions at $\sqrt{s_{\mathrm{NN}}} = 5.02$ TeV. JHEP **10**, 161 (2018). https://doi.org/10.1007/JHEP10(2018)161. arXiv:1805.05145
1289. ALICE Collaboration, First measurement of jet mass in Pb-Pb and p-Pb collisions at the LHC. Phys. Lett. B **776**, 249 (2018). https://doi.org/10.1016/j.physletb.2017.11.044. arXiv:1702.00804
1290. H.A. Andrews et al., Novel tools and observables for jet physics in heavy-ion collisions. J. Phys. G **47**, 065102 (2020). https://doi.org/10.1088/1361-6471/ab7cbc. arXiv:1808.03689
1291. V. Vaidya, Effective field theory for jet substructure in heavy ion collisions, arXiv:2010.00028
1292. K. Datta, A.J. Larkoski, Novel jet observables from machine learning. JHEP **03**, 086 (2018). https://doi.org/10.1007/JHEP03(2018)086. arXiv:1710.01305
1293. K. Datta, A. Larkoski, B. Nachman, Automating the construction of jet observables with machine learning. Phys. Rev. D **100**, 095016 (2019). https://doi.org/10.1103/PhysRevD.100.095016. arXiv:1902.07180
1294. J. Collado, J.N. Howard, T. Faucett, T. Tong, P. Baldi, D. Whiteson, Learning to identify electrons, arXiv:2011.01984
1295. T. Faucett, J. Thaler, D. Whiteson, Mapping machine-learned physics into a human-readable space, arXiv:2010.11998
1296. F.C.C. Collaboration, FCC physics opportunities: future circular collider conceptual design report volume 1. Eur. Phys. J. C **79**, 474 (2019). https://doi.org/10.1140/epjc/s10052-019-6904-3
1297. FCC Collaboration, HE-LHC: the high-energy Large Hadron Collider: future circular collider conceptual design report volume 4.Eur. Phys. J. ST **228**, 1109 (2019). https://doi.org/10.1140/epjst/e2019-900088-6
1298. F.C.C. Collaboration, FCC-hh: the hadron collider: future circular collider conceptual design report volume 3. Eur. Phys. J. ST **228**, 755 (2019). https://doi.org/10.1140/epjst/e2019-900087-0

1299. E. Coleman, M. Freytsis, A. Hinzmann, M. Narain, J. Thaler, N. Tran et al., The importance of calorimetry for highly-boosted jet substructure. JINST **13**, T01003 (2018). https://doi.org/10.1088/1748-0221/13/01/T01003. arXiv:1709.08705

1300. C.-H. Yeh, S. Chekanov, A. Kotwal, J. Proudfoot, S. Sen, N. Tran et al., Studies of granularity of a hadronic calorimeter for tens-of-TeV jets at a 100 TeV *pp* collider. JINST **14**, P05008 (2019). https://doi.org/10.1088/1748-0221/14/05/P05008. arXiv:1901.11146

Index

R. Kogler, *Advances in Jet Substructure at the LHC*, Springer Tracts in Modern Physics 284, https://doi.org/10.1007/978-3-030-72858-8

The manufacturer's authorised representative in the EU is Springer Nature Customer Service Centre GmbH, Europaplatz 3, 69115 Heidelberg, Germany. If you have any concerns regarding our products, please contact ProductSafety@springernature.com

Printed and bound by CPI Group (UK) Ltd, Croydon, CR0 4YY
30/09/2026
02225287-0009